The "BOBBED" WIRE VI BIBLE

By JACK GLOVER

Cow Puddle Press
Sunset, Texas

COVER AND INSIDE ILLUSTRATIONS
BY AL MARTIN NAPOLETANO, NAVATO, CALIFORNIA.

WIRE ILLUSTRATIONS BY GARY ODLE, BOWIE, TEXAS
AND JUNE MARQUIS, PONDER, TEXAS,
& ARD SMEDLEY, SAN ANTONIO, TEXAS

ADDITIONAL COPIES OF THIS BOOK AVAILABLE FROM
JACK GLOVER
SUNSET TRADING POST
SUNSET, TEXAS 76270
PHONE 817-872-2027

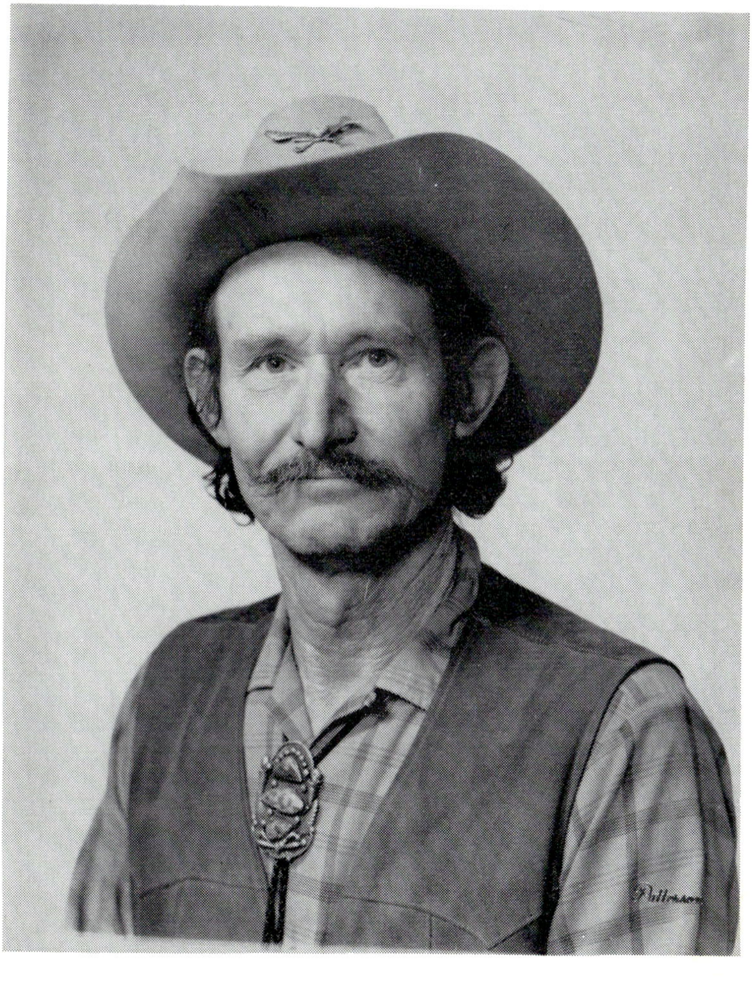

I dedicate this book to
barbed wire collectors everywhere

JACK GLOVER

PREFACE

Howdy,

From time to time I put out another book as enough new finds in wire show up to justify it. I think this book should hold for several years as I have all in it up to a new find a couple of weeks ago. I was asked by several to put out a supplement to go with my book on the new finds. By the majority, they didn't want a supplement as they are easily lost and troublesome to keep with a book. Due to the low cost of my book, I think it best to revise. It's more costly to me to do this, but if it's what the collector wants, this I'll do.

A lot of wires are controversial, like the Glidden and Rose barbs. Also the Sunderland. Some say it's this, other's that. This is the kind of interest that gets to the history of the patents. I don't believe everyone will ever agree on all wire, but the checking and double checking will get to the basic history of wire someday.

I started to put my wire in this edition in single line, double line, ribbons and so on. The collectors didn't want this as they were used to my numbers as such set up in my book and I decided to run it as is. There are enough books on wire in this manner. Collectors told me to leave it as is.

Our foreign members are finding patents in the patent offices in the countries in which they are located and this really helps. Some wire which I thought carried U.S. Patents, were in reality, shipped into this country and carried foreign patents. The people in the Northeast are collecting more and interest is picking up. They are finding new patents, also.

As I asked before, you folks please notify me of a new find whenever you locate it and I would like to put in a future Edition. Some wires were used in early days, discontinued, and then put out later for several different uses. You can find new and old in these wires. I hope you enjoy this book and use it to its best advantage. This is what I put it out for. Take care of your previous Editions, they are getting more valuable all the time. After eight years of study and seven previous Editions, I believe this book will hold for several years. Happy Hunting.

JACK GLOVER
Sunset Trading Post
Sunset, Texas 76270

A Short History of Barbed Wire

The beginning for the need of barbed wire was seen in the year of 1867 as a few men applied, that year, for patents pertaining to restraining wire with sharp points to turn and to hold livestock.

The names of the men who applied that year were Alphonso Dabb, Lucien B. Smith and William D. Hunt. In 1868, Michael Kelly patented his "thorny fence." The next few years found Lyman P. Judson, Jacob Haish, the Merrill brothers, I. L. Ellwood and Henry M. Rose patenting wire. These men could see the vast need for a fence that would hold livestock in a vicinity where neither poles nor rock were available, a fence that would be easy to erect and would last for a long period of time.

The millions of acres of the Plains States where fences would be needed were upper most in these men's minds and they set about to get their share of the business.

The first wires these men made were made on coffee mills, grind stones and some were even made in blacksmith shops.

Some of the early wires had signal plates and wooden blocks and such so the livestock could see the wire. Livestock raised in pole corrals had a tendency to run through wire as though it did not exist. This caused damage to the livestock, so tags and plates were installed to help stop the problem.

From 1867 until 1880, hundreds of wires were patented. Some men claimed patent infringements and there were quite a few court hearings. There is no telling how many hundreds of miles of Kelly, Scutt and Sunderland wire was sold as their wires are still found in nearly every part of the Nation.

There were some men who had better backing and better advertising than others and by 1885, Joseph Glidden was the chief wire manufacturer in the United States. This was accomplished by the purchase of other wire plants, franchises and dealerships. Haish, Washburn, Scutt and other wire manufacturers had made their fortunes before Glidden set out to corner the wire market.

The men mentioned were all living in Illinois, Iowa and surrounding areas, and they could see the need for wire since they lived so near the Great Plains.

The earliest wire patent on record, I suppose, is the Merriweather wire patented at New Braunfels, Texas, in 1853. This wire, however, was not a barbed wire, rather, it was a smooth wire.

As World War One came up, there was wire manufactured to stop or slow up troops on the battlefields. As time progressed and steel companies began putting wires on the market, two and four point wires became most common. The military wire currently being used for stockades and compounds is the most vicious of all previous wires.

So, through the many years that wire has existed, there have been many different types, starting from a coffee mill and growing into one of the truly large businesses of the world.

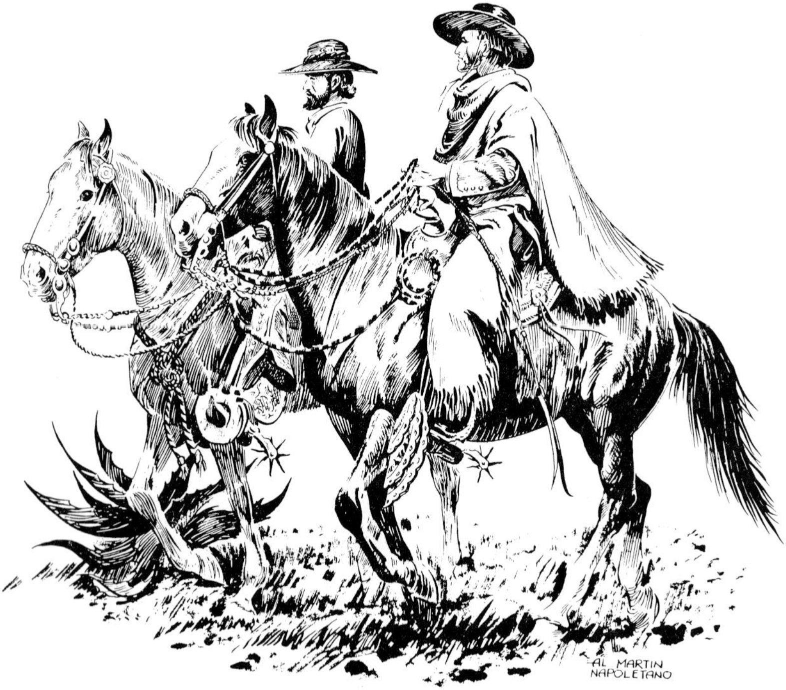
AL MARTIN
NAPOLETANO

"THE IMPACT OF WIRE IN THE WEST"

The rancher and stockman, homesteader and nester, had their troubles when the fencing began. It set friend against friend, neighbor against neighbor, every time a shipment of wire came to town. What eventually happened, it set the free range grazer against anyone who fenced his place. The free rangers could see all the miles and miles of range, but couldn't see why anyone would want to fence it. The people migrating West was homesteading and taking up claims, putting the squeeze on the large free grass man and sooner or later it was bound to cause trouble. It came when the water holes and better bottom land was fenced. Fence cutting wars was wide spread in Texas, beginning in the seventies and continuing into the eighties. The large free grass ranchers had a lot of men working for them and made it hard on the smaller settler. Night riders were out to scare settlers into moving off. The riders would usually leave a coffin on the porch of a family they wanted moved out. If this didn't scare him into leaving, more than likely, they would burn him out. These men cut fences, sometimes every wire between each post. Cattle would be pushed in and eat all the families crops or graze his meadows. Law and order was at a standstill, as local law might side with the free grazers. Usually the free grazers were the old settlers of the area and the settlers coming in were the new comers and were often held in contempt.

Many a man has been killed over a fence line. After the fencing of property came into existence through surveys, a lot of neighbors got into arguments over where the survey or section line was to go. Neither being in accord over where it was supposed to be. Some land owners dropped back from the survey line and each built his own fence and maintained it, leaving a 10 or 15 foot lane between them. These lanes were referred to as "devils lane." After the Hereford came into Texas to replace the Longhorn, the smaller spreads came into being. The Hereford was not as hardy as the Longhorn. This breed didn't range as far and had to have more water than the Longhorn. You could say the free range went out with the Longhorn as they both disappeared about the same time. When the XIT Ranch in the Panhandle fenced in their land, it was generally accepted that barb wire was a way of life on the range from then on.

THE JUSTIN COMPANIES

BOOTS · BELTS · LEATHERGOODS

P.O. BOX 548 · FORT WORTH, TEXAS 76101 · TELEPHONE (817) 332-4385

February 26, 1980

There is very little to say about Jack Glover that hasn't already been said when it comes to bobbed wire. He has long been regarded as the undisputed authority on the subject.

Jack's years of searching, collecting and researching have earned him a spot among the select few in the "bobbed wire hall of fame". That's why his name always comes up when collectors get together.

You're sure to find almost any piece of information about bobbed wire between the covers of this book.

Sincerely,

John Justin

1

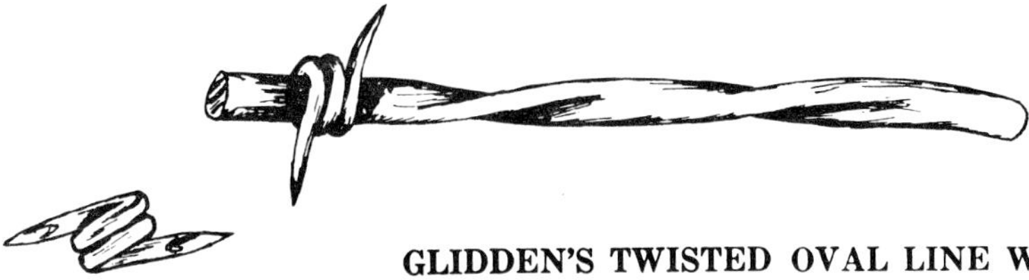

GLIDDEN'S TWISTED OVAL LINE WIRE

PAT. AUG. 22, 1876

2

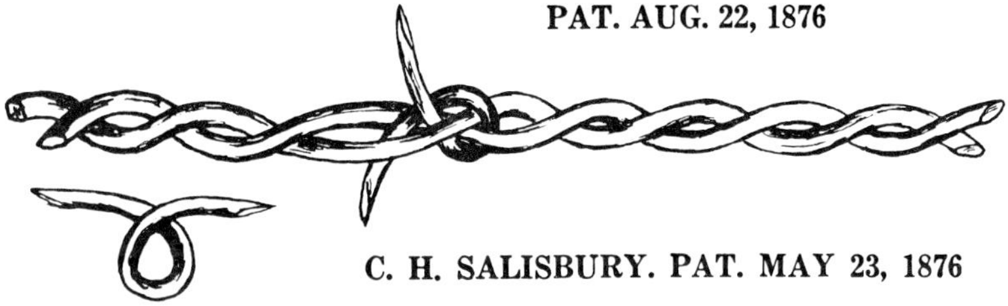

C. H. SALISBURY. PAT. MAY 23, 1876

3

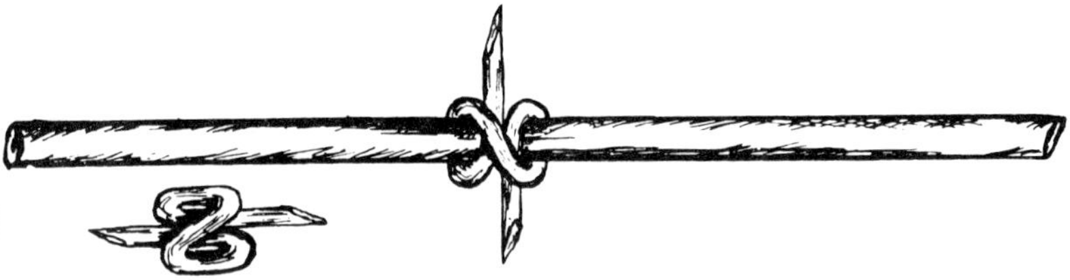

R. EMERSON. PAT. APRIL 25, 1876

4

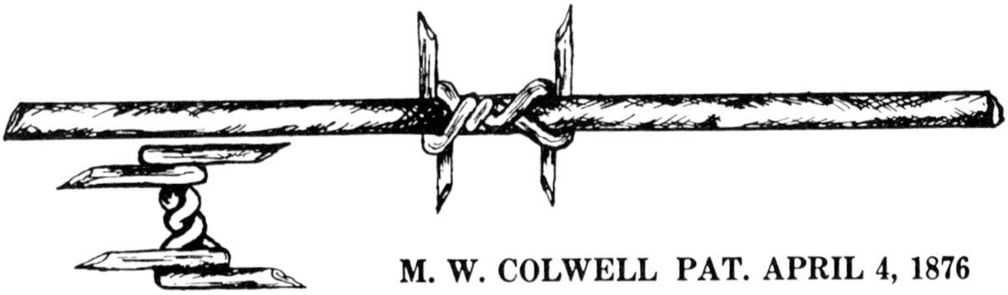

M. W. COLWELL PAT. APRIL 4, 1876

5

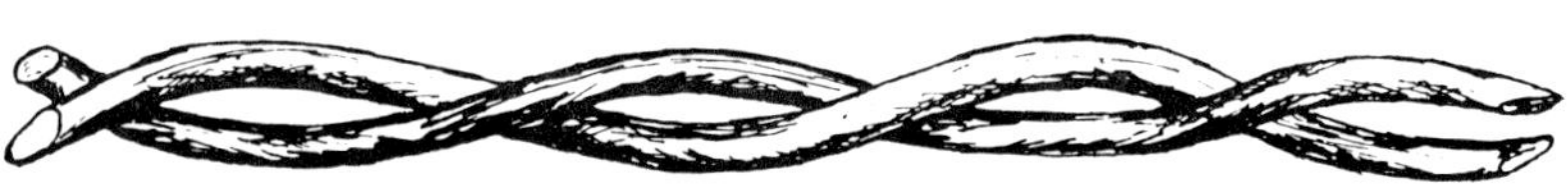

TWISTED TWO STRAND HORSE WIRE

PAT. J. B. CLEAVELAND, MARCH 20, 1894

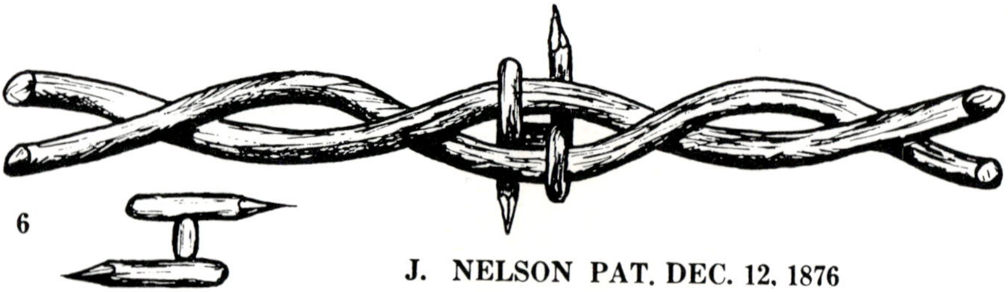

6

J. NELSON PAT. DEC. 12, 1876

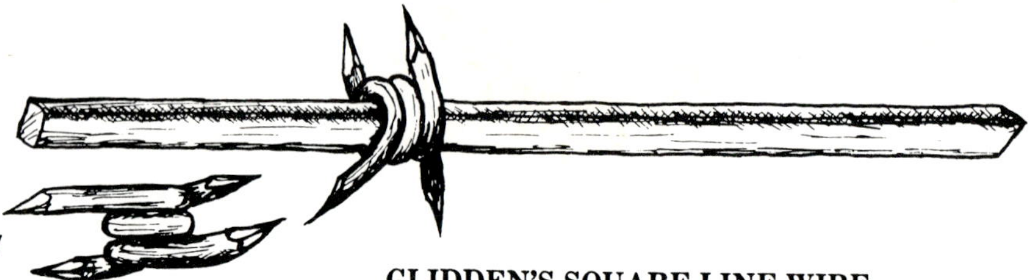

7

GLIDDEN'S SQUARE LINE WIRE

PAT. J. F. GLIDDEN, FEB. 8, 1876

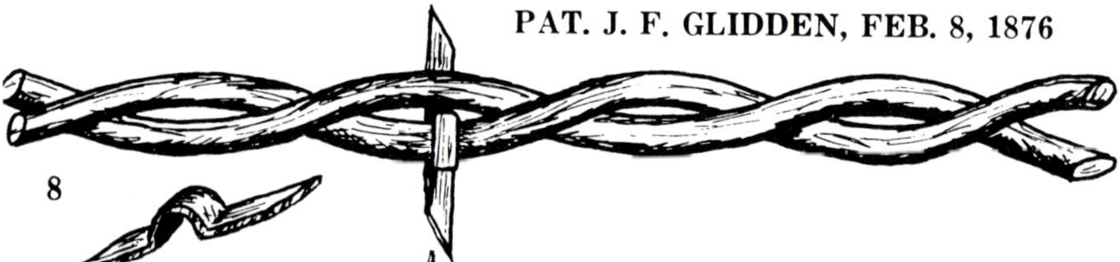

8

E. M. CRANDAL PAT. SEPT. 27, 1881

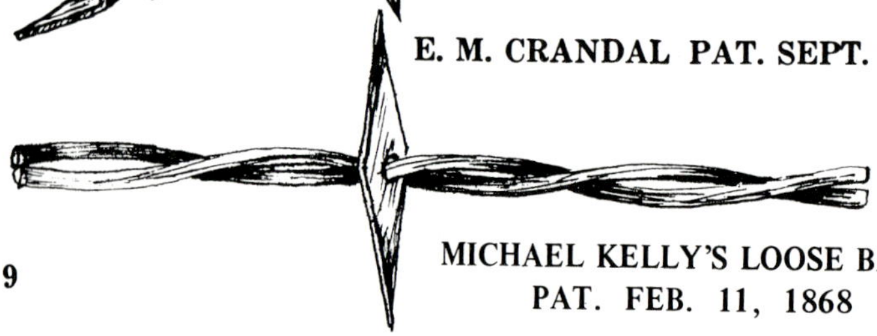

9

MICHAEL KELLY'S LOOSE BARB
PAT. FEB. 11, 1868

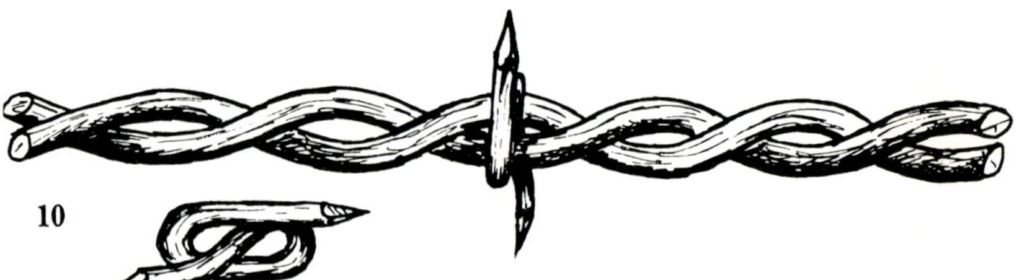

10

H. B. SCUTT PAT. SEPT. 21, 1880

11

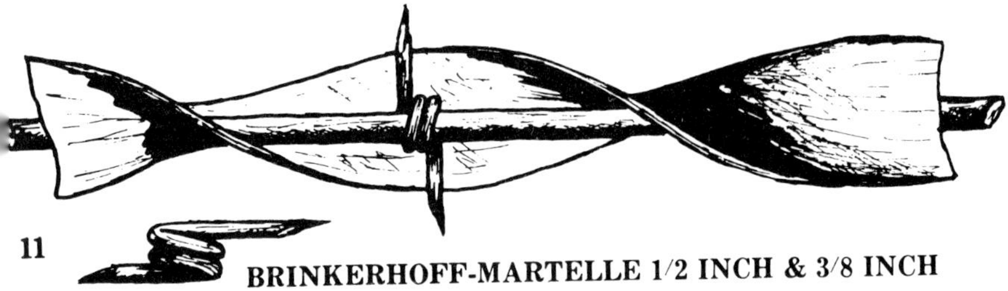

BRINKERHOFF-MARTELLE 1/2 INCH & 3/8 INCH
RIBBON PAT. AUG. 11, 1885

12

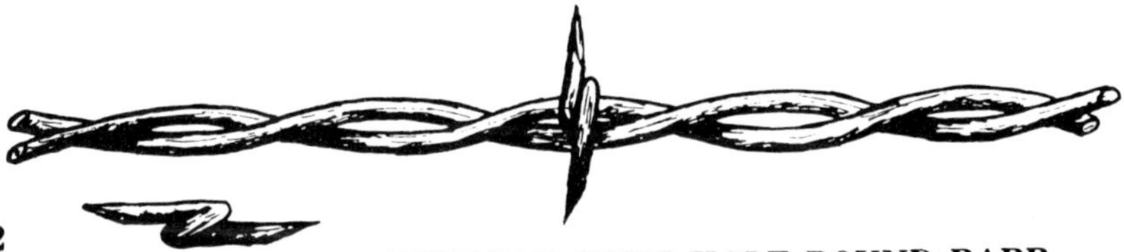

WAUKEGAN WIRE HALF ROUND BARB

13

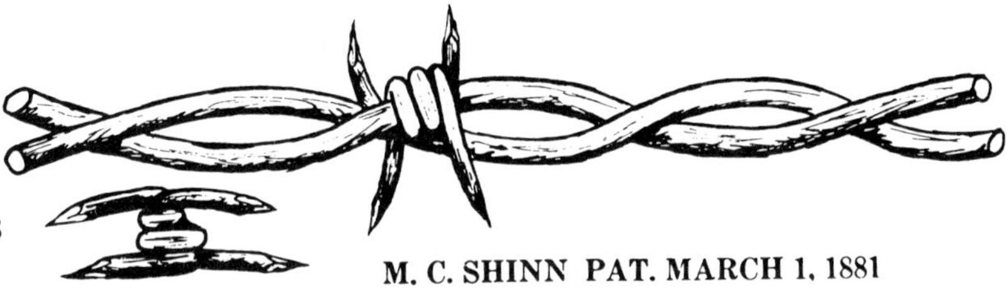

M. C. SHINN PAT. MARCH 1, 1881

14

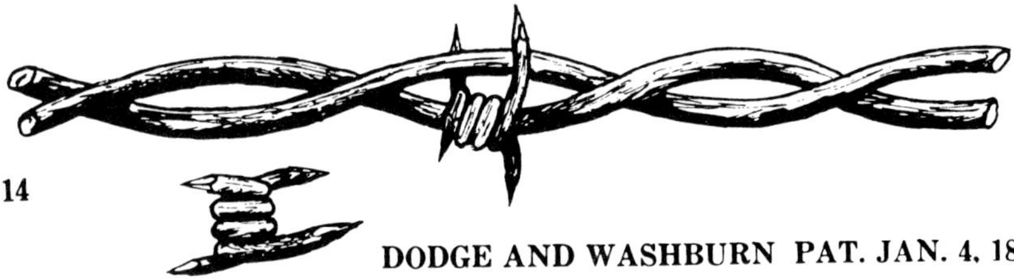

DODGE AND WASHBURN PAT. JAN. 4, 1882

15

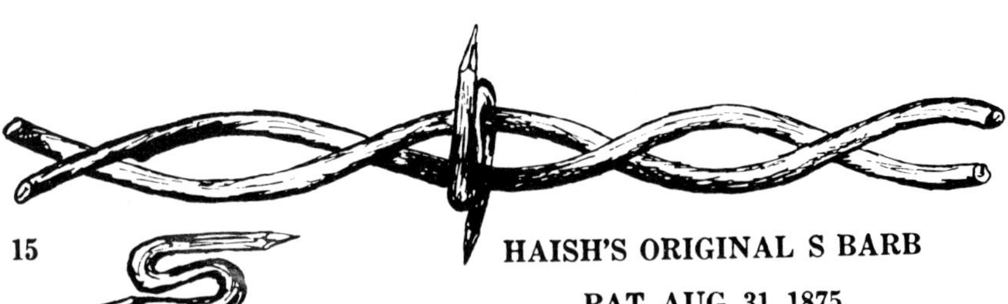

HAISH'S ORIGINAL S BARB
PAT. AUG. 31, 1875

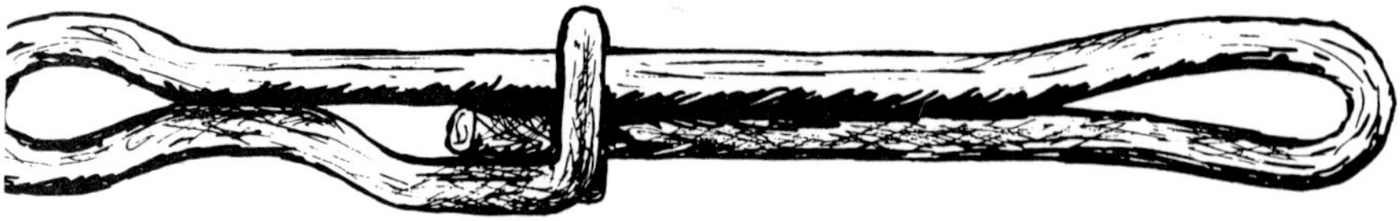

16

BARBLESS LINK WIRE, TWELVE INCH LINKS

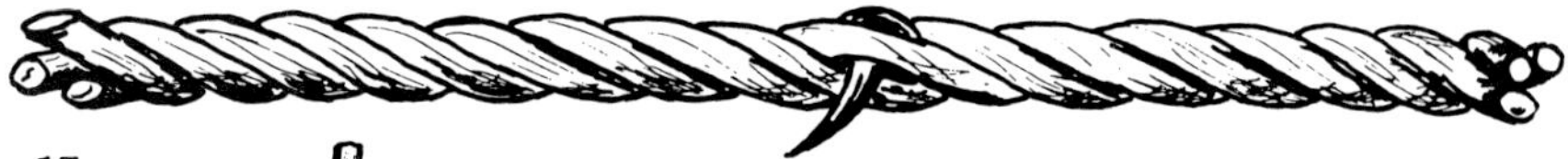

17

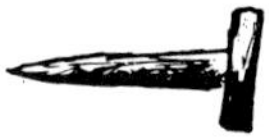

H. M. UNDERWOOD TACK WIRE

PAT. AUG. 6, 1878

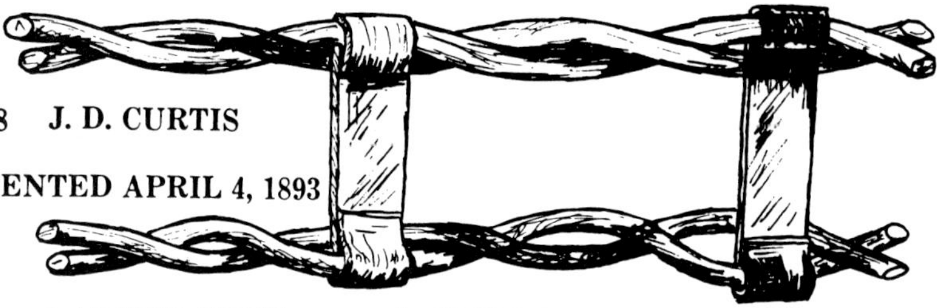

18 J. D. CURTIS

PATENTED APRIL 4, 1893

LADDER WIRE CLIPS THREE INCHES APART

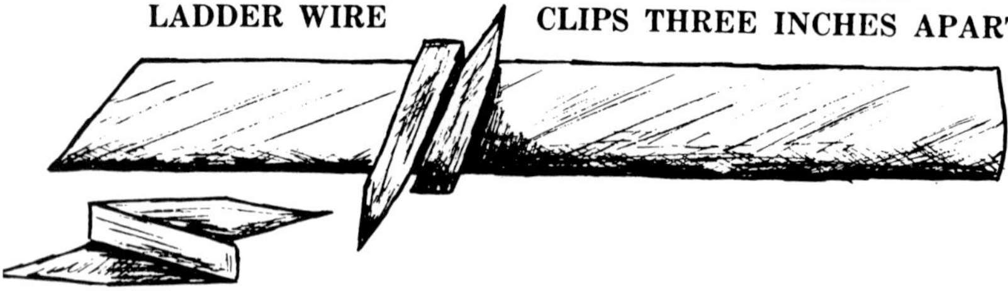

19 FORD'S FLAT RIBBON, PATENTED JUNE 9, 1875

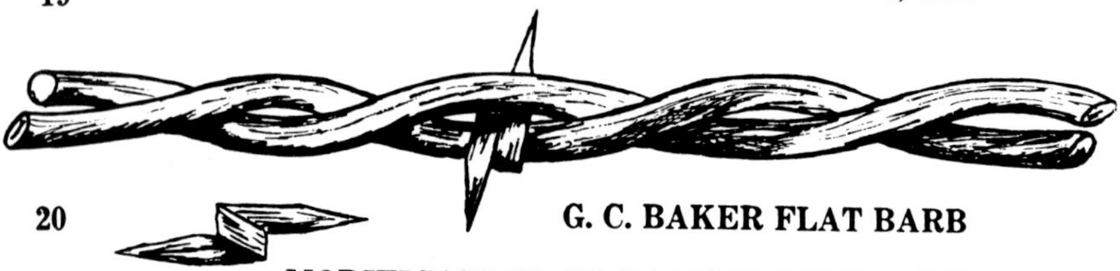

20 G. C. BAKER FLAT BARB

MODIFICATION OF PATENT FEBRUARY 27, 1883

21

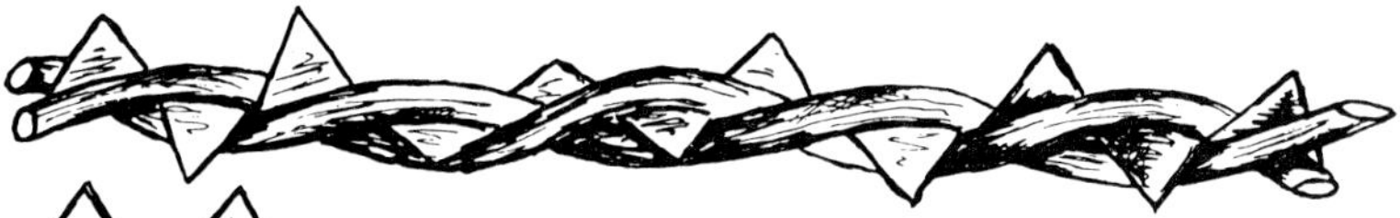

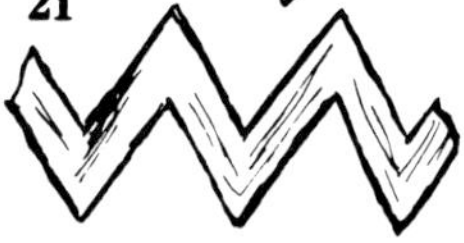

CRANDAL'S CHAMPION PAT. NOV. 4, 1879

22

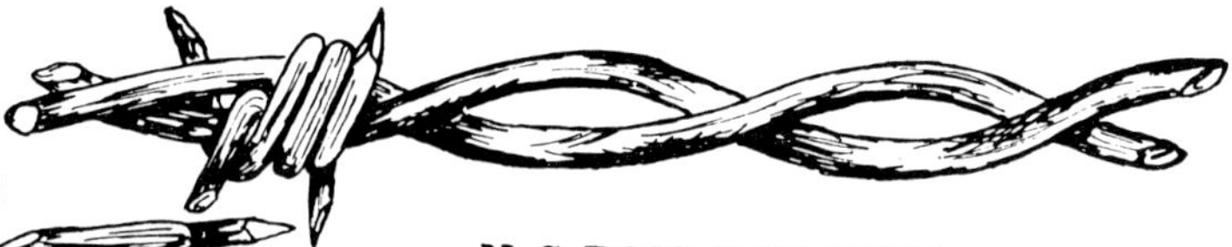

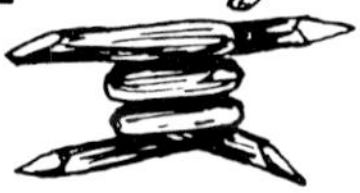

N. G. ROSS PAT. JUNE 10, 1879

23

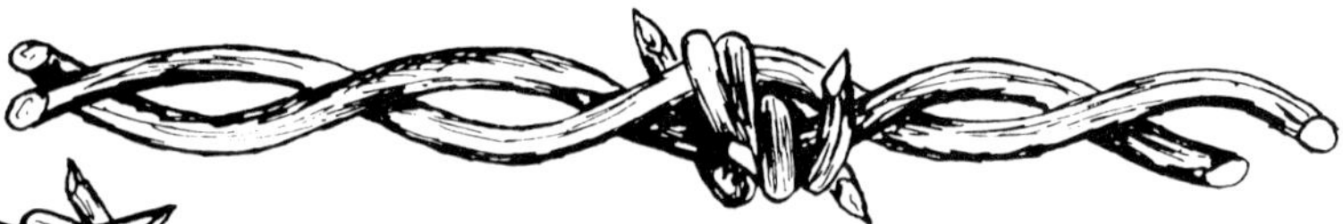

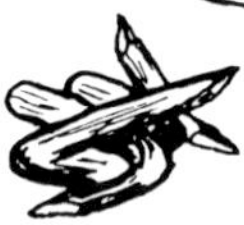

N. G. ROSS SMALL GAUGE & BARB

PAT. JUNE 10, 1879

24

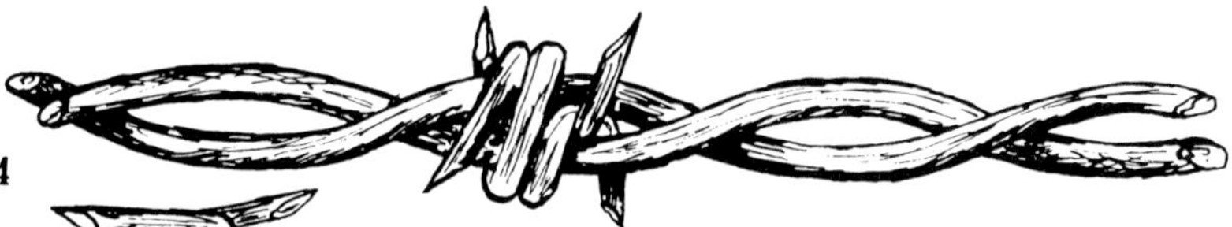

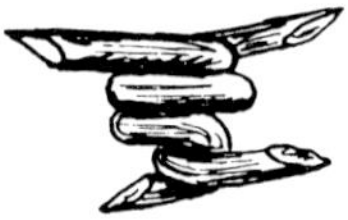

A. S. BURNELL HOOKED FOUR POINT
PATENTED JUNE 19, 1877

25

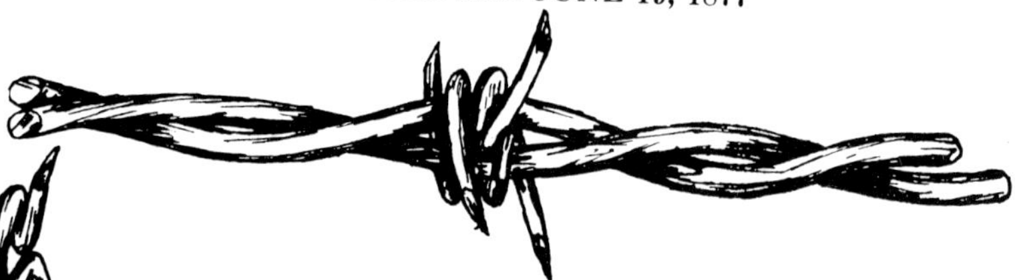

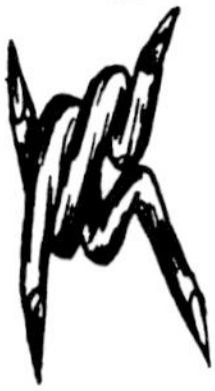

BURNELL FOUR POINT PAT. JUNE 19, 1877

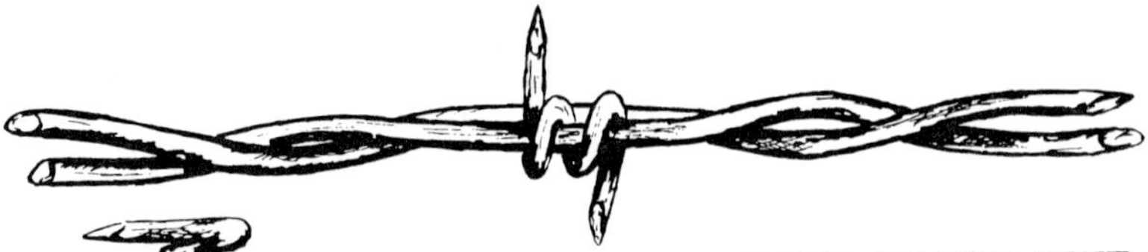

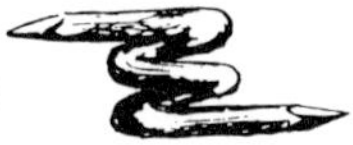

26 L. E. SUNDERLAND BARB ON TWO LINE WIRE

PAT. AUG. 12, 1884

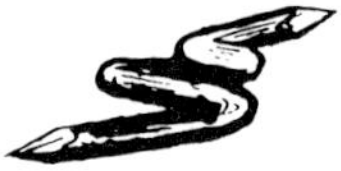

27 J. BROTHERTON ROUND BARB

BARB IS REVERSE TWIST. PAT. SEPT. 3, 1887

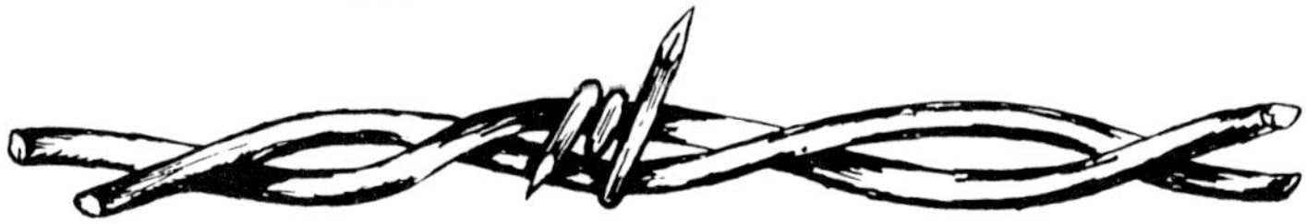

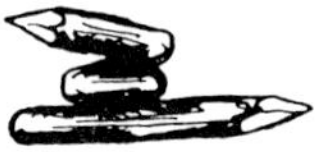

PAT. J. BROTHERTON, SEPT. 3, 1887

28 NOTE CENTER OF BARB

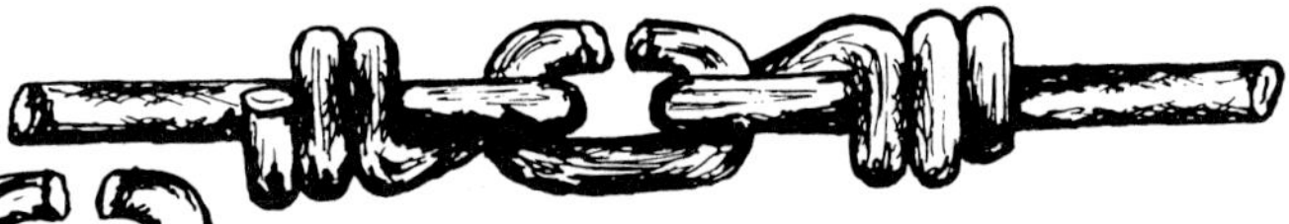

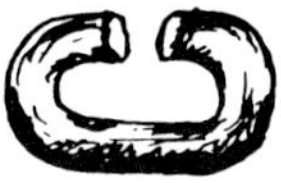

PLANTER TRIP WIRE

USED FOR FENCING LATER

29

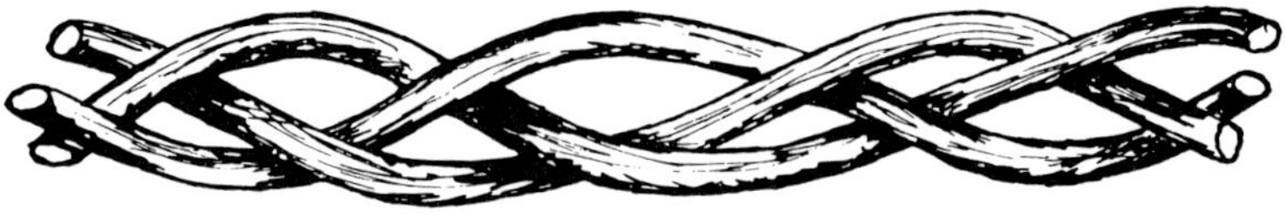

30

OTHNIEL PRESTON BRAIDED WIRE PAT. OCT. 18, 1881

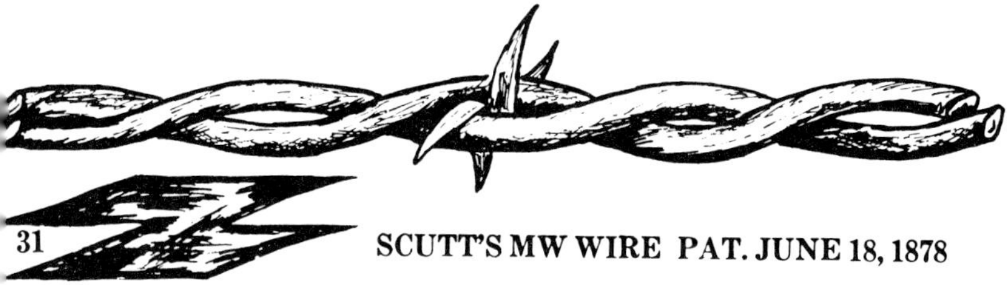

31

SCUTT'S MW WIRE PAT. JUNE 18, 1878

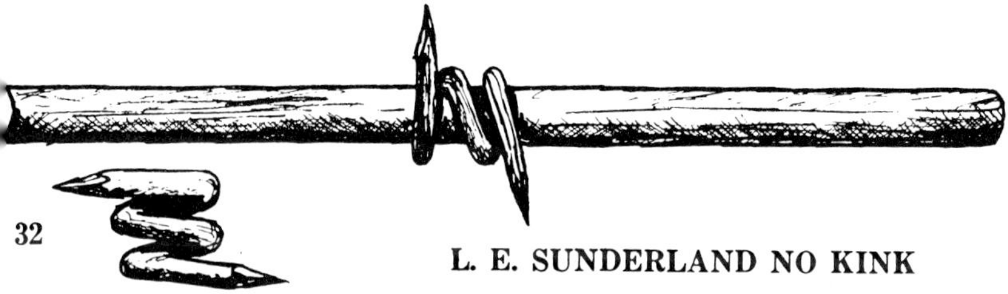

32

L. E. SUNDERLAND NO KINK

PAT. AUG. 12, 1884

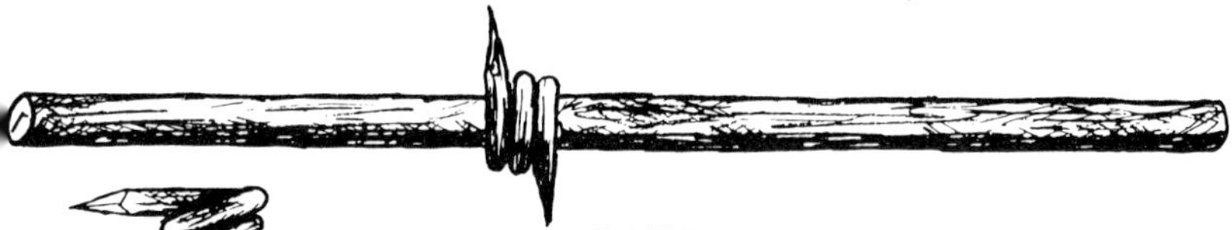

33

BAKER'S SINGLE STRAND

PAT. G. C. BAKER, FEB. 27, 1883

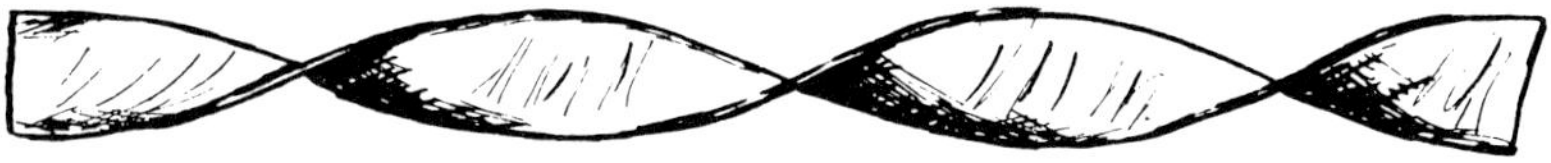

34

TWISTED RIBBON WIRE FOUR WIDTHS KNOWN.

PAT. J. BRINKERHOFF, MAY 17, 1881

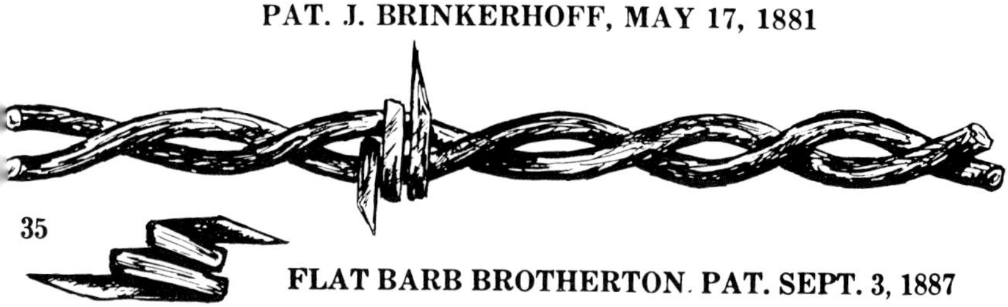

35

FLAT BARB BROTHERTON. PAT. SEPT. 3, 1887

36

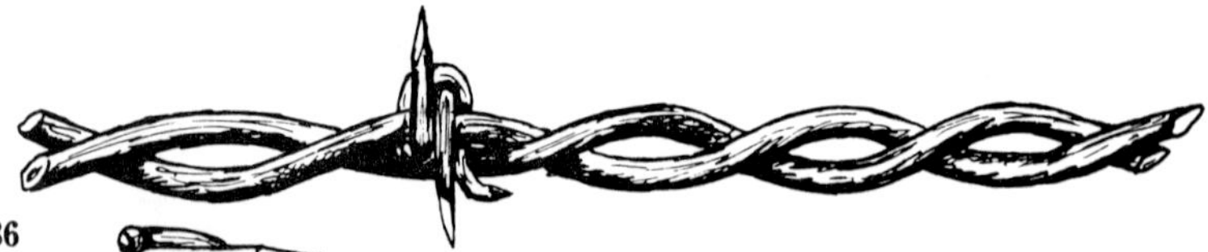

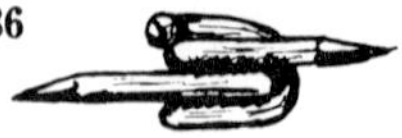

A. J. UPHAM'S BENT STAPLE BARB
PATENTED JULY 18, 1882

37

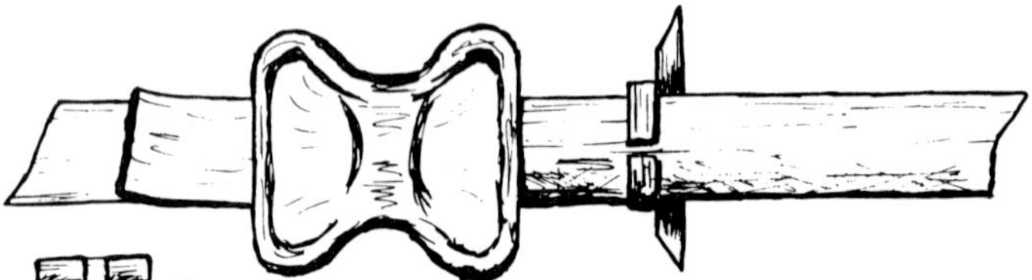

BRINKERHOFF RIVETED SPLICER
PAT. J. BRINKERHOFF, MAY 17, 1881

BOTTOM VIEW

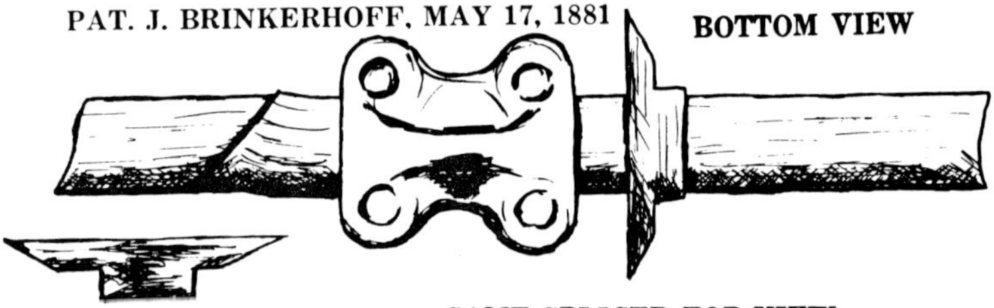

SAME SPLICER TOP VIEW

38

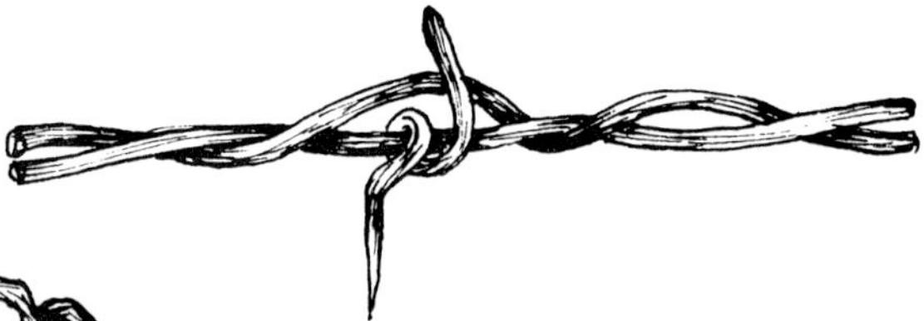

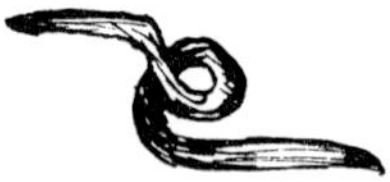

TWO LINE, TWO POINT "SNAKE BARB"

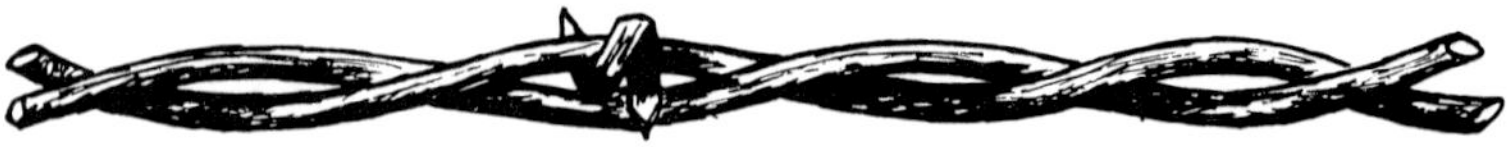

39

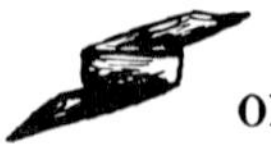

BAKER BARB
ONE LINE WIRE 12 GAUGE THE OTHER 16 GAUGE
PAT. G. C. BAKER, FEB. 27, 1883

40

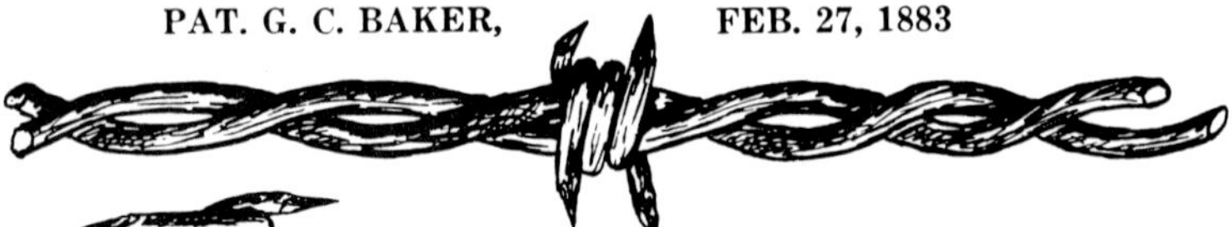

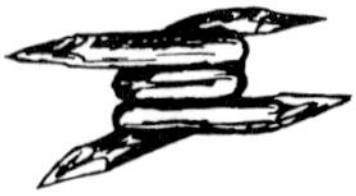

BOTH BARBS AROUND BOTH LINE WIRES
"TWO AROUND TWO"
PAT. DODGE & WASHBURN, JAN. 4, 1882

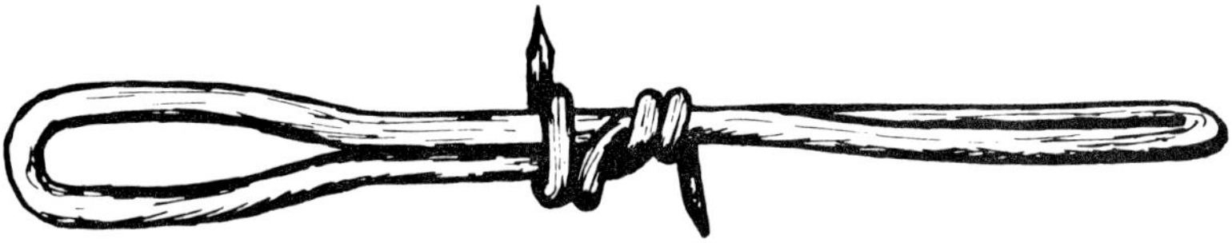

41

CRANDAL'S TELEGRAPH SPLICE

PAT. MAY 24, 1881

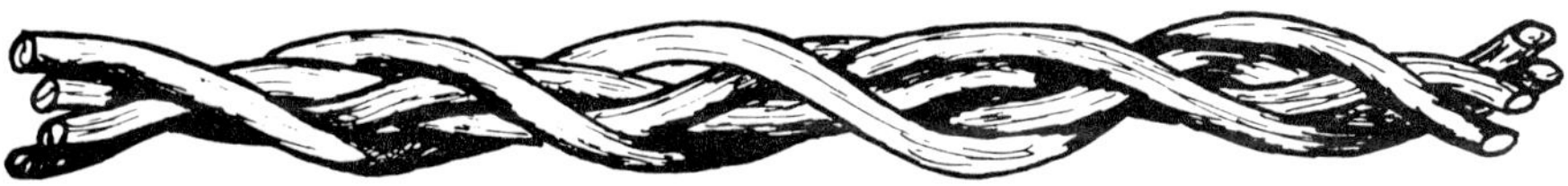

42

FOUR WIRE ORNAMENTAL YARD FENCE WIRE

TWISTED TWO AROUND TWO

PAT. J. B. CLEAVELAND, MAR. 20, 1894

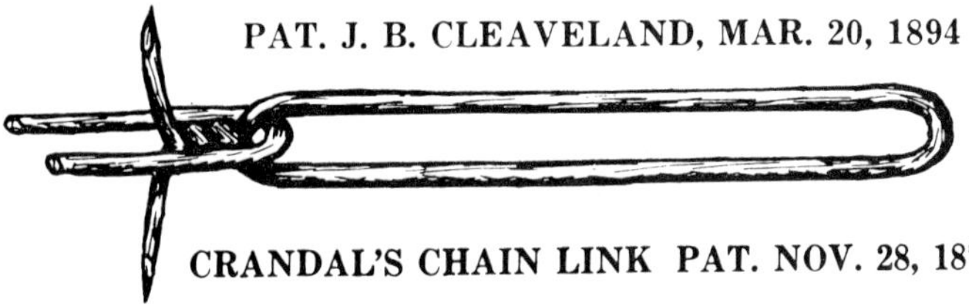

43

CRANDAL'S CHAIN LINK PAT. NOV. 28, 1876

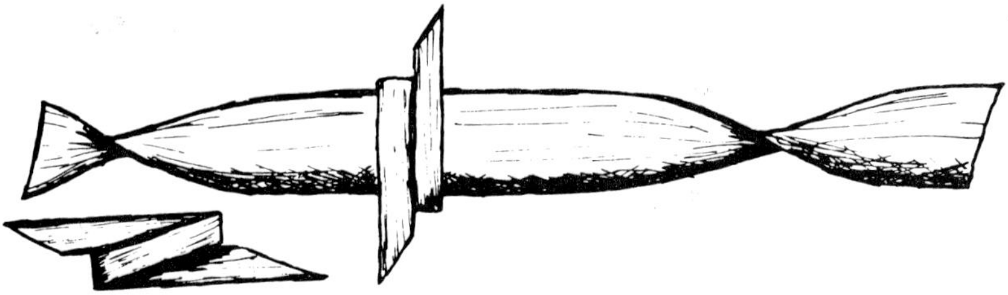

44

PAT. BY F. D. FORD JUNE 9, 1875

45

J. W. HARBAUGH'S SMOOTH RIBBON WIRE

PAT. JUNE 7, 1881

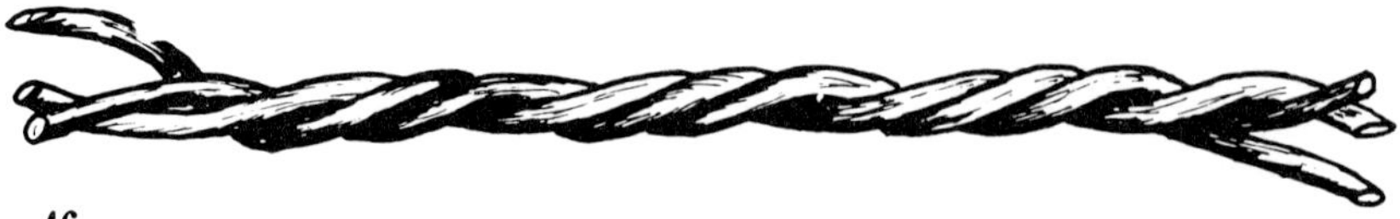

46

BARBLESS THREE LINE TWIST

PAT. J. B. CLEAVELAND, MAR. 20, 1894

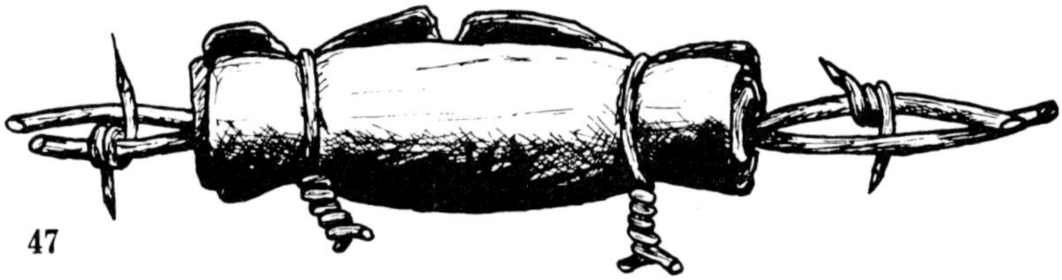

47

HOME MADE INNERTUBE INSULATOR ON TOP

FENCE WIRE USED AS TELEPHONE LINE

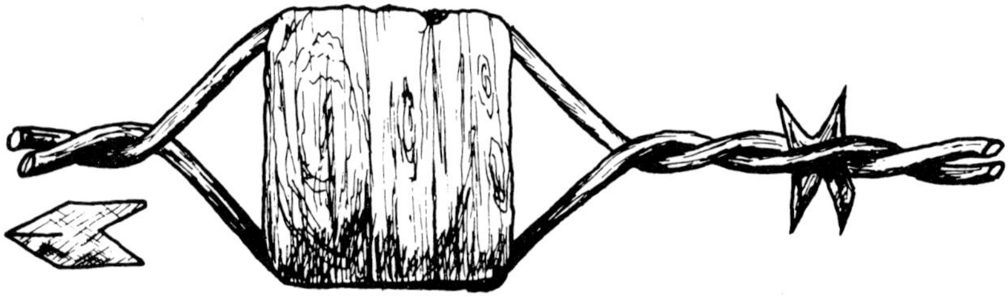

48

H. B. SCUTT'S WOODEN BLOCK
PATENTED FEBRUARY 10, 1880

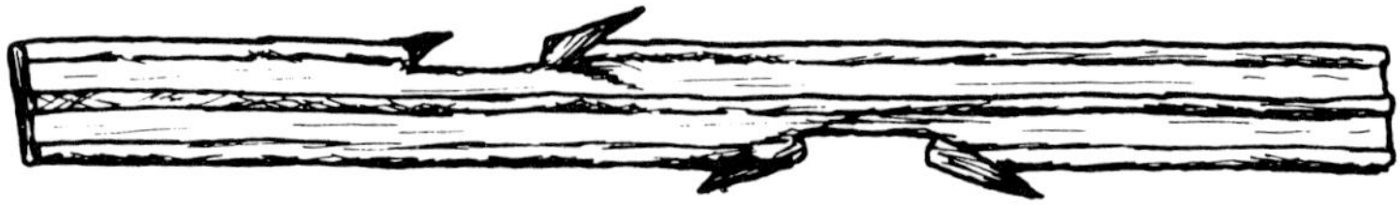

49

J .W. HARBAUGH'S TORN RIBBON
PATENTED JUNE 7, 1881

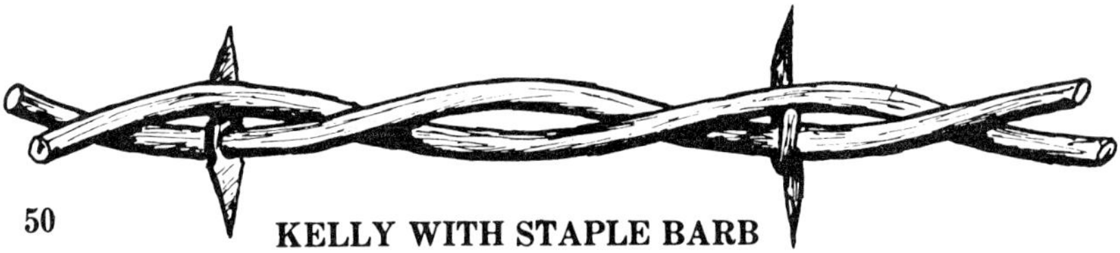

50

KELLY WITH STAPLE BARB

PAT. FEB. 11, 1868

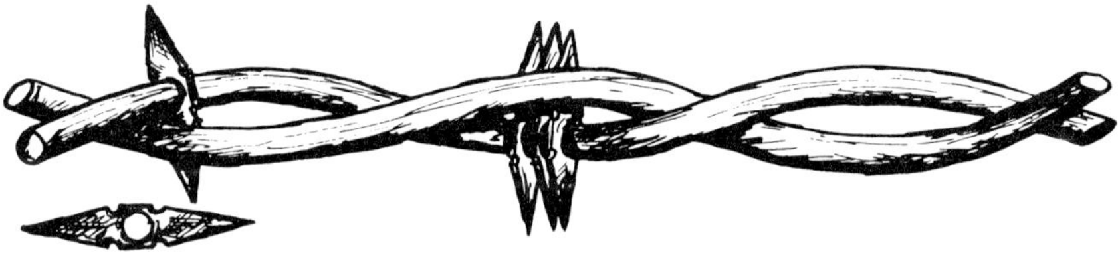

51

FACTORY IRREGULAR KELLY

PAT. FEB. 11, 1868

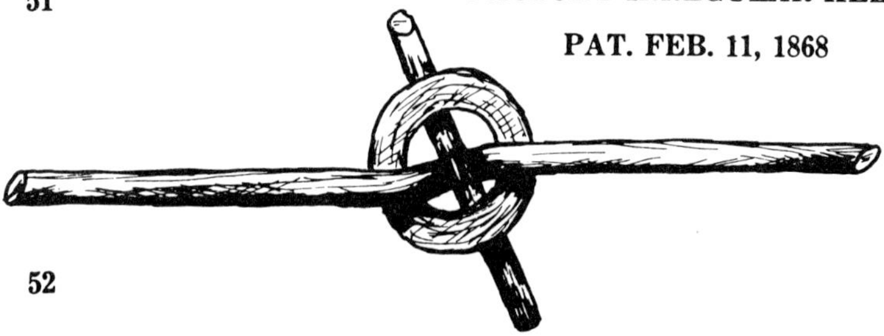

52

FENCE STAY: BENT WASHER HAMMERED FLAT

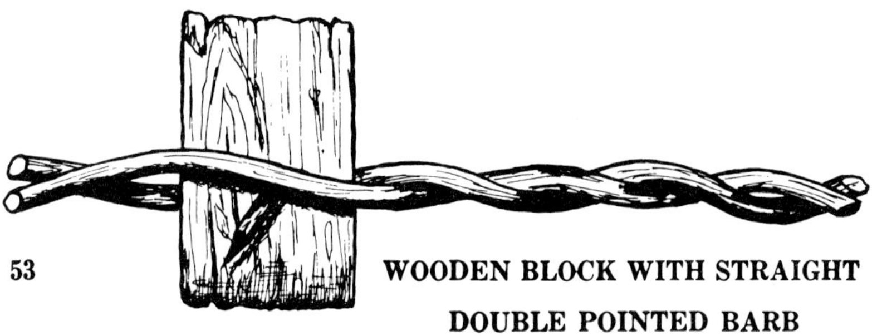

53

WOODEN BLOCK WITH STRAIGHT DOUBLE POINTED BARB

A. C. HULBERT, PATENTED APRIL 15, 1884

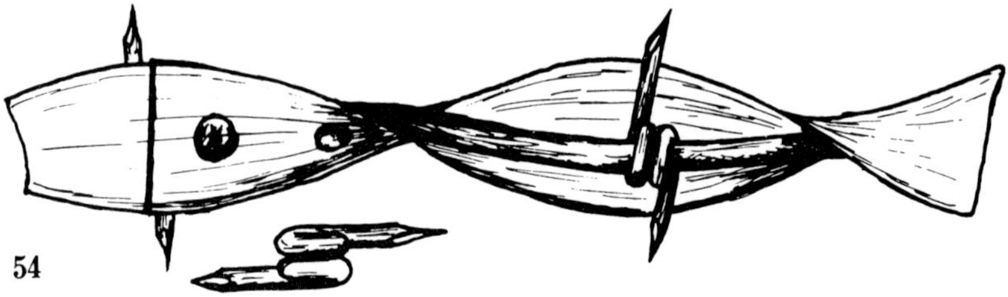

54

BRINKS MARTELLE FACTORY SPLICE

PAT. AUG. 11, 1885

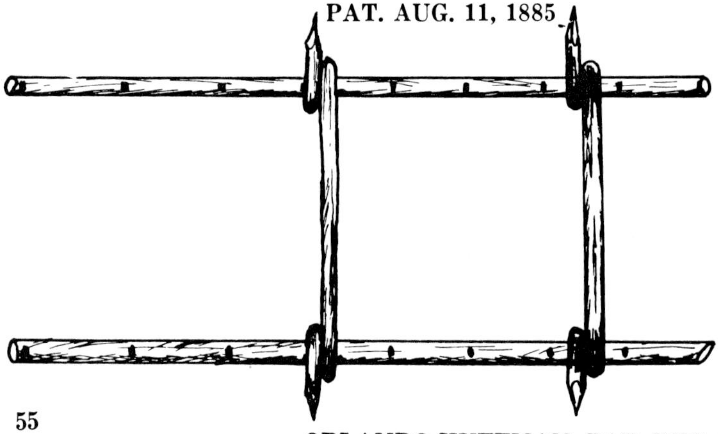

55

ORLANDO HUFFMAN. PAT. DEC. 9, 1890

COMMONLY CALLED

DECKERS TWO INCH SPREAD

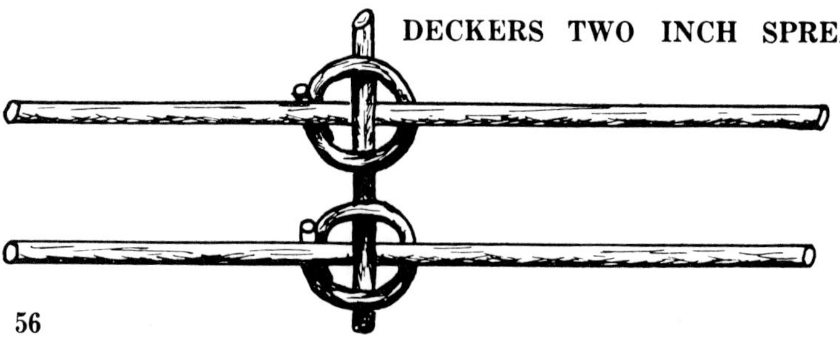

56

HOG WIRE WITH RING JOINT CLAMPS

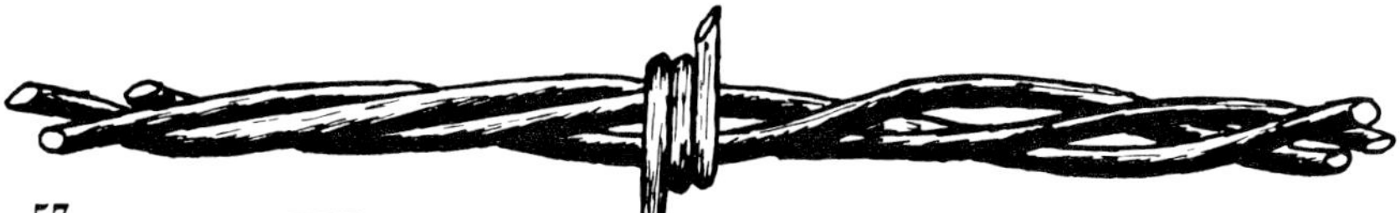

57 

PAT. J. F. GLIDDEN, NOV. 24, 1874
THREE LINE WITH GLIDDEN BARB
AROUND ALL THREE

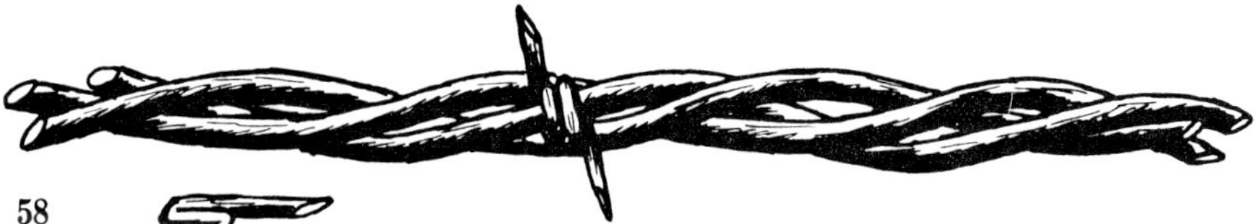

58 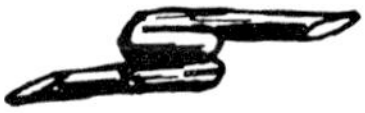

THREE LINE WITH GLIDDEN TYPE BARB
PAT. J. F. GLIDDEN, NOV. 24, 1874 RIGHT TWIST

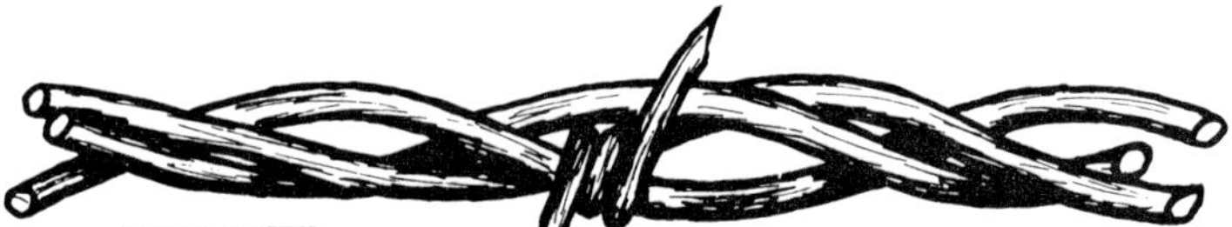

59 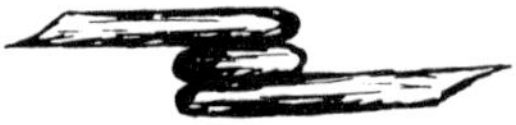

PAT. J. F. GLIDDEN, NOV. 24, 1874
THREE LINE WITH GLIDDEN TYPE BARB
LEFT TWIST

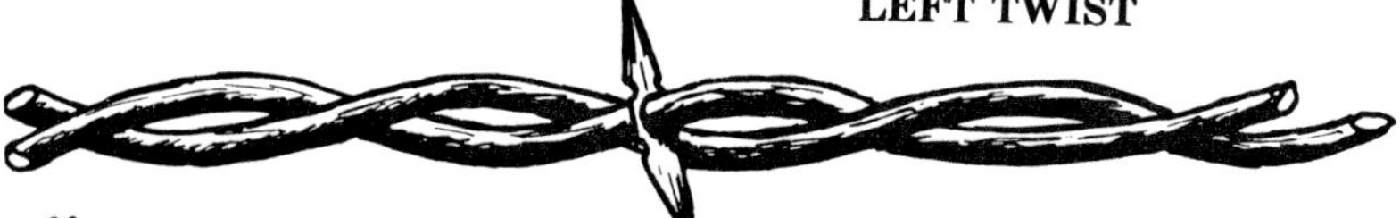

60 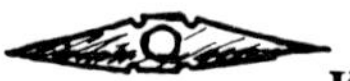

KELLY'S "DIAMOND POINT" RIGHT TWIST
PAT. FEB. 11, 1868
CLAMPED ON LINE WIRE

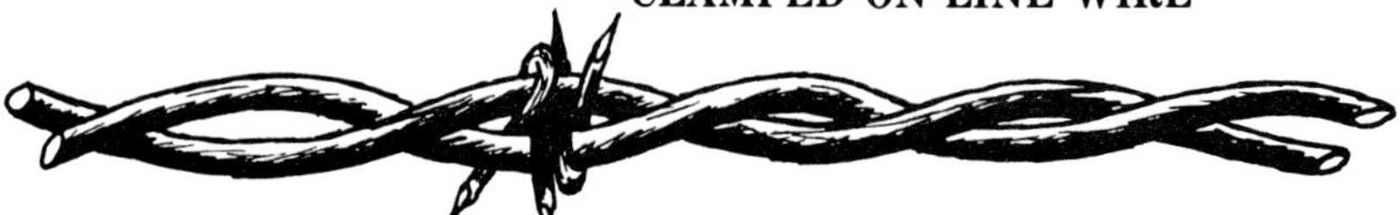

61 

S. M. STEVENS. PAT. DEC. 16, 1879

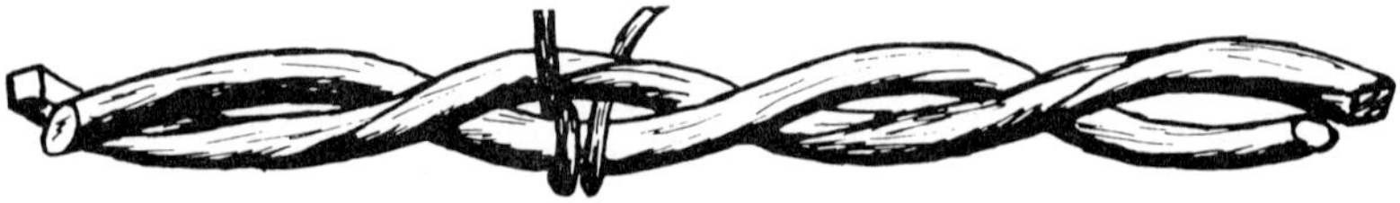

62

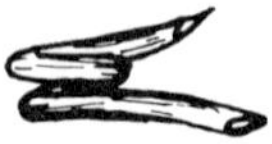

UNION PACIFIC R. R. WIRE
ONE ROUND, ONE SQUARE LINE WIRE
PAT. J. F. GLIDDEN, NOV. 24, 1874

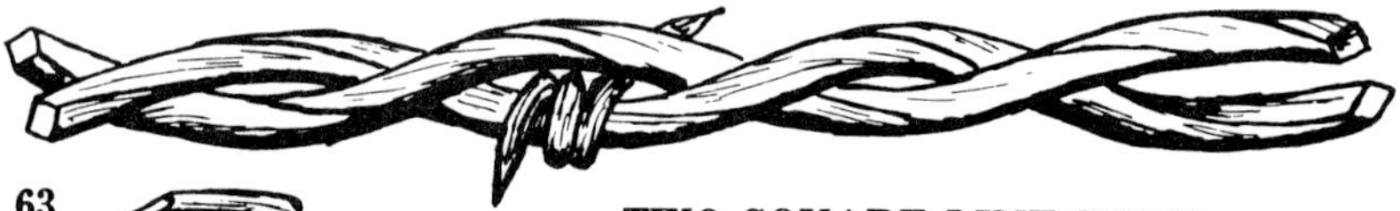

63

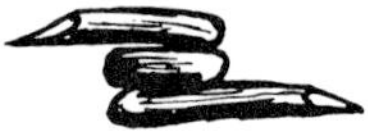

TWO SQUARE LINE WIRE
PAT. J. F. GLIDDEN, NOV. 24, 1874

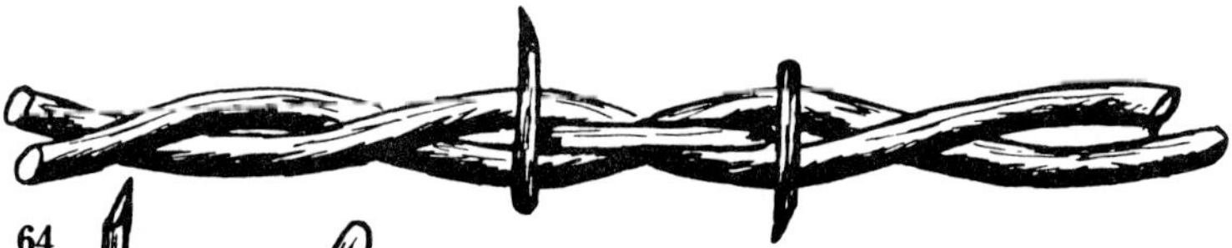

64

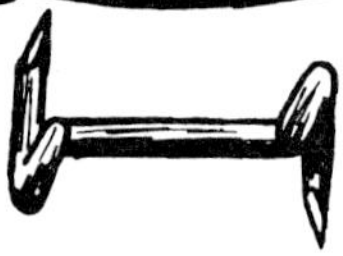

NADELHOFFER'S TWIST
PAT. JULY 22, 1884

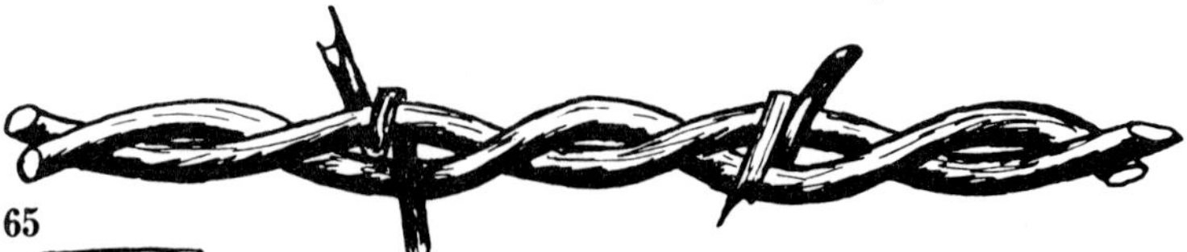

65

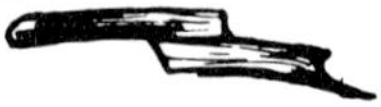

BAKER BARB ODD TYPE
PAT. G. C. BAKER, FEB. 27, 1883

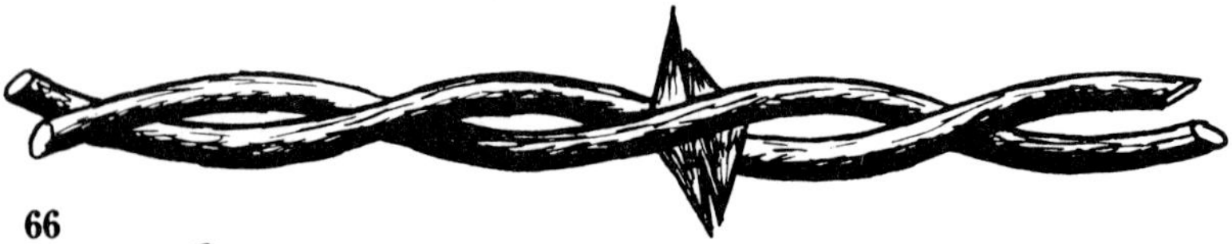

66

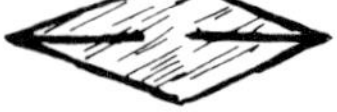

H. N. FRENTRESS SPLIT DIAMOND
PAT. DEC. 14, 1875

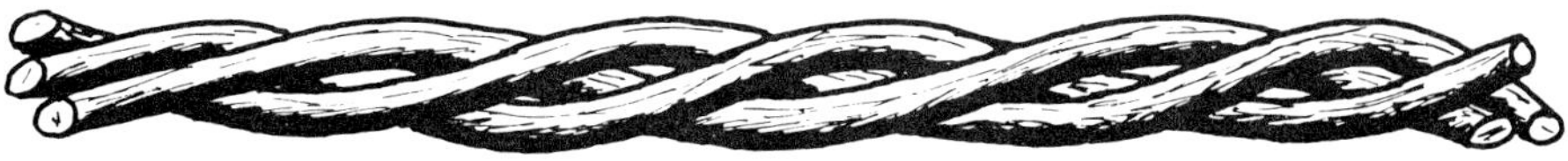

67

HEAVY GAUGE THREE LINE TWIST

PAT. J. B. CLEAVELAND, MAR. 20, 1894

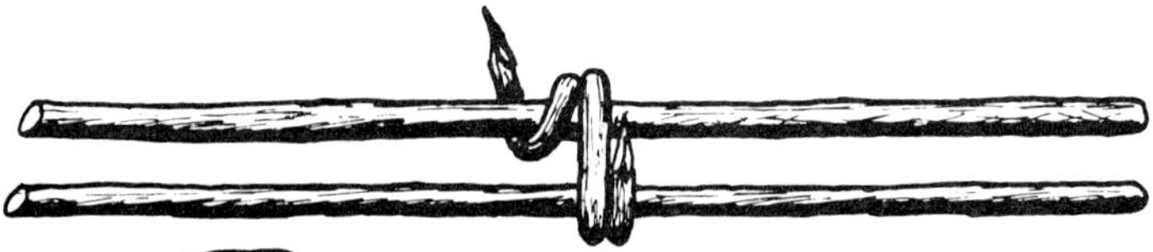

68

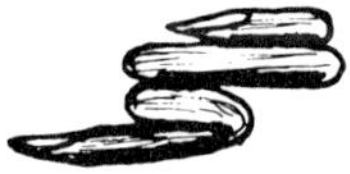

FLATTENED PARALLEL WIRE

O. HUFFMAN, PATENTED DECEMBER 31, 1889

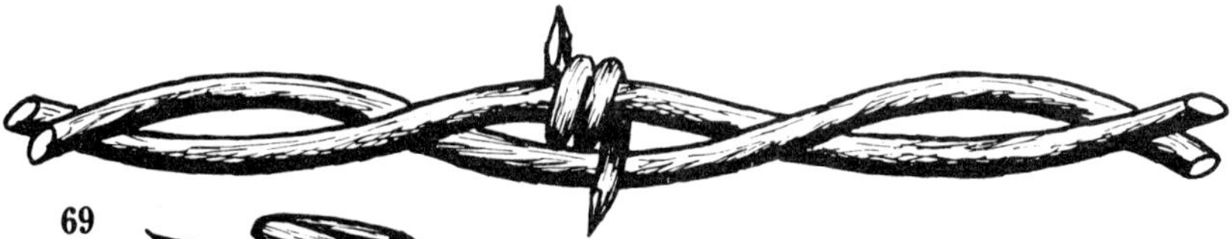

69

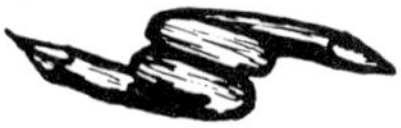

J. F. GLIDDEN WINNER PAT. NOV. 24, 1874

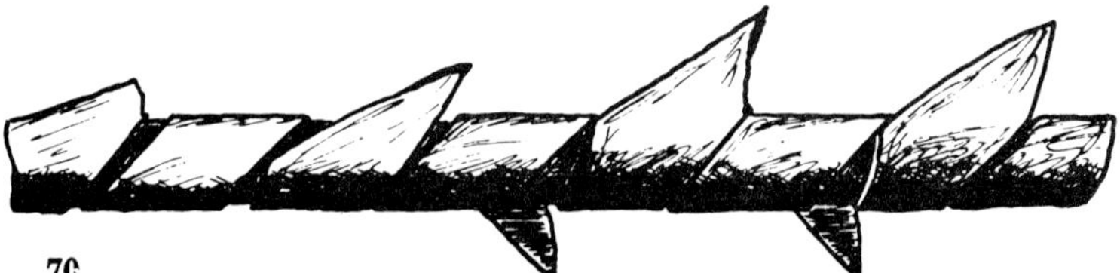

70

PAT. JAN. 22, 1878

CAREY'S MACHINE PATENT

"GREENBRIAR" OR "WRAP AROUND"

JOHN HOLLNER, ITHICA, NEB.

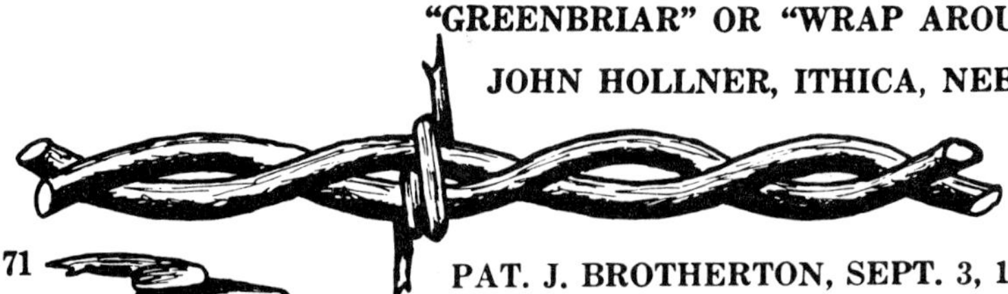

71

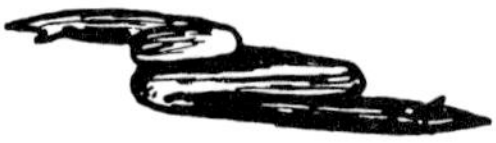

PAT. J. BROTHERTON, SEPT. 3, 1887

FORKED TONGUE BROTHERTON BARB

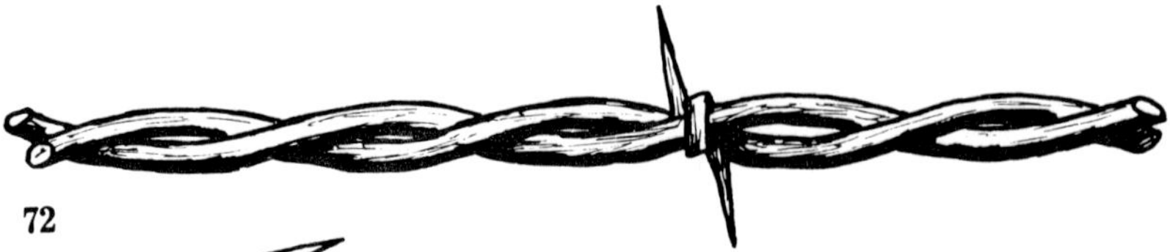

72

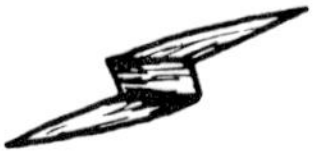

BAKER PERFECT FLAT BARB
IN VARIOUS LENGTHS & SPACING
PAT. G. C. BAKER, FEB. 27, 1883

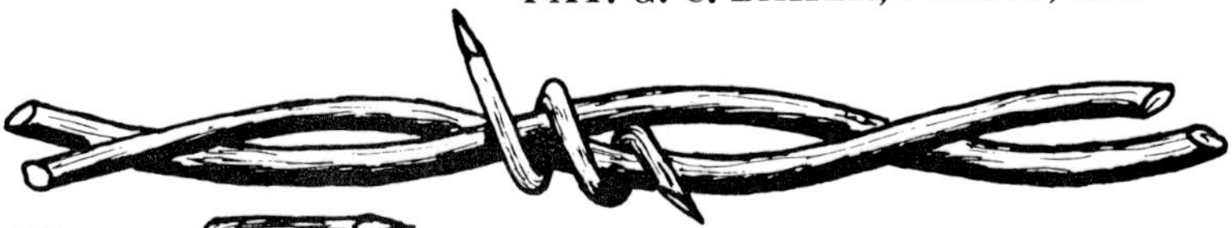

73

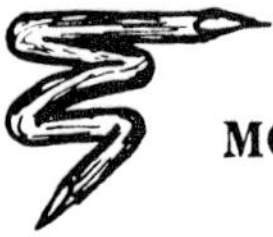

MODERN WIRE WITH LOOSE TWIST BARB

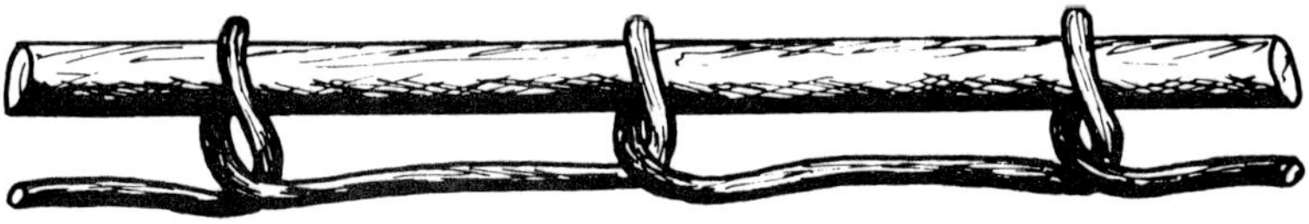

74

SECTION OF FENCE STAY

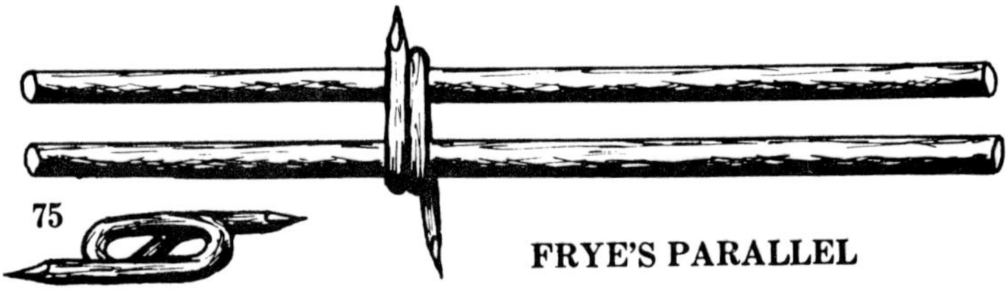

75

FRYE'S PARALLEL

H. H. FRYE PATENTED MAY 28, 1878

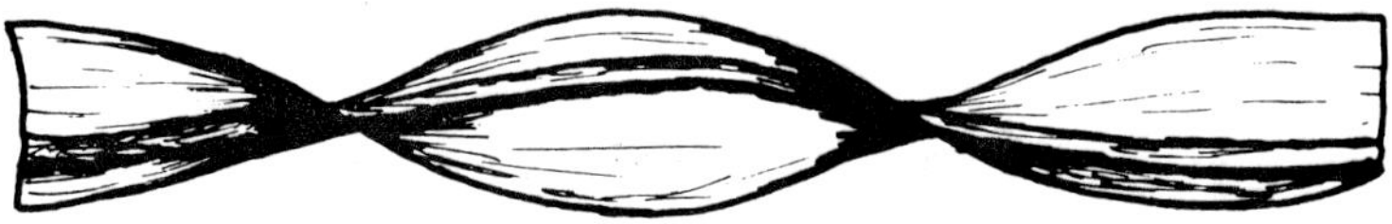

76 T. V. ALLIS BARBLESS. NARROW, MEDIUM AND WIDE WIDTHS. RIBBON WIRE. PAT. JUNE 26, 1881 CENTER RIDGE ON BOTH SIDES.

77

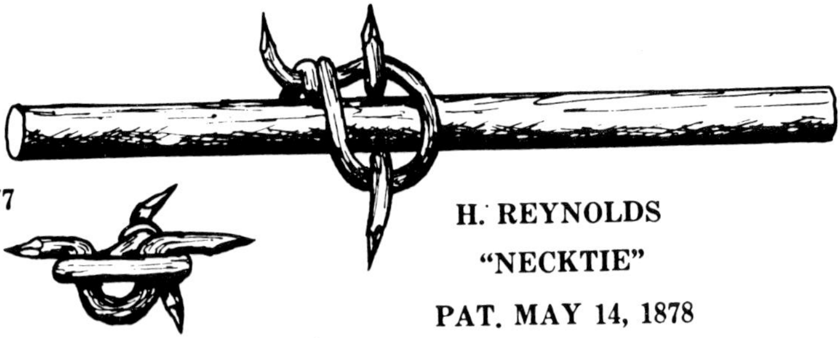

H. REYNOLDS

"NECKTIE"

PAT. MAY 14, 1878

78

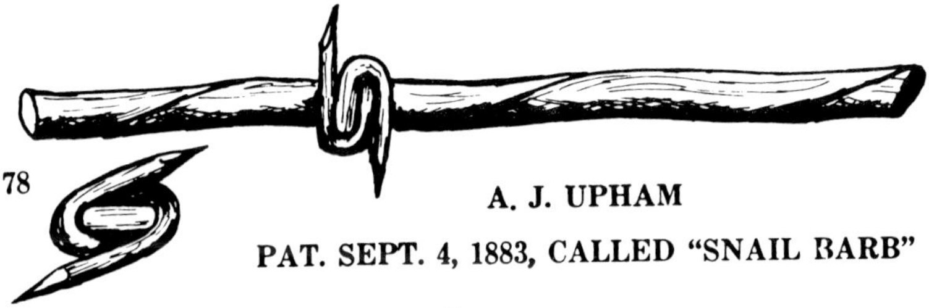

A. J. UPHAM

PAT. SEPT. 4, 1883, CALLED "SNAIL BARB"

79

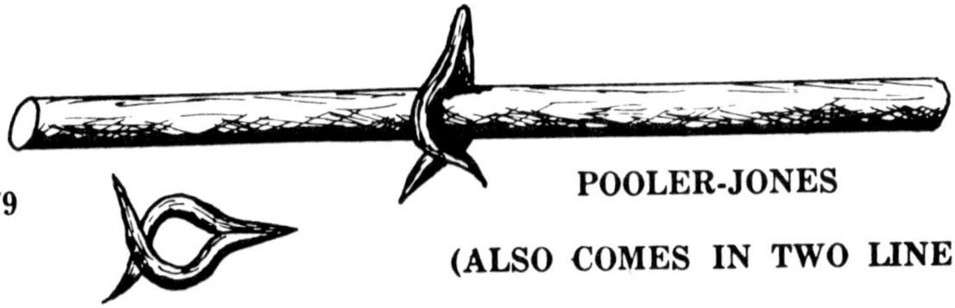

POOLER-JONES

(ALSO COMES IN TWO LINE)

PAT. AUGUST 29, 1876 SOLD BY POUND

80

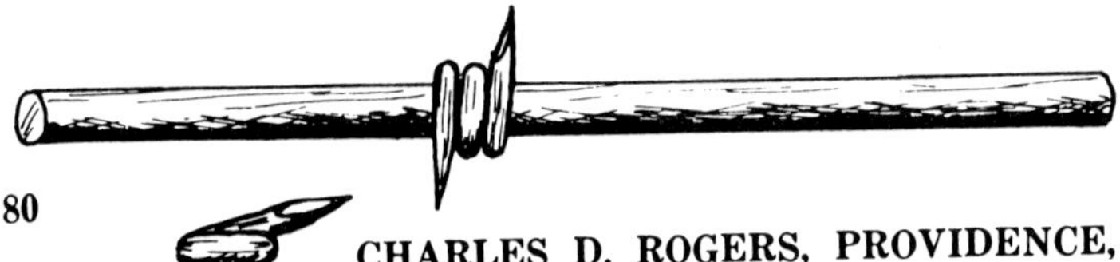

CHARLES D. ROGERS, PROVIDENCE, RHODE ISLAND

PAT. JAN. 10, 1888

81

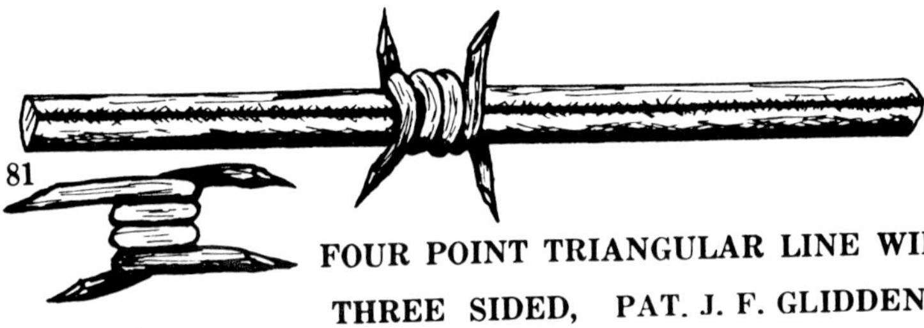

FOUR POINT TRIANGULAR LINE WIRE

THREE SIDED, PAT. J. F. GLIDDEN,

NOV. 24, 1874

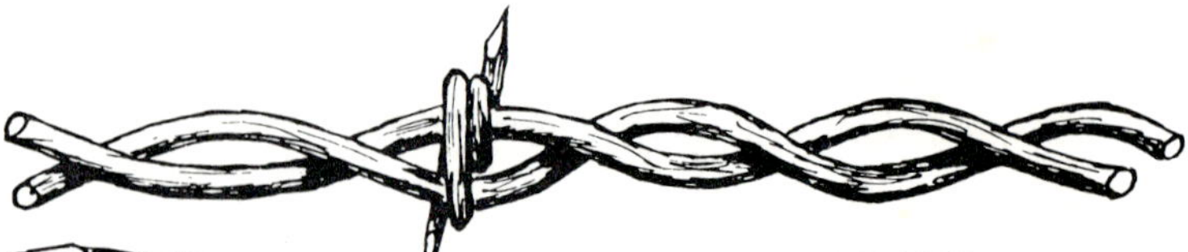

82 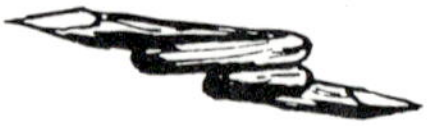

BROTHERTON BARB

PAT. SEPT. 3, 1887

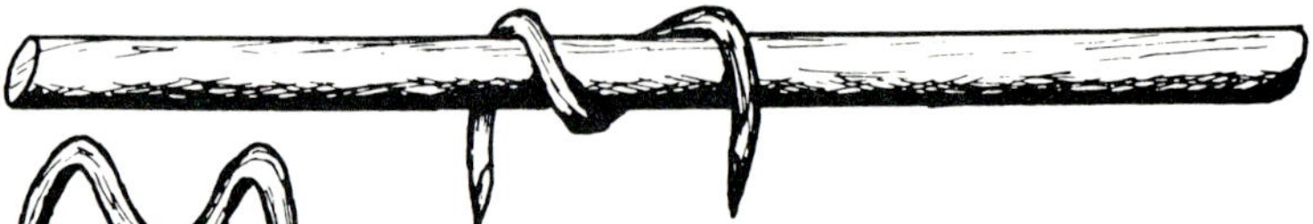

83 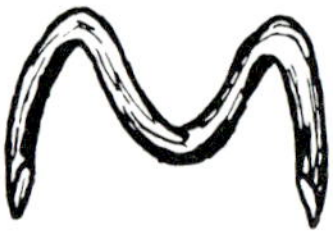

SINGLE LINE WIDE WRAP BARB
M. DALEY, PATENTED OCTOBER 29, 1878

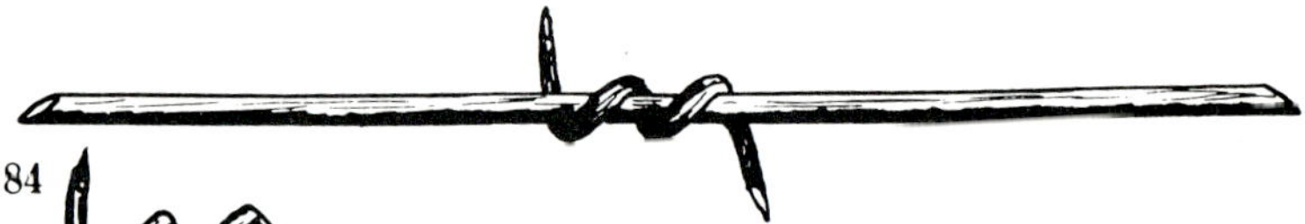

84 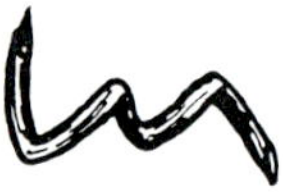

BARBED ELECTRIC FENCE WIRE

PAT. R. S. DOEN, OCT. 20, 1959

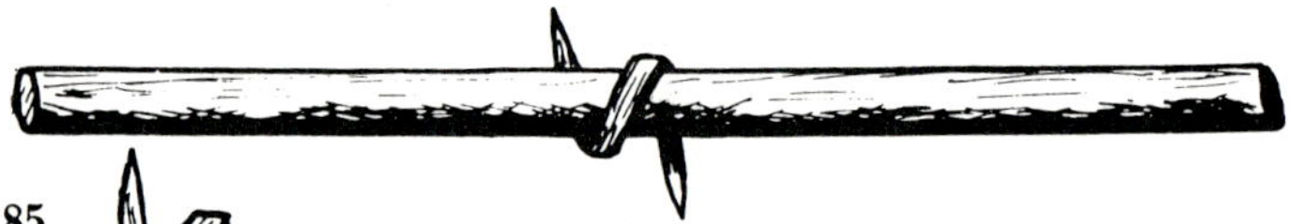

85 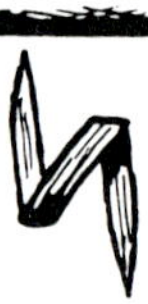

BAKER PERFECT BARB

ON SINGLE SMALL LINE WIRE

PAT. G. C. BAKER, FEB. 27, 1883

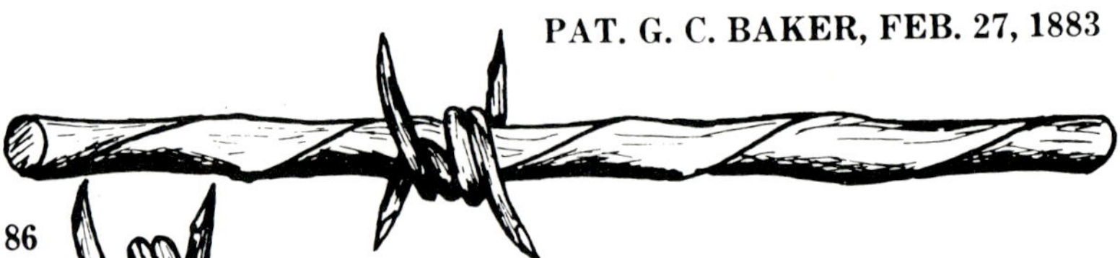

86 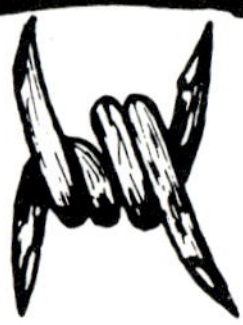

TWIST OVAL FOUR POINT LOCK BARB
M. C. SHINN, PATENTED MARCH 1, 1881

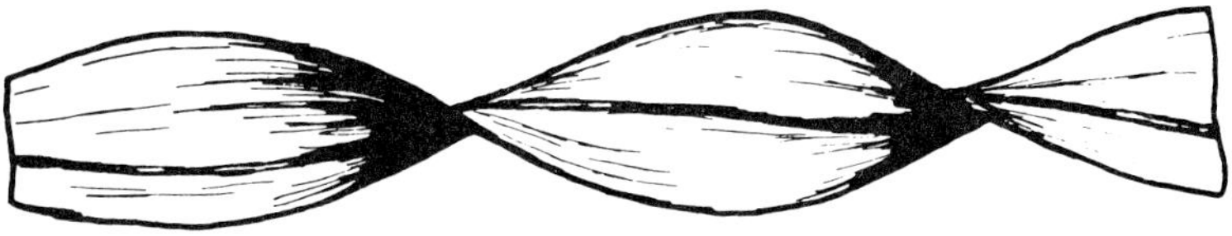

87 CRIMPED CENTER RIBBON WIRE
USED AS FENCING

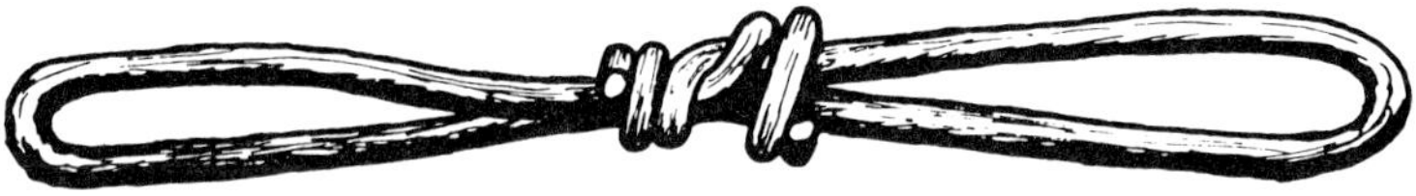

88 CADY MACHINE PATENT
BARBLESS CHAIN LINK

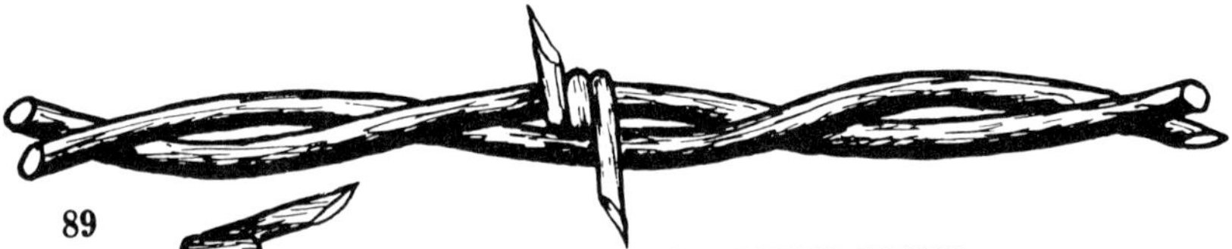

89

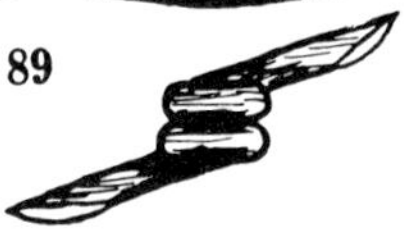

MODERN WIRE
WITH HALF ROUND GLIDDEN BARB
PAT. J. F. GLIDDEN, NOV. 24, 1874

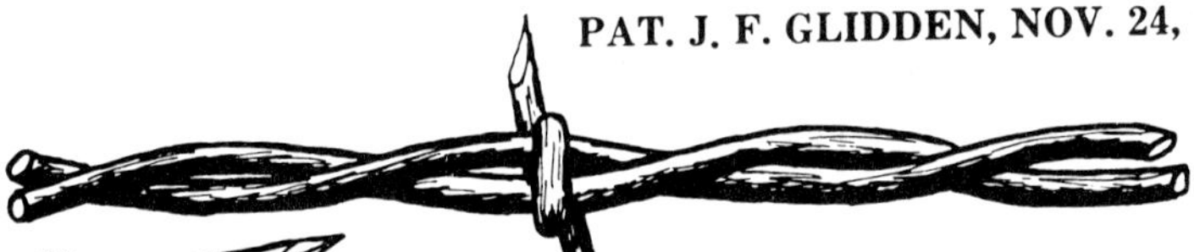

90

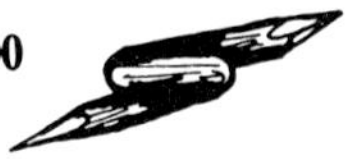

PAT. A. J. UPHAM, AUG. 26, 1876
"1 AROUND 2"

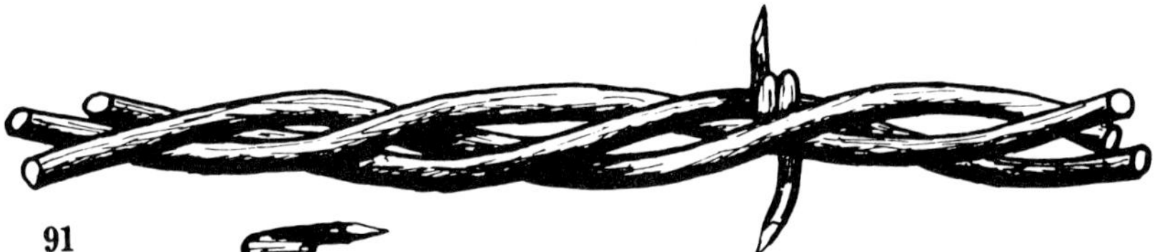

91

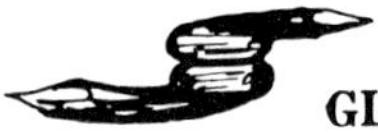

GLIDDEN TYPE BARB, TWO LINES TWISTED
THEN TWISTED WITH THE THIRD
PAT. J. F. GLIDDEN, NOV. 24, 1874

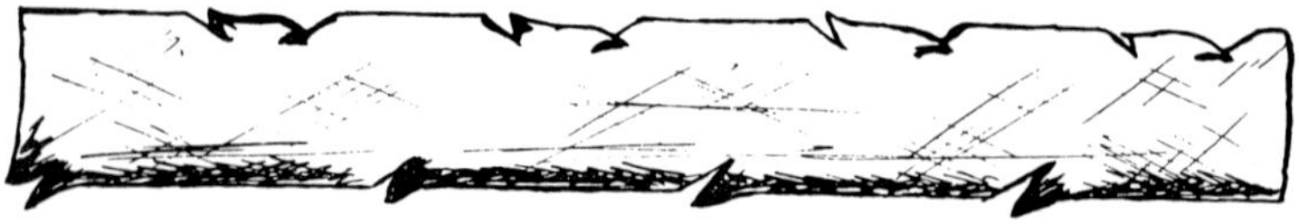

92

L. P. JUDSON NOTCHED RIBBON

PAT. AUG. 15, 1871

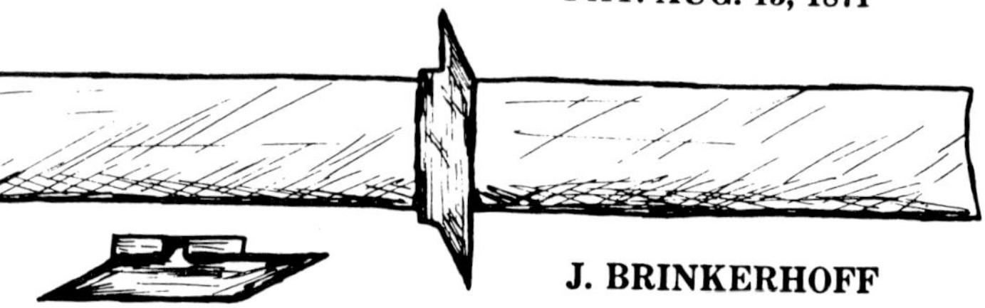

93

J. BRINKERHOFF

PAT. MAY 17, 1881

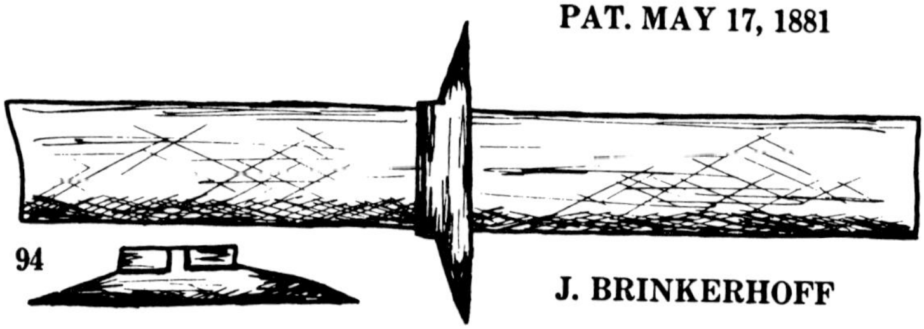

94

J. BRINKERHOFF

PAT. MAY 17, 1881

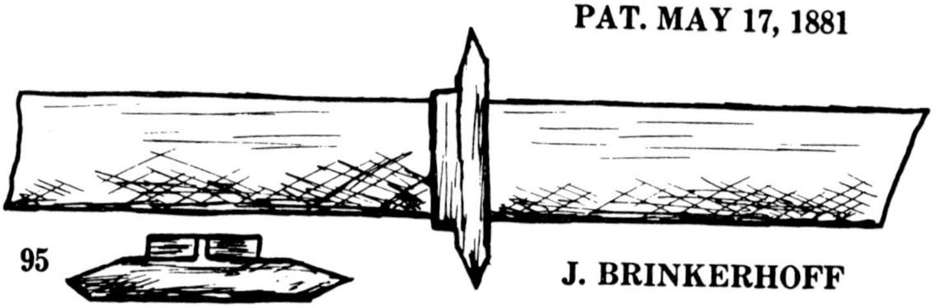

95

J. BRINKERHOFF

PAT. MAY 17, 1881

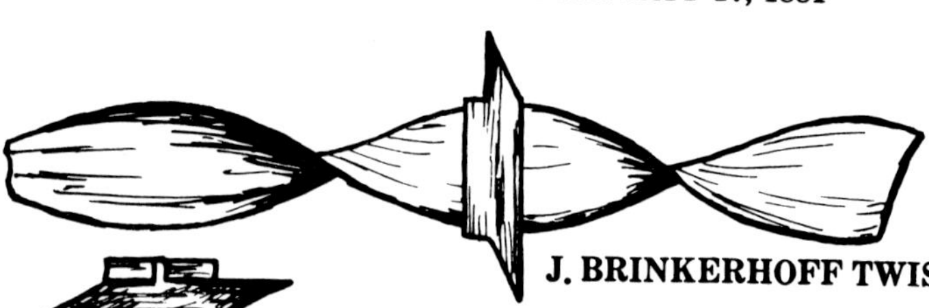

96

J. BRINKERHOFF TWISTED

PAT. APRIL 8, 1879

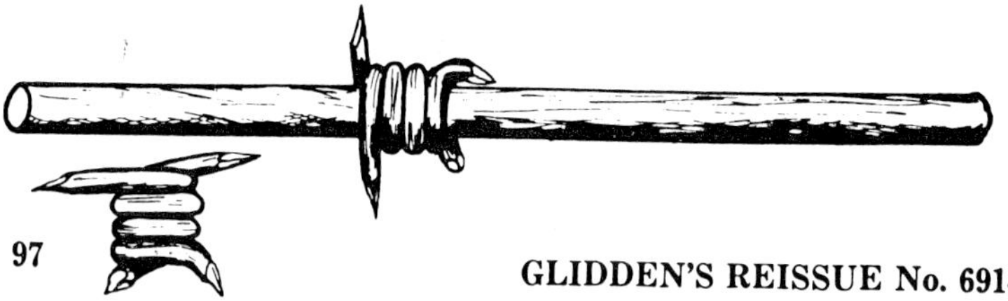

97

GLIDDEN'S REISSUE No. 6914

PAT. J. F. GLIDDEN, FEB. 8, 1876

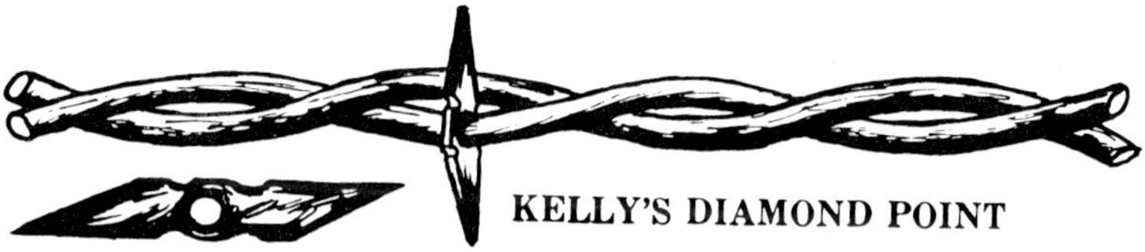

98

KELLY'S DIAMOND POINT

WITH CRIMPED ON BARB

PAT. FEB. 11, 1868

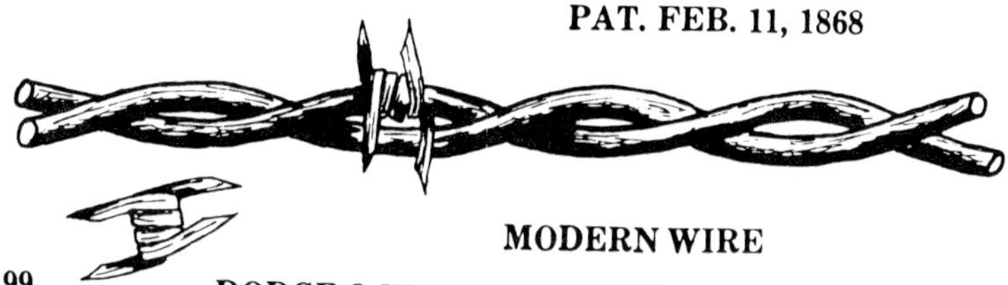

99

MODERN WIRE

DODGE & WASHBURN TYPE HALF ROUND BARB

PAT. JAN. 4, 1872

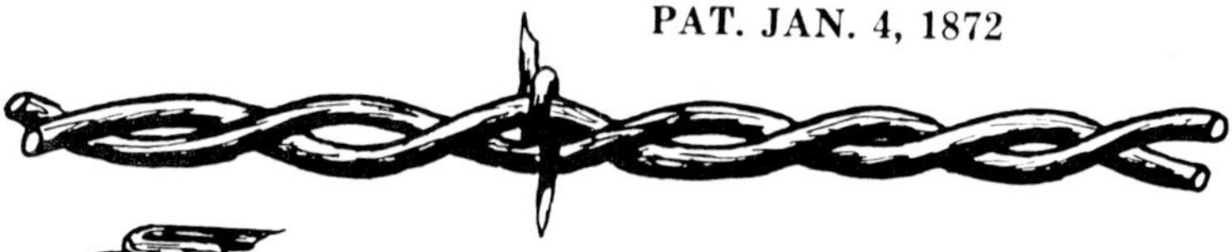

100

A. J. UPHAM STAPLE BARB ON TWO LINES

PAT. AUG. 29, 1876

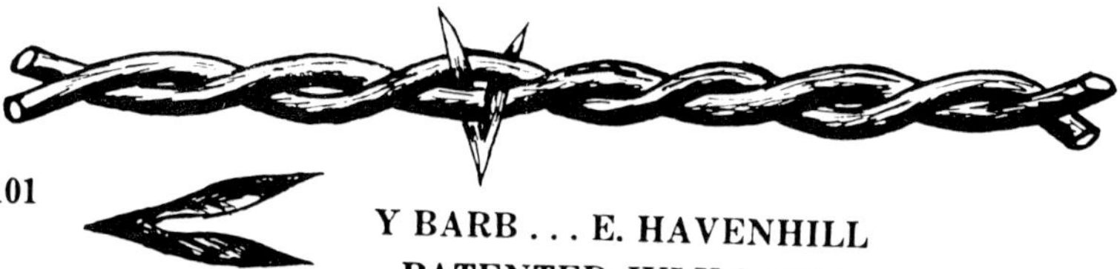

101

Y BARB . . . E. HAVENHILL

PATENTED JULY 2, 1878

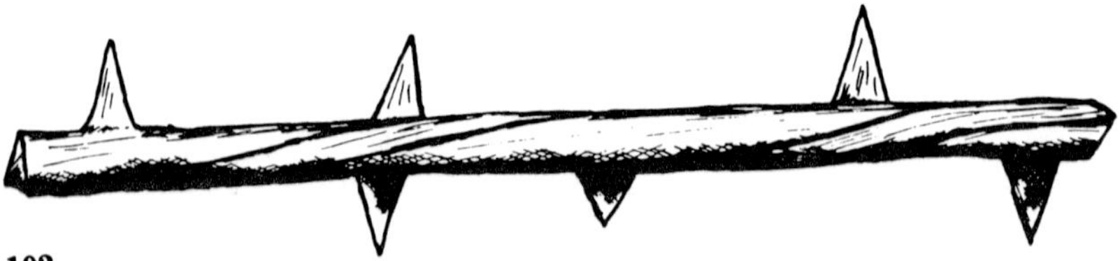

102

E. SIMS, PATENTED MAY 30, 1876

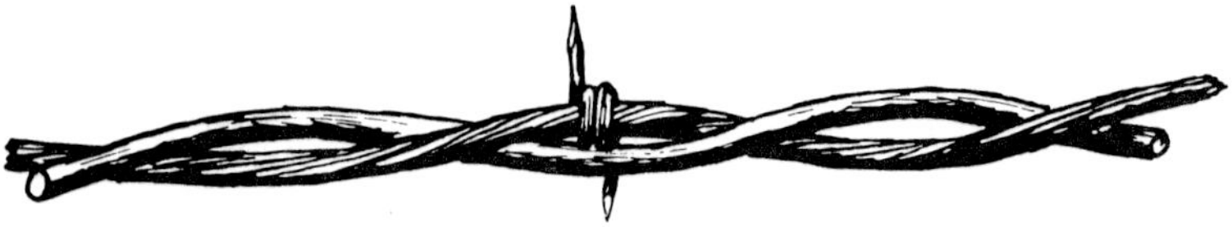

103 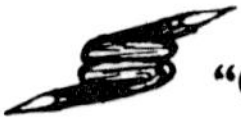 "CABLE WIRE" SEVEN STRAND CABLE TWISTE

WITH LINE WIRE. BARB IS ON CABLE

PAT. J. F. GLIDDEN, NOV. 24, 1874

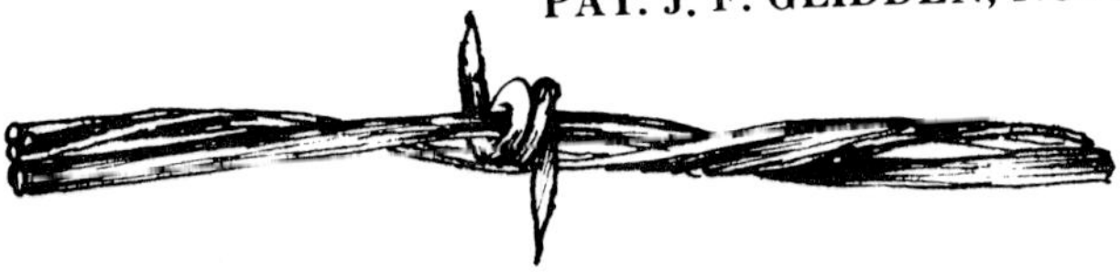

104

J. GLIDDENS' THREE STRAND CABLE WIRE

PAT. FEB. 8, 1876, BARB ON LARGE LINE WIRE

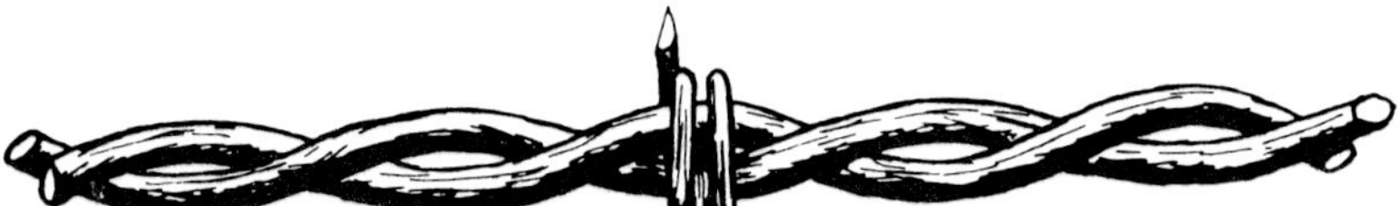

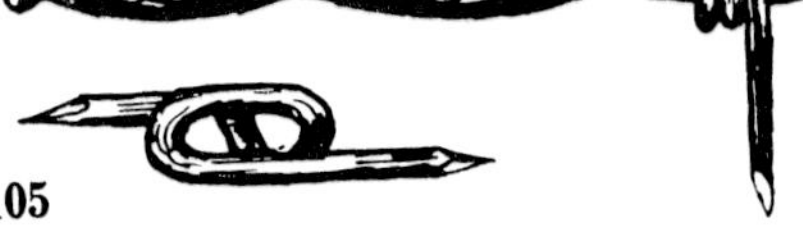

FRYE'S TWIST

105

PATENTED MAY 28, 1878

H. H. FRYE

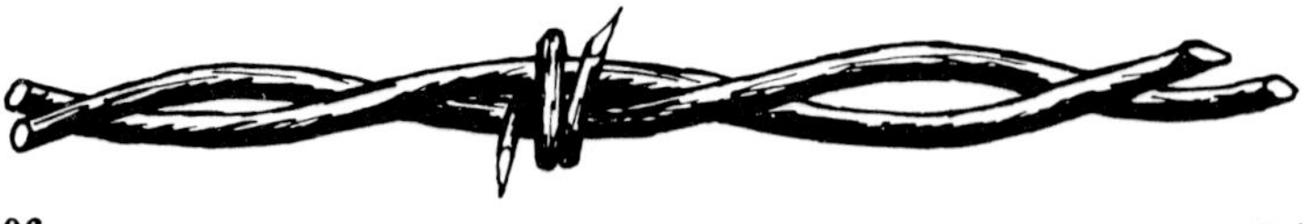

106 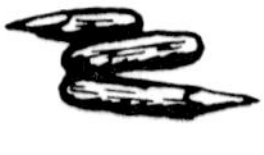 BROTHERTON $2\frac{1}{4}$ INCH BETWEEN BARBS

"HOG WIRE" PAT. SEPT. 3, 1878

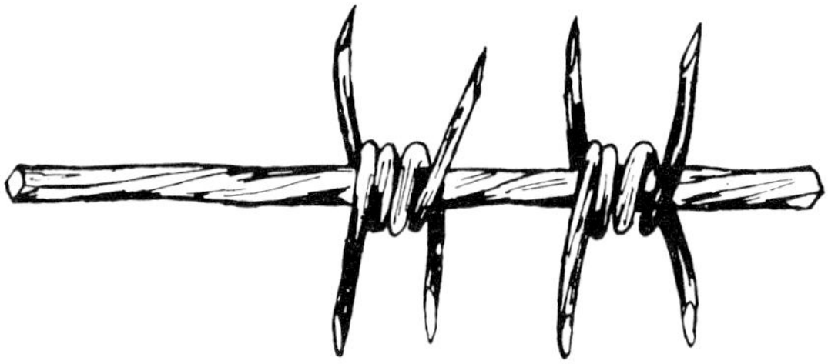

107

GERMAN W.W. I ENTANGLEMENT WIRE

SQUARE TWISTED LINE

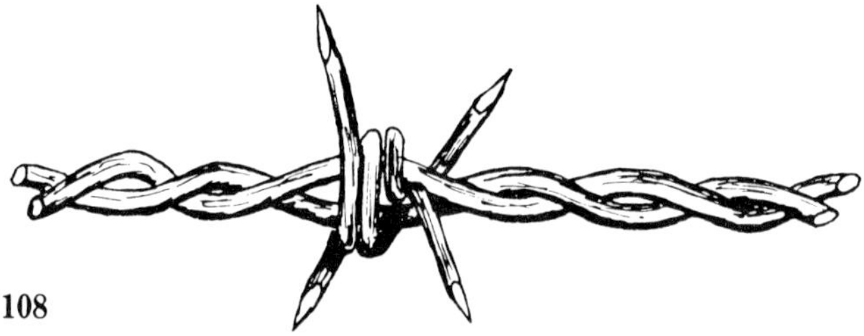

108

EXTRA LARGE KOREAN WAR WIRE

RESEMBLES BURNELL'S FOUR POINT

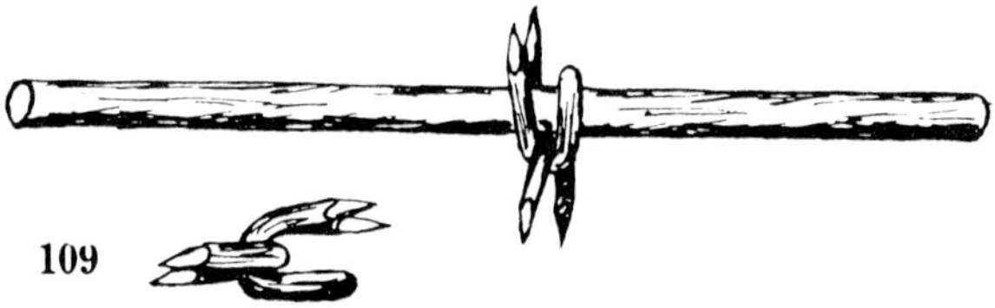

109

LEWIS T. LARSON CALLED WING

PATENTED FEBRUARY 26, 1878

110

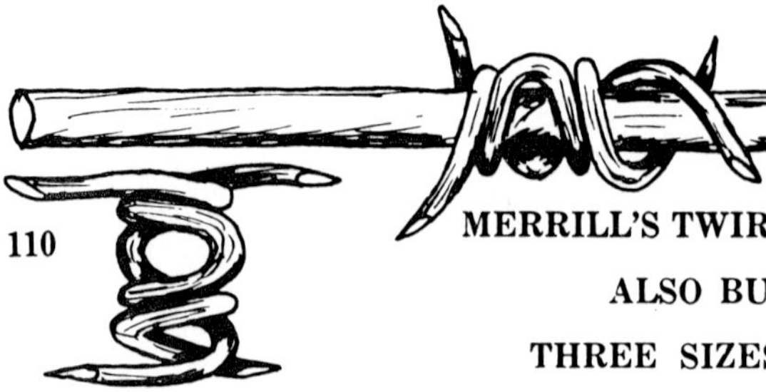

MERRILL'S TWIRL:CALLED HOLDFAST

ALSO BUFFALO WIRE

THREE SIZES OF LINE WIRE

PAT. DEC. 26, 1876

111

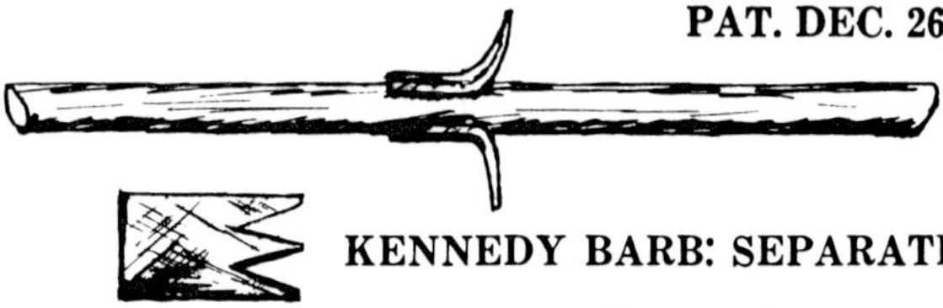

KENNEDY BARB: SEPARATE BARBS TO B

INSTALLED ON SMOOTH WIRE FENCE

PAT. AUG. 11, 1878

112

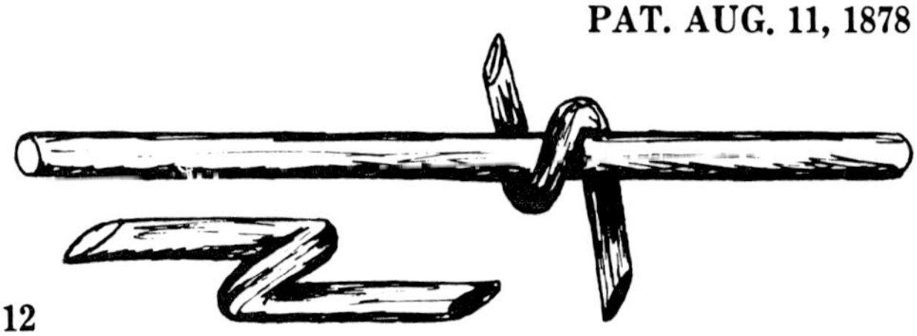

PROBABLY HOME MADE STAPLE WIRE

LARGE, MEDIUM & SMALL

113

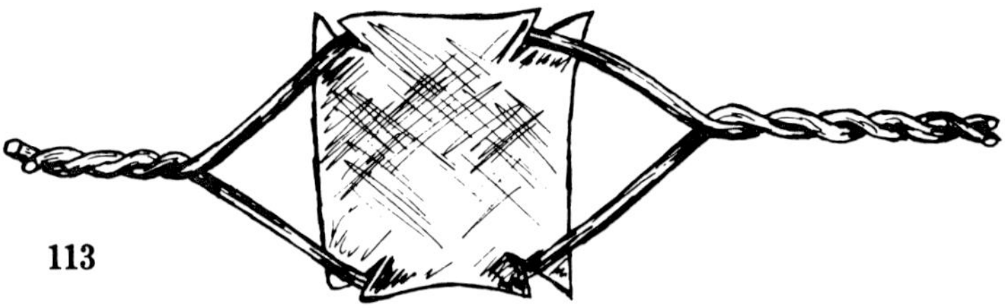

STUBBE PLATE. LARGE & SMALL

PAT. OCT. 23, 1883

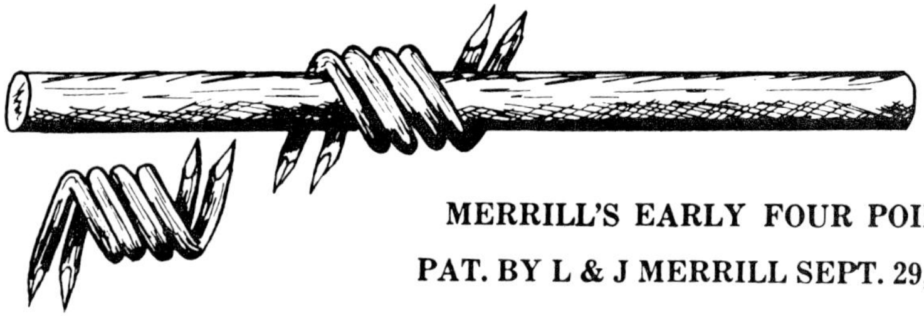

MERRILL'S EARLY FOUR POINT

PAT. BY L & J MERRILL SEPT. 29, 1871

114

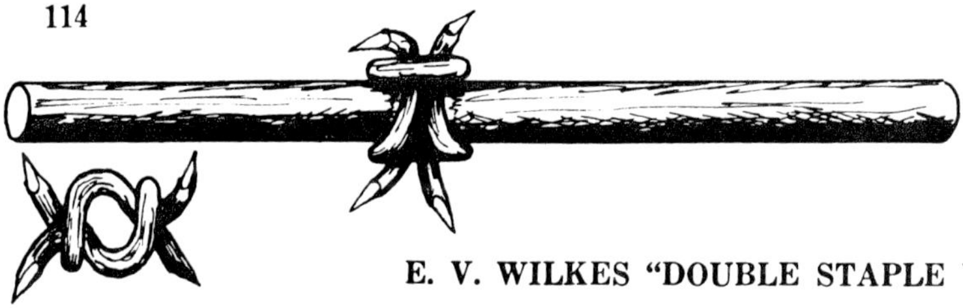

E. V. WILKES "DOUBLE STAPLE WIRE"

PAT. JUNE 17, 1879

115

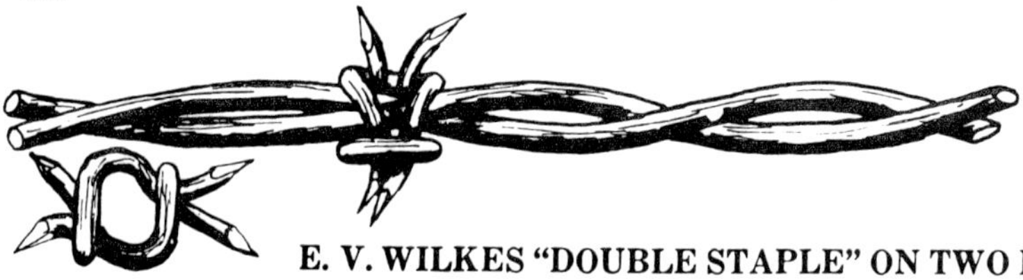

E. V. WILKES "DOUBLE STAPLE" ON TWO LINES

PAT. JUNE 17, 1879

116

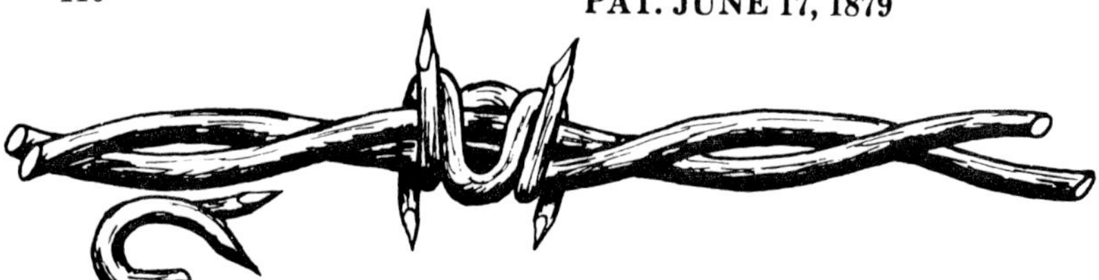

MERRILL'S TWIRL

CALLED HOLDFAST ON TWO LINES

PAT. DEC. 26, 1876

117

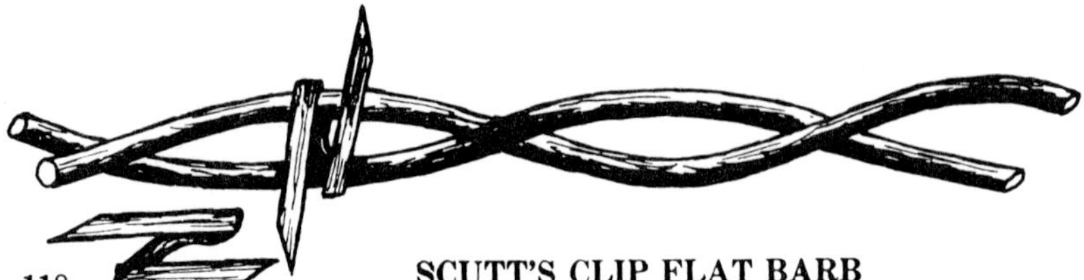

118

SCUTT'S CLIP FLAT BARB
HIRAM B. SCUTT
PATENTED DECEMBER 22, 1885

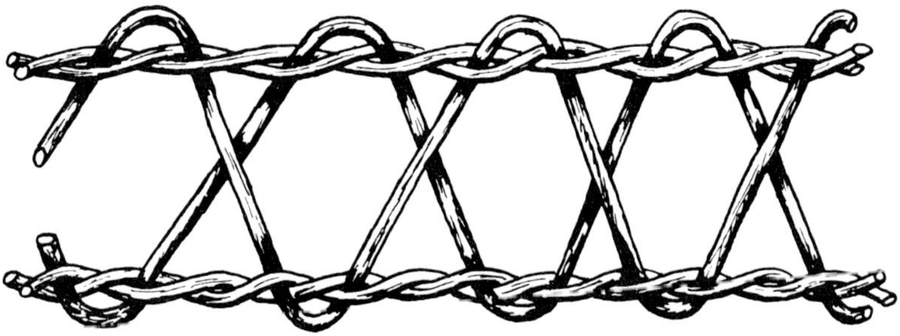

119

ORNAMENTAL FENCE WIRE

PAT. W. E. HATHAWAY — A. B. WOODARD

AUG. 12, 1890

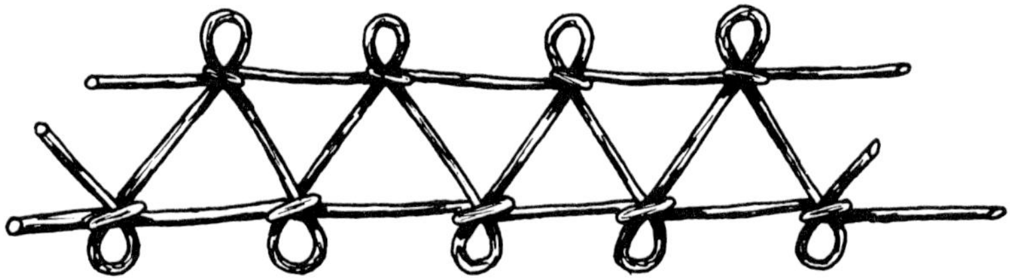

120

ORNAMENTAL FENCE WIRE

PAT. F. J. INGRAHAM, FEB. 16, 1892

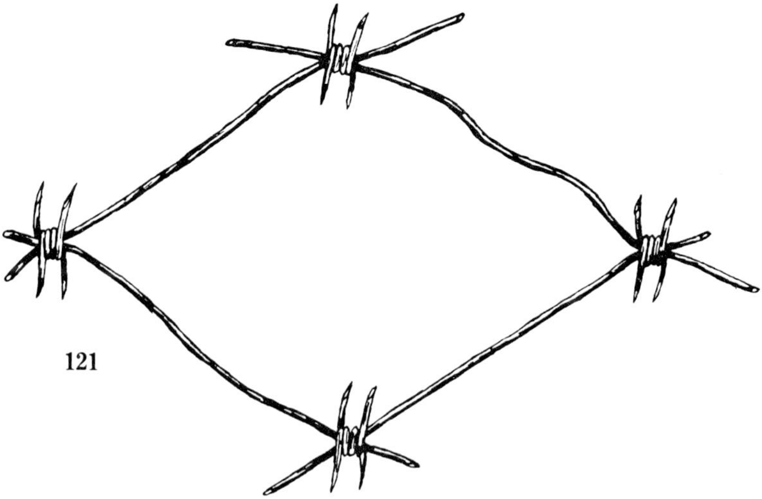

121

BARBED NET WIRE ONLY 12 IN. HIGH USED FOR

HOG PASTURES

PAT. DODGE & WASHBURN

JAN. 4, 1882

122 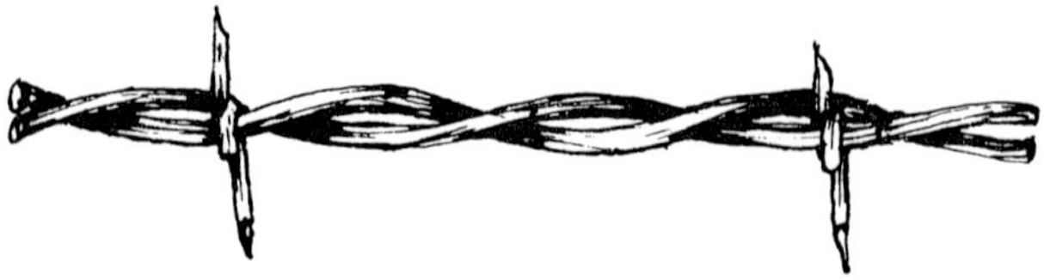

PURCHES MILES STAPLE BARB ON BOTH LINES
PAT. OCT. 8, 1878

123 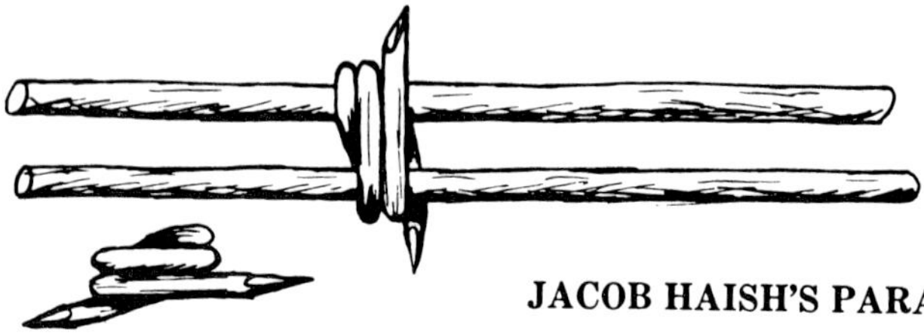

JACOB HAISH'S PARALLEL

PAT. AUG. 31, 1885

124 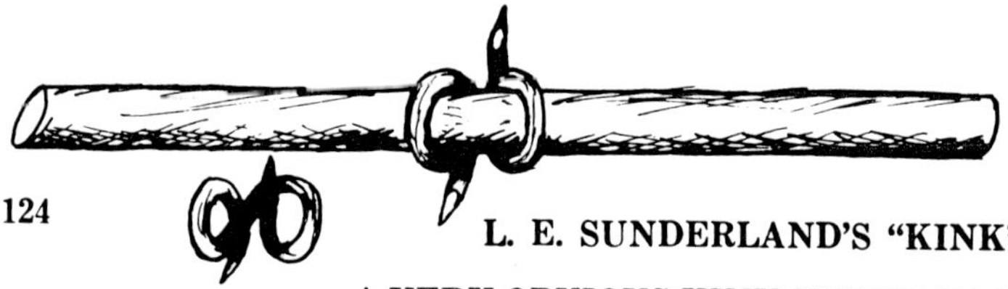

L. E. SUNDERLAND'S "KINK"

A VERY OBVIOUS KINK UNDER EACH BARB

PAT. AUG. 12, 1884

125 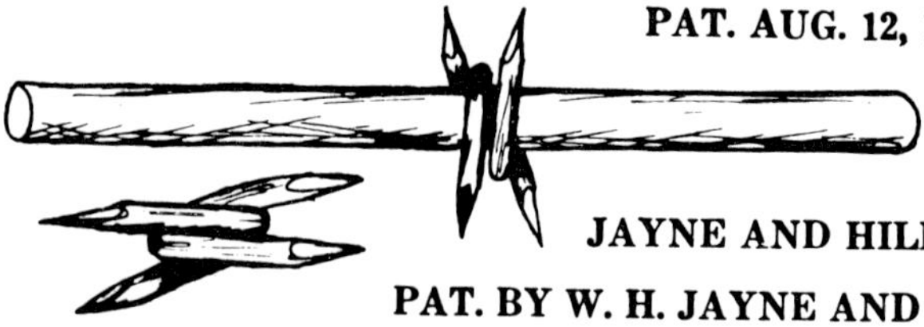

JAYNE AND HILL

PAT. BY W. H. JAYNE AND J. H. HILL

APRIL 11, 1876

126 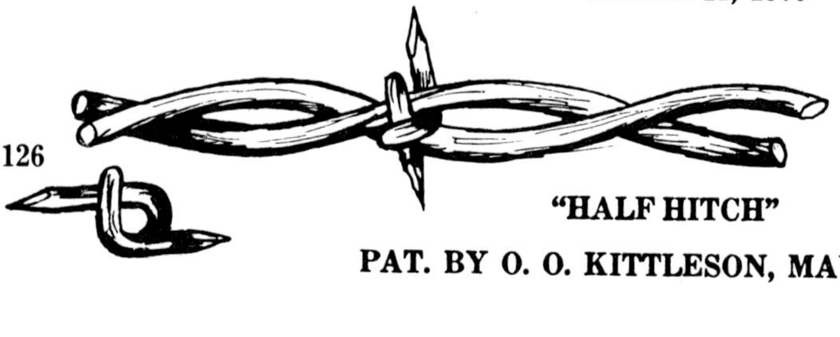

"HALF HITCH"

PAT. BY O. O. KITTLESON, MAY 7, 1878

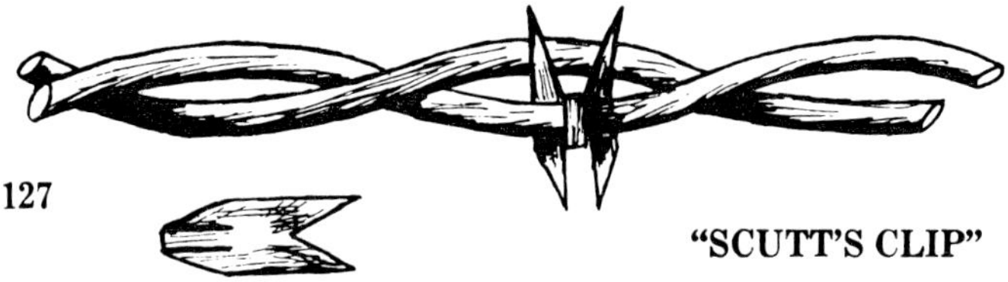

127

"SCUTT'S CLIP"

PAT. BY H. B. SCUTT JUNE 18, 1878

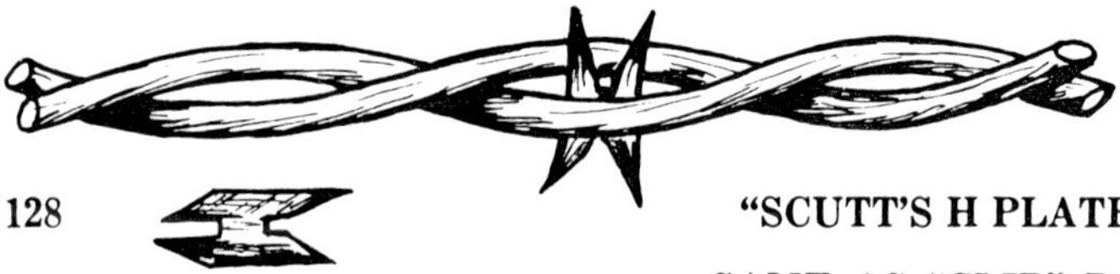

128

"SCUTT'S H PLATE"

SAME AS "CLIP" PAT.

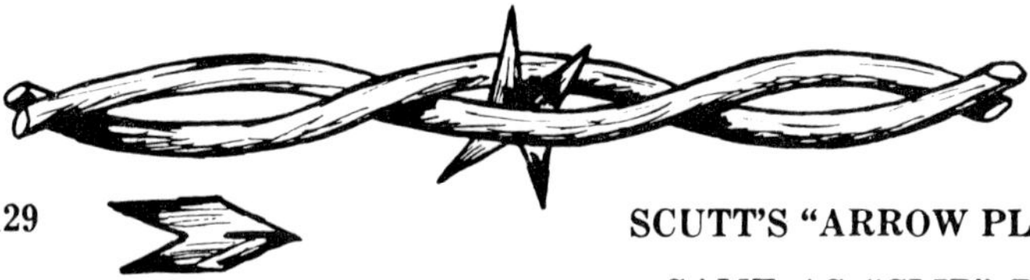

129

SCUTT'S "ARROW PLATE"

SAME AS "CLIP" PAT.

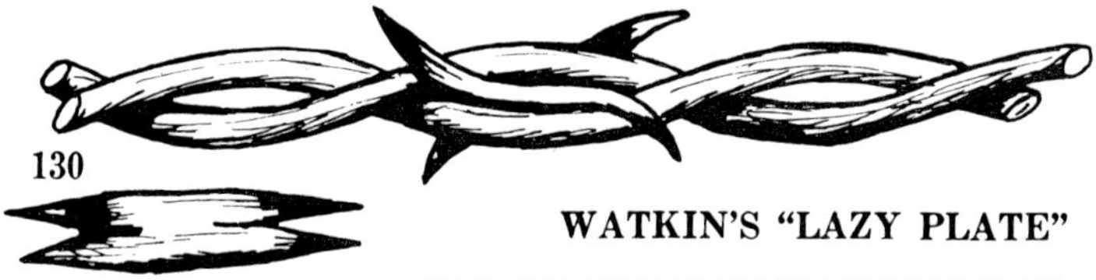

130

WATKIN'S "LAZY PLATE"

PAT. BY WILLIAM WATKINS NOV. 21, 1876

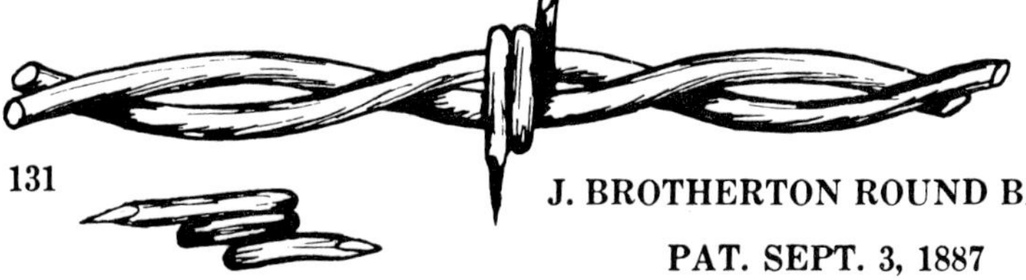

131

J. BROTHERTON ROUND BARB

PAT. SEPT. 3, 1887

132

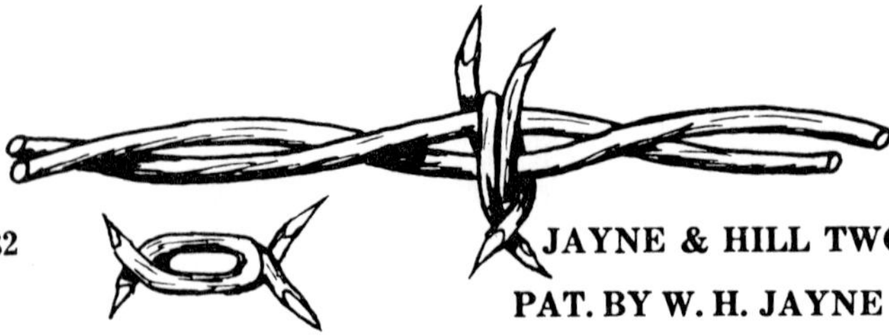

JAYNE & HILL TWO STRAND
PAT. BY W. H. JAYNE & J. H. HILL
APRIL 11, 1876. NO BARBS "BETWEEN"

133

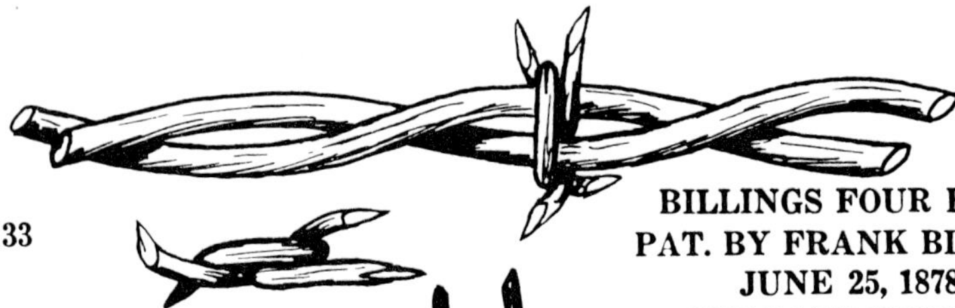

BILLINGS FOUR POINT
PAT. BY FRANK BILLINGS
JUNE 25, 1878
ONE BARB "BETWEEN"

134

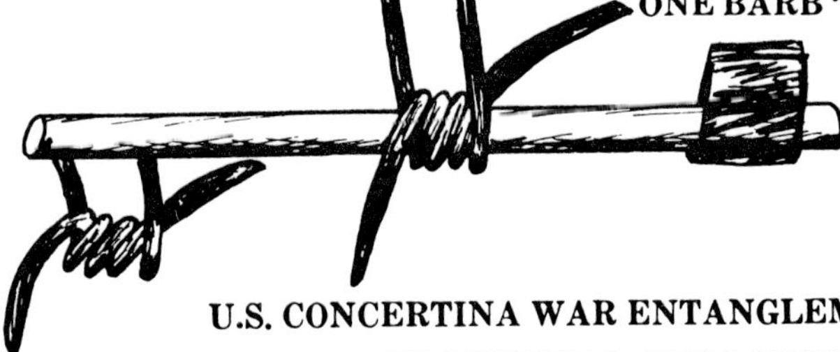

U.S. CONCERTINA WAR ENTANGLEMENT WIRE.
HARD STEEL LINE WAR. STILL IN PRODUCTION.

135

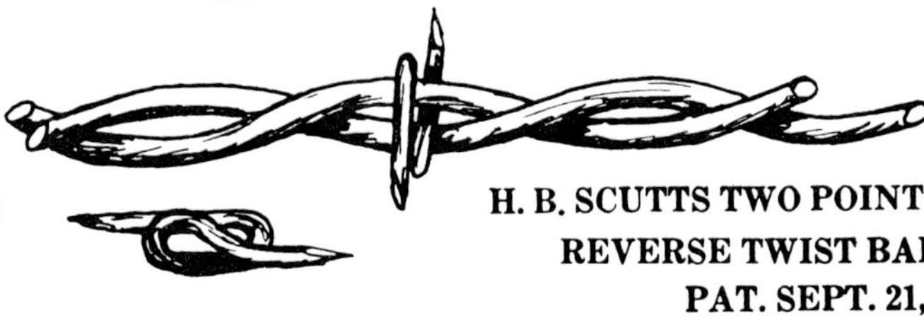

H. B. SCUTTS TWO POINT
REVERSE TWIST BARB
PAT. SEPT. 21, 1880

136

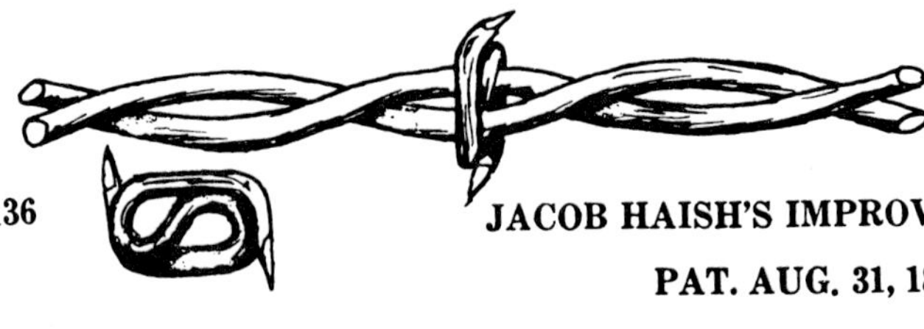

JACOB HAISH'S IMPROVED S BARB
PAT. AUG. 31, 1875

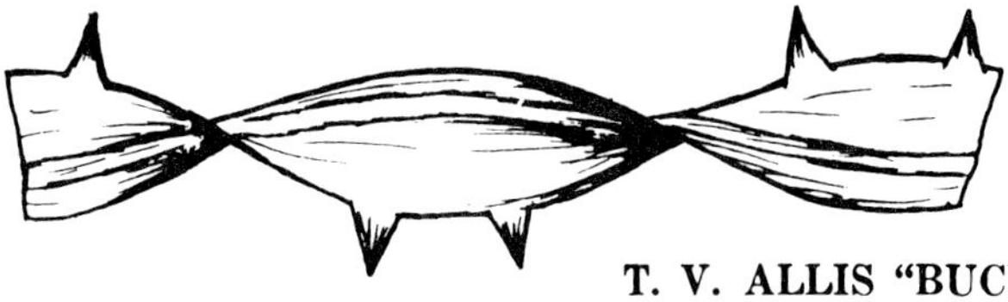

137

T. V. ALLIS "BUCKTHORN"

WIDE, MEDIUM, NARROW WIDTHS

PAT. JULY 26, 1881

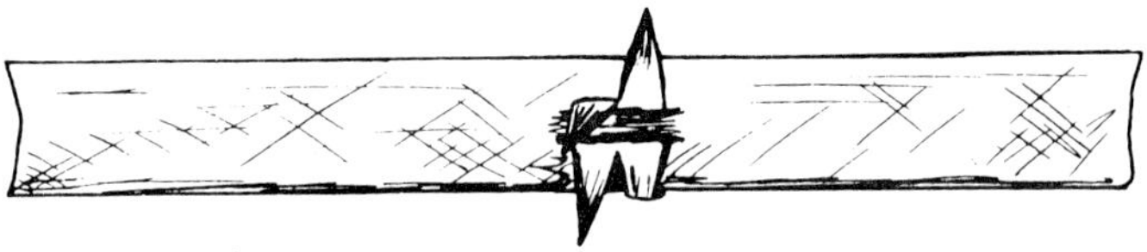

138 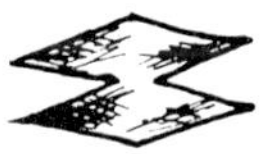

I. L. ELLWOOD RIBBON WIRE

PAT. FEB. 24, 1874

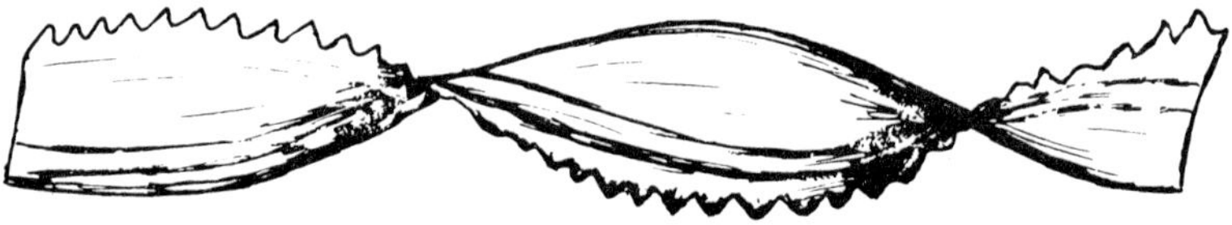

139

PAT. T. V. ALLIS

JULY 26, 1881

SAW TOOTH RIBBON WIRE

TEETH VARY IN SIZE AND DEPTH

PROBABLY ALLIS MANUFACTURE

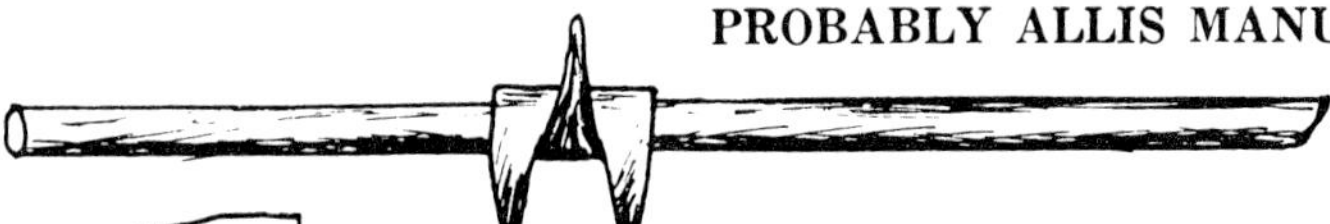

140 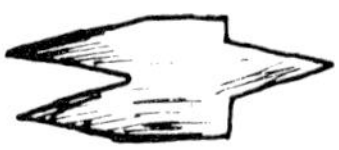

MILLIS KNICKERBOCKER SEPARATE BARB

SOLD BY POUND

PAT. DEC. 12, 1876

141

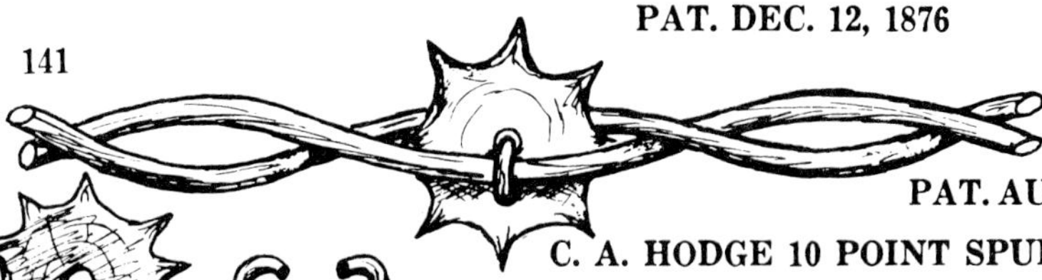

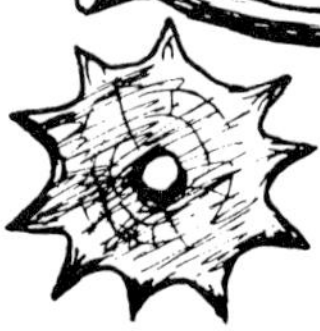

PAT. AUG. 2, 1887

C. A. HODGE 10 POINT SPUR ROWEL

MIGHT COME IN 8 POINT

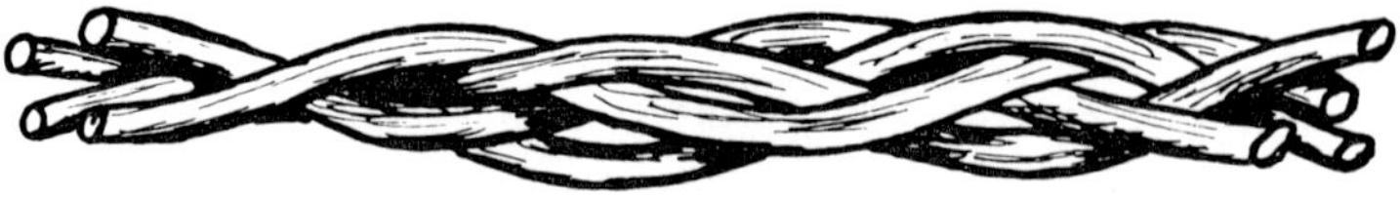

142

"FLEMISH FOUR BRAID"

PAT. J. B. CLEAVELAND, MAR. 20, 1894

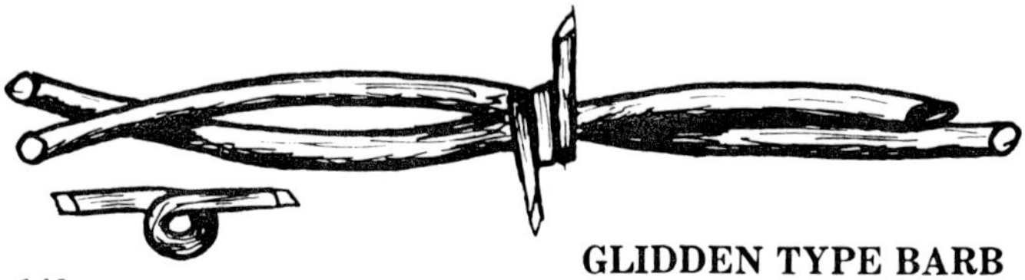

143

GLIDDEN TYPE BARB

SQUARE WIRE USED FOR BARB

PAT. J. F. GLIDDEN, NOV. 24, 1874

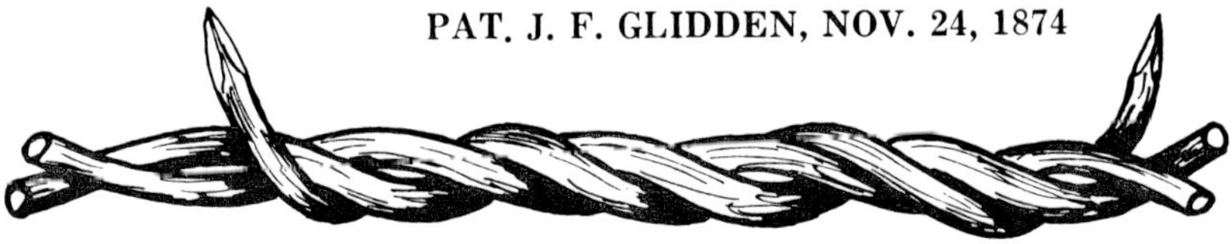

144

M. P. MIGHELL "2 AWHILE 3 AWHILE"

PAT. FEB. 5, 1878

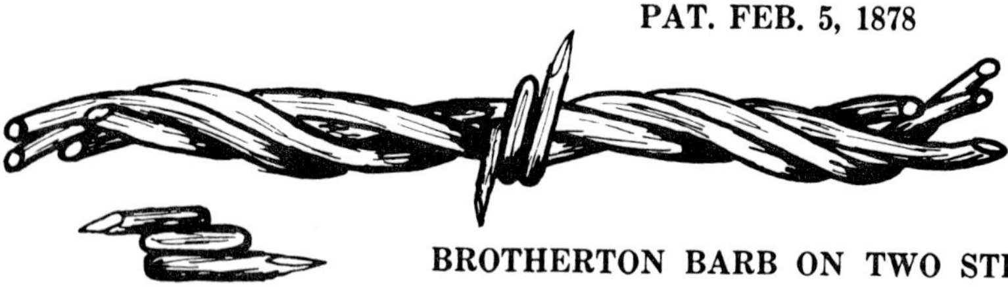

145

BROTHERTON BARB ON TWO STRAND TWISTED WITH TWO SMOOTH STRAND

PAT. J. BROTHERTON, SEPT. 3, 1887

146

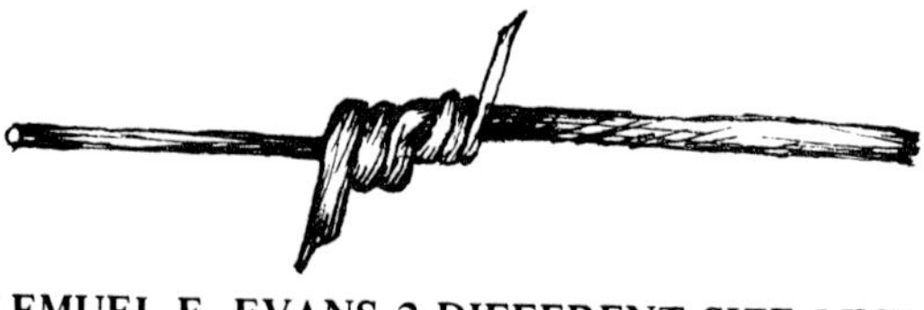

LEMUEL E. EVANS 2 DIFFERENT SIZE LINE THAT ALTERNATE. PAT. OCT. 24, 1876

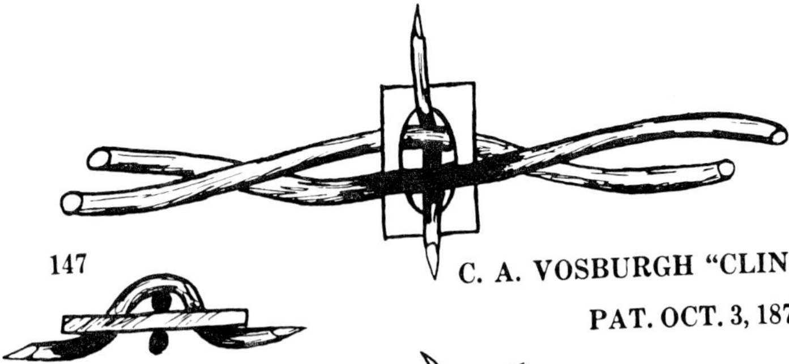

147

C. A. VOSBURGH "CLINCH WIRE"
PAT. OCT. 3, 1876

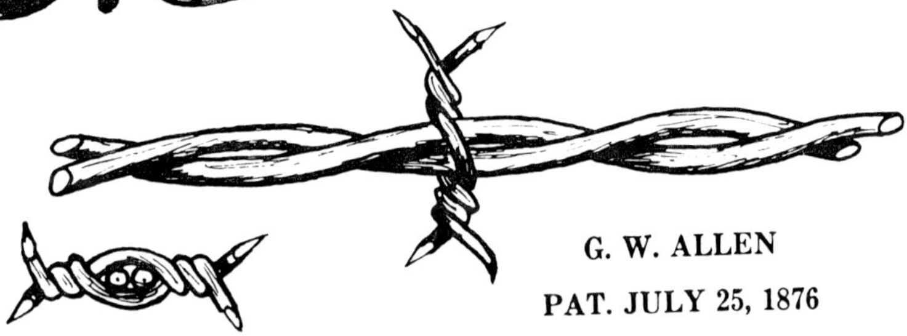

G. W. ALLEN
PAT. JULY 25, 1876

148

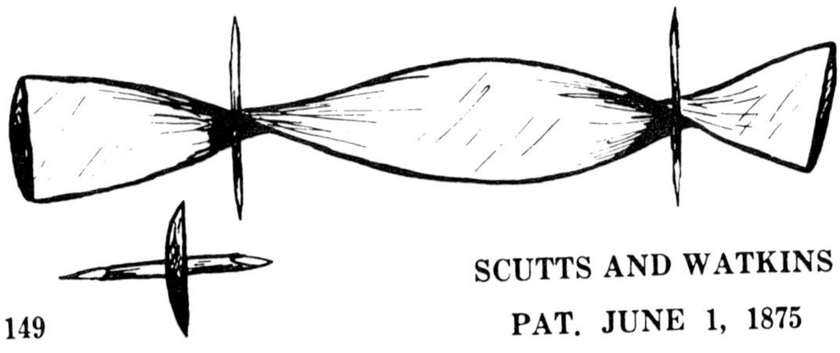

SCUTTS AND WATKINS
PAT. JUNE 1, 1875

149

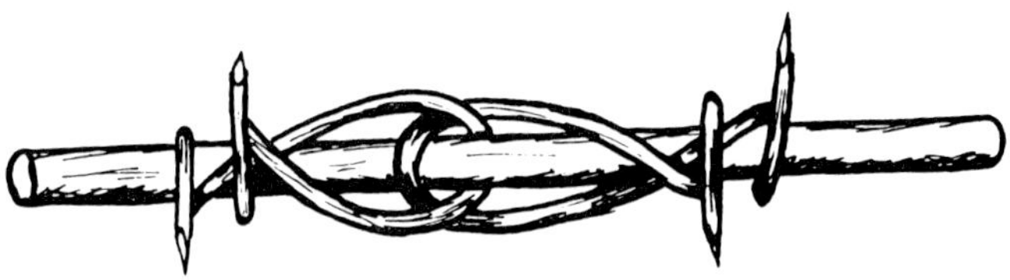

BARBED WIRE FROM OLD MEXICO

150

151

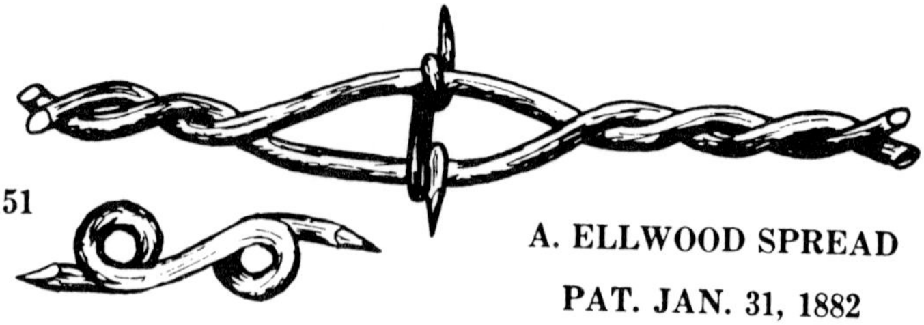

A. ELLWOOD SPREAD

PAT. JAN. 31, 1882

152

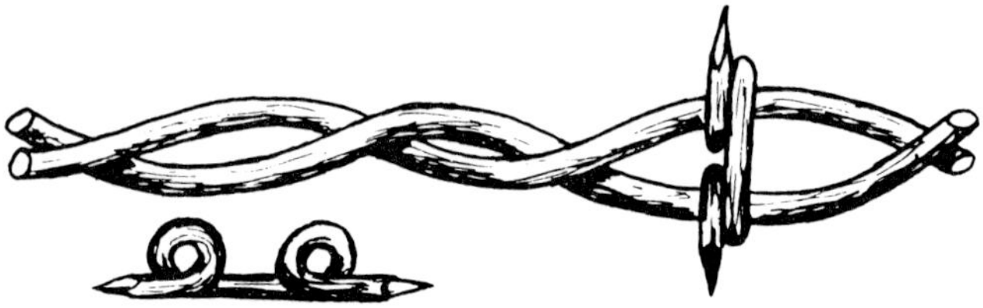

A. C. DECKER'S SPREAD

PAT. JUNE 3, 1884

153

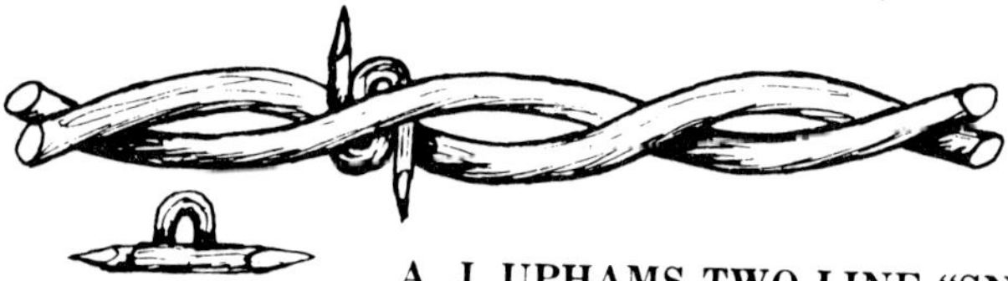

A. J. UPHAMS TWO LINE "SNAIL BARB"

PAT. SEPT. 4, 1883

154

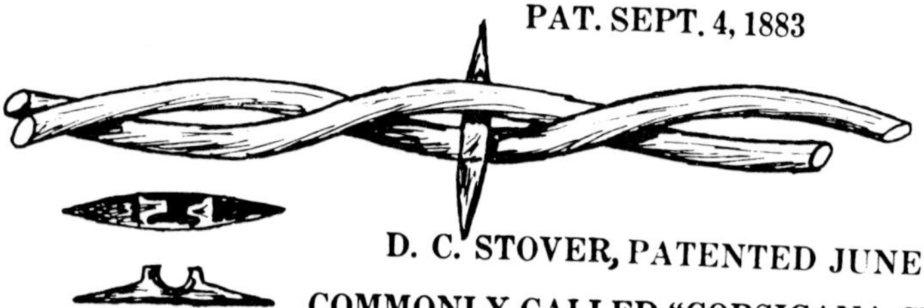

D. C. STOVER, PATENTED JUNE 29, 1875

COMMONLY CALLED "CORSICANA CLIP"

155

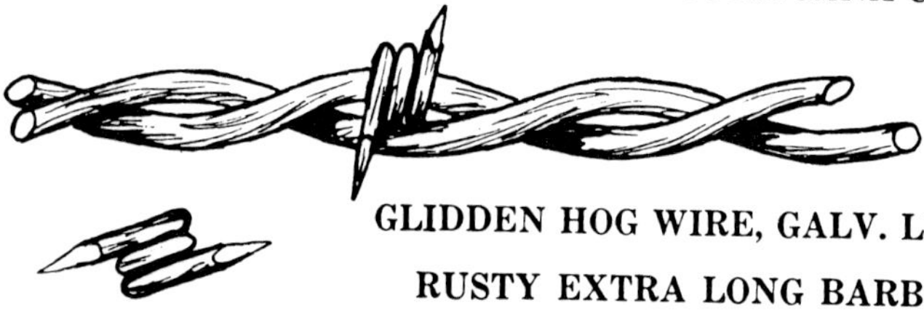

GLIDDEN HOG WIRE, GALV. LINE

RUSTY EXTRA LONG BARBS.

PAT. J. F. GLIDDEN, NOV. 24, 1874

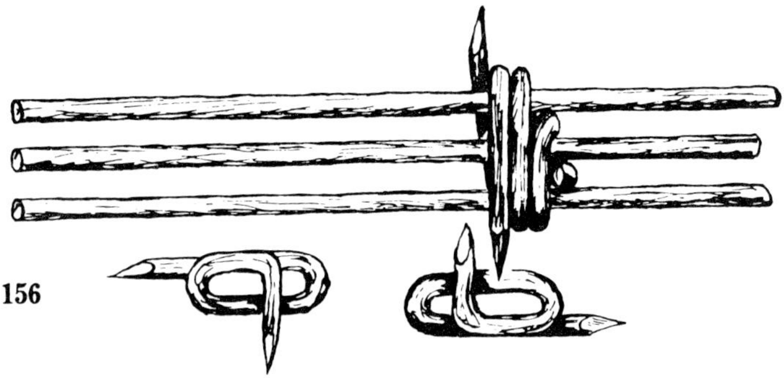

156

CALLED CLINE'S RAIL

J. B. CLINE PATENTED DECEMBER 25, 1883

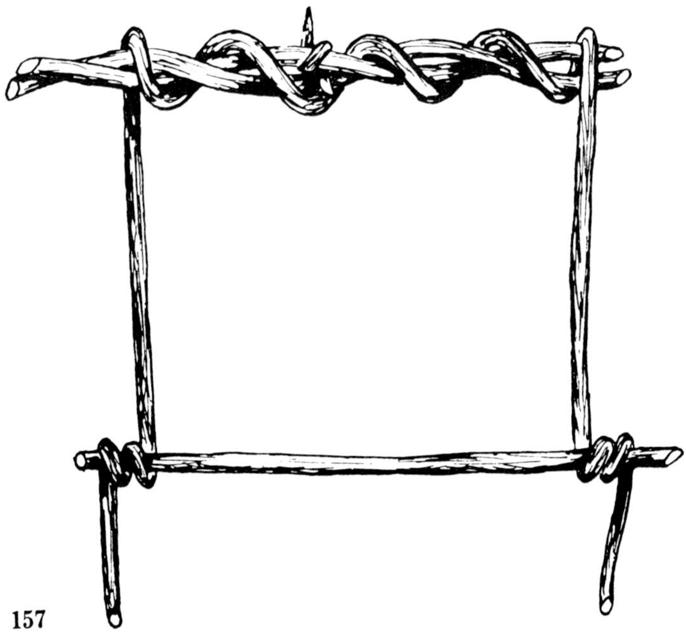

157

HOG WIRE WITH BARB WIRE TOP STRAND

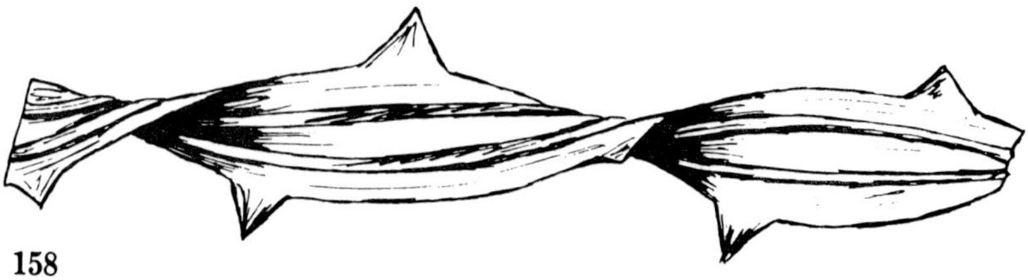

158

PLASTIC ARMORED FENCE WIRE MODERN RESEMBLES BUCKTHORN EXCEPT BARBS ON BOTH SIDES. HAS A TWO WIRE CORE

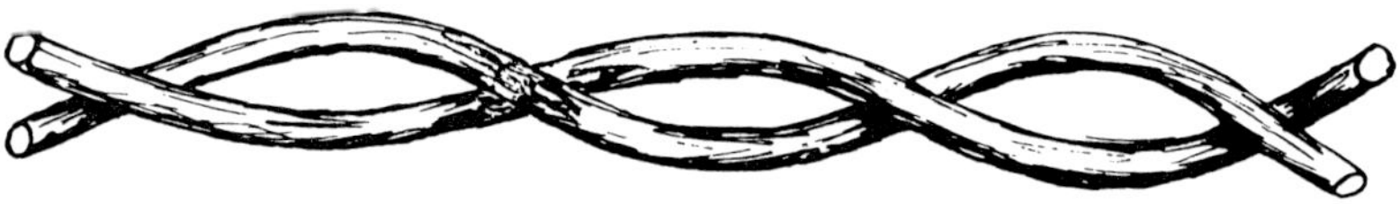

159

LEFT HAND TWIST
TWISTED COUNTER CLOCKWISE

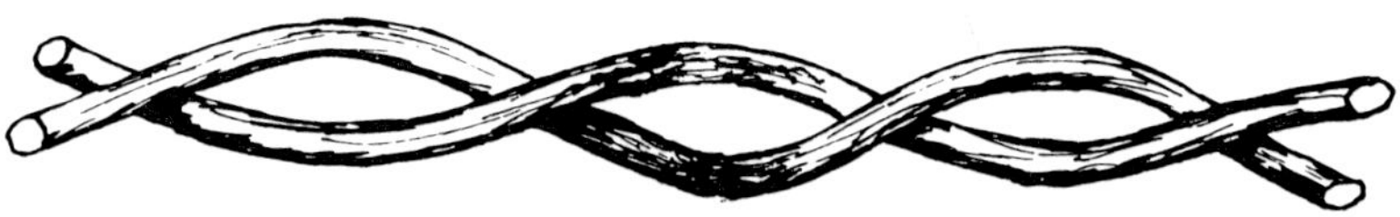

160

RIGHT HAND TWIST. TWISTED CLOCKWISE.

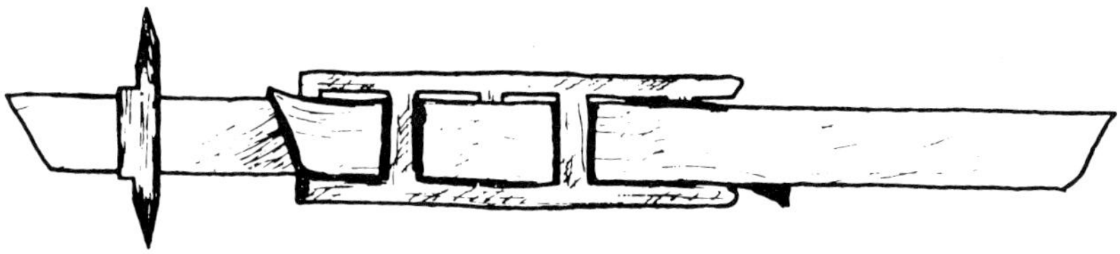

161 BRINKERHOFF BELT BUCKLE SPLICER

PAT. J. BRINKERHOFF, MAY 17, 1881

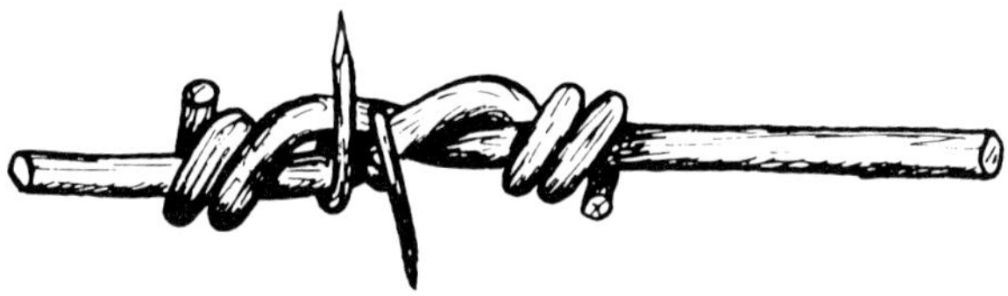

162

ONE LINE WIRE FACTORY SPLICE

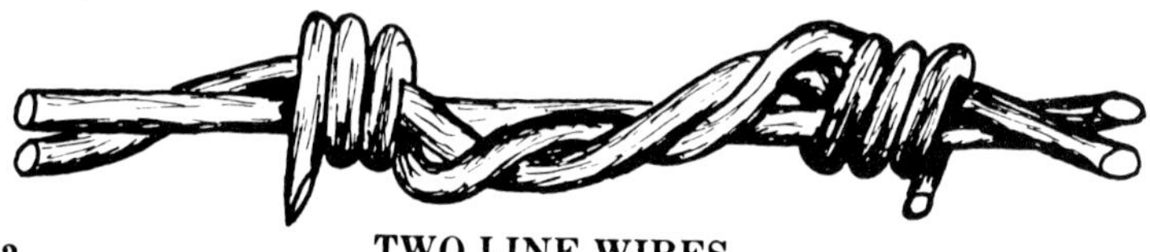

163 TWO LINE WIRES,

ONE WIRE FACTORY SPLICED

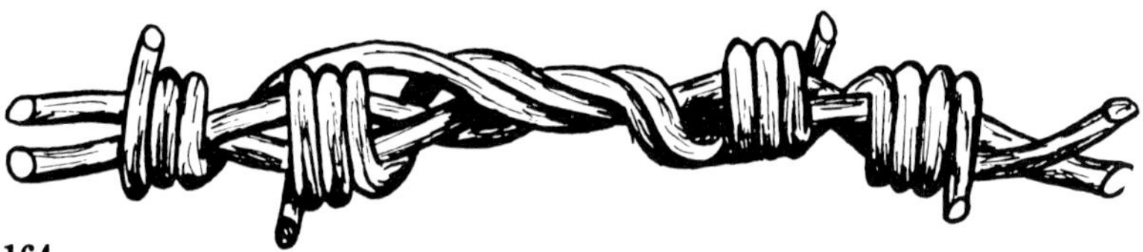

164

TWO LINE WIRES, BOTH FACTORY SPLICED

165 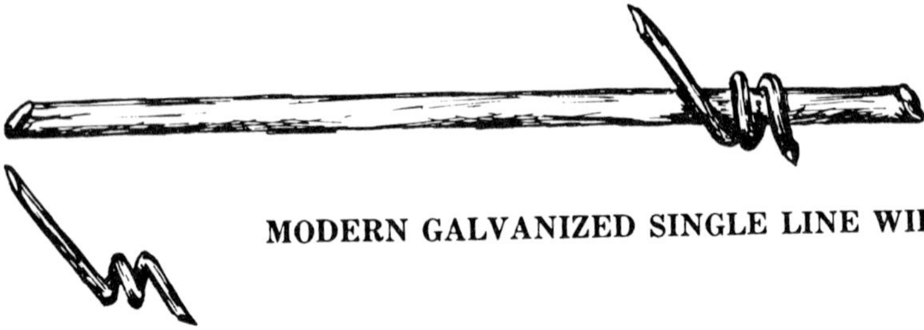

MODERN GALVANIZED SINGLE LINE WIRE

166 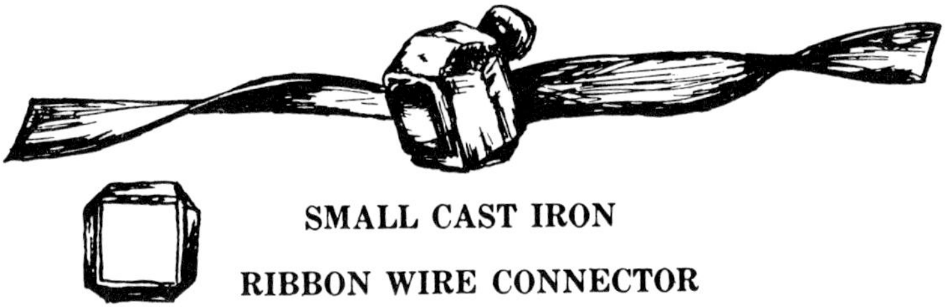

SMALL CAST IRON

RIBBON WIRE CONNECTOR

PAT. J. BRINKERHOFF, MAY 17, 1881

167 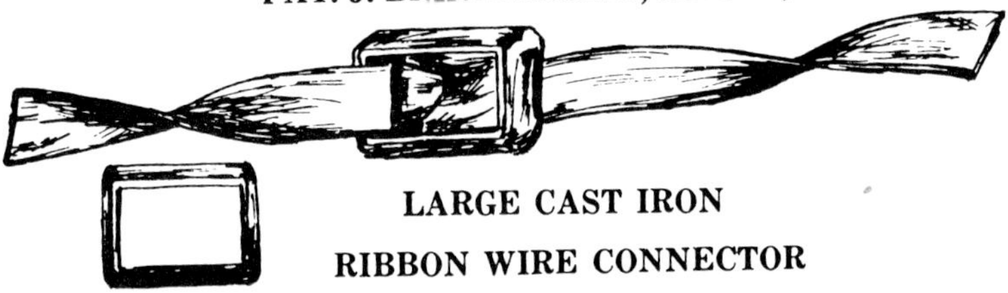

LARGE CAST IRON

RIBBON WIRE CONNECTOR

168 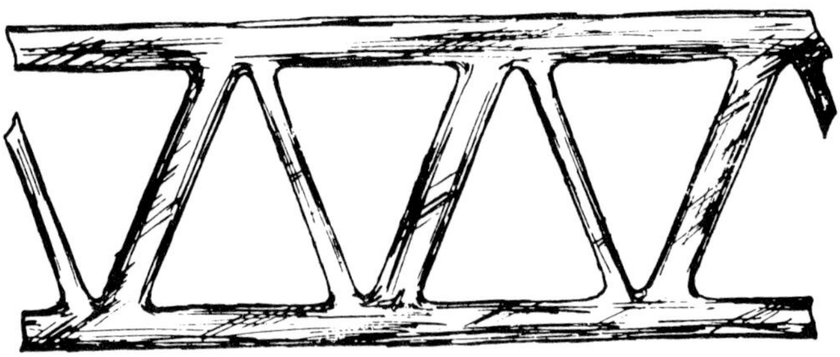

KILMER'S "WINDOW" WIRE

PATENTED MAY 12, 1885

169 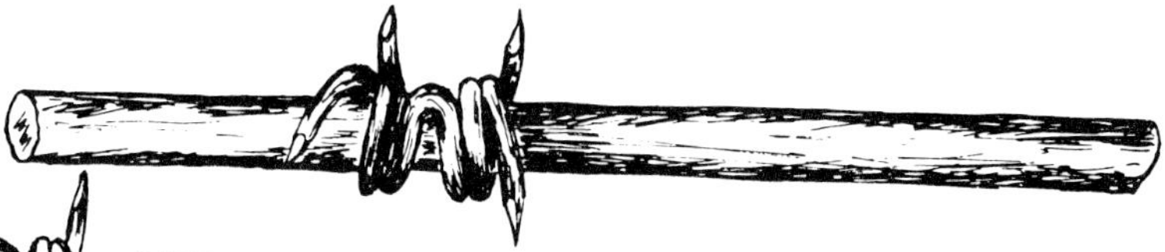

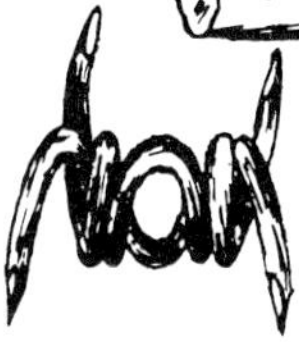

MERRILL'S TWIRL LARGE FLATTENED LINE WIRE

PATENTED DECEMBER 26, 1876

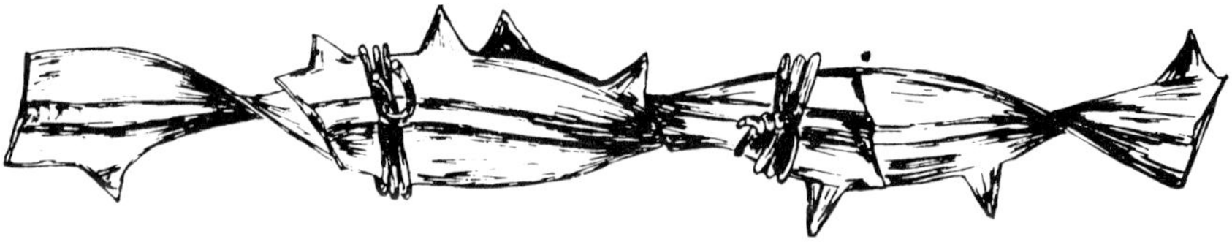

170

BUCKTHORN GALVANIZED FACTORY SPLICE

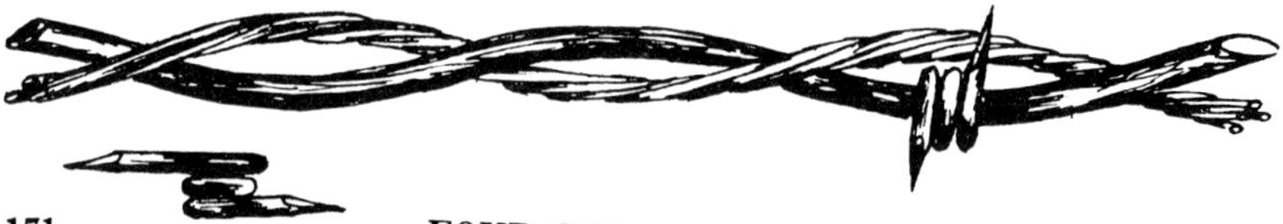

171

FOUR STRAND CABLE WIRE

PAT. J. F. GLIDDEN, NOV. 24, 1874

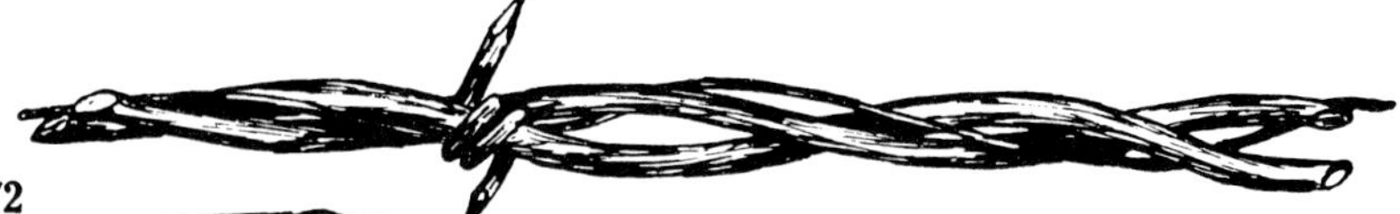

172 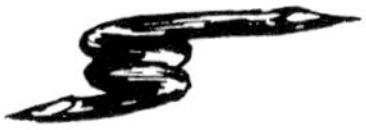

GLIDDEN THREE LINE

WITH ONE LINE EXTRA SMALL

PAT. J. F. GLIDDEN, NOV. 24, 1874

173 

M. M. MACK (ALTERNATE BARBS CUT)

PATENTED MAY 4, 1875

174

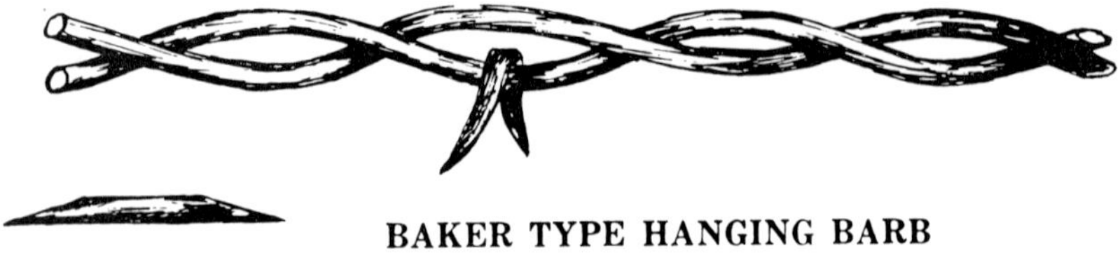

BAKER TYPE HANGING BARB

PAT. G. C. BAKER, FEB. 27, 1883

175

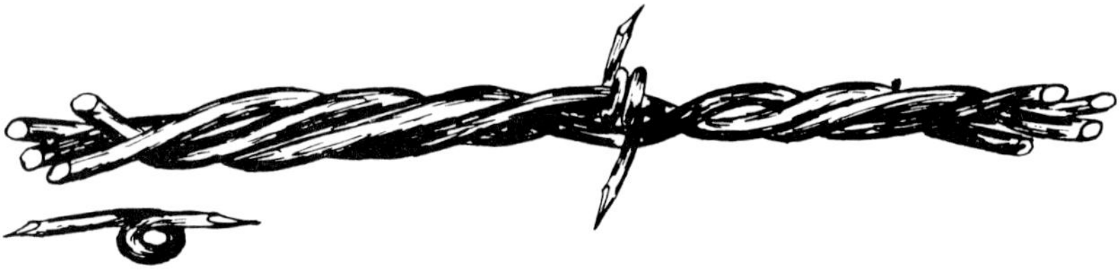

GLIDDEN FOUR LINE WIRE

PAT. NOV. 24, 1874

176

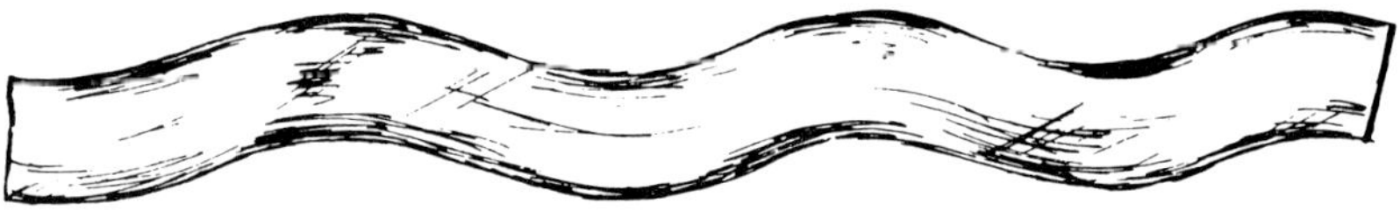

SNAKE WIRE PATENTED BY W. H. MERIWETHER

(NEW BRAUNFELS, TEXAS) NOVEMBER 8, 1853

177

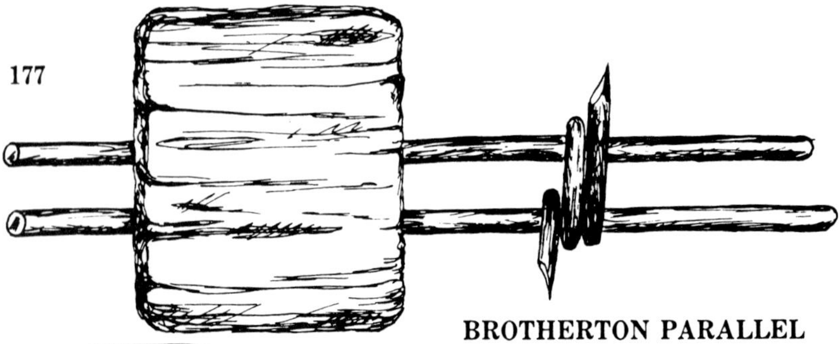

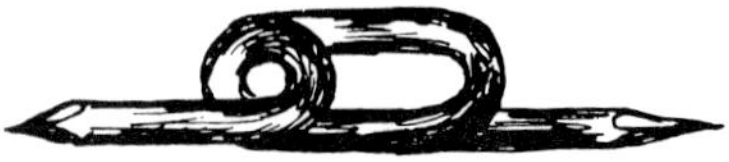

BROTHERTON PARALLEL

STAPLED ON WOODEN BLOCK

PATENTED SEPTEMBER 3, 1878

178

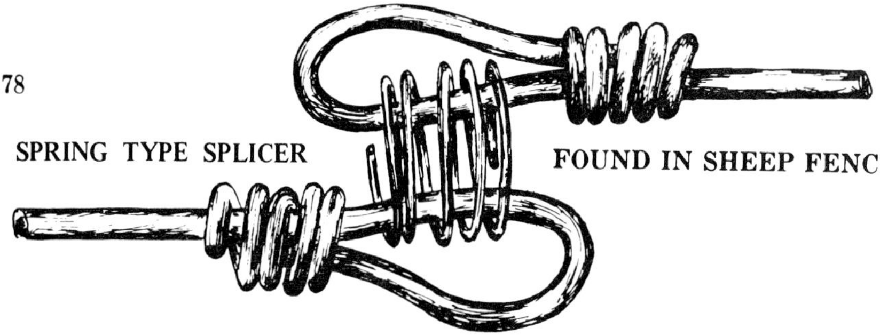

SPRING TYPE SPLICER FOUND IN SHEEP FENCE

179

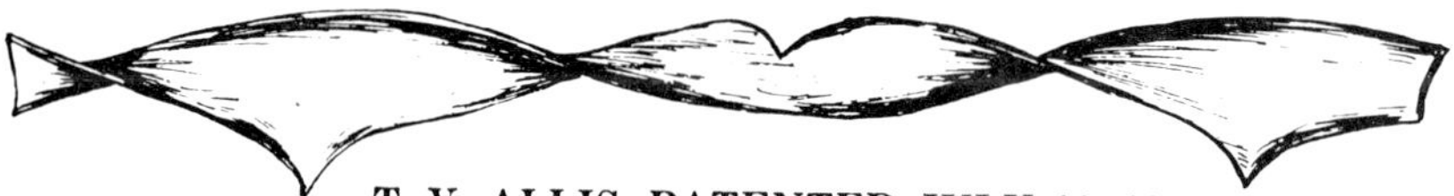

T. V. ALLIS, PATENTED JULY 11, 1893

180

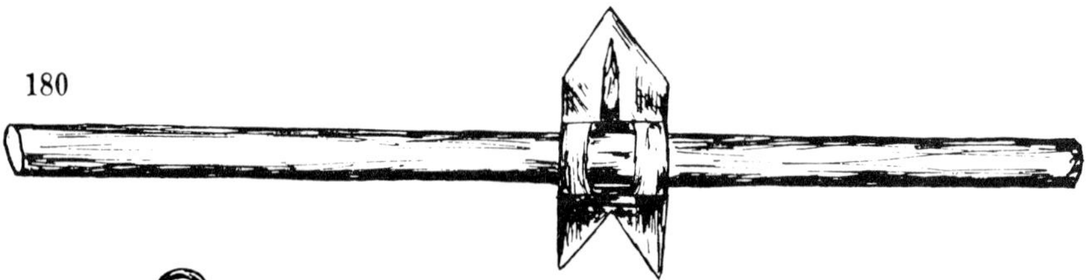

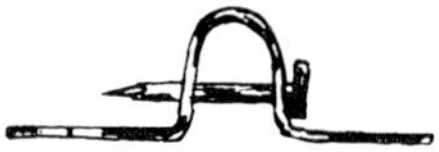

M. S. HARSHA, PATENTED JULY 4, 1876

181

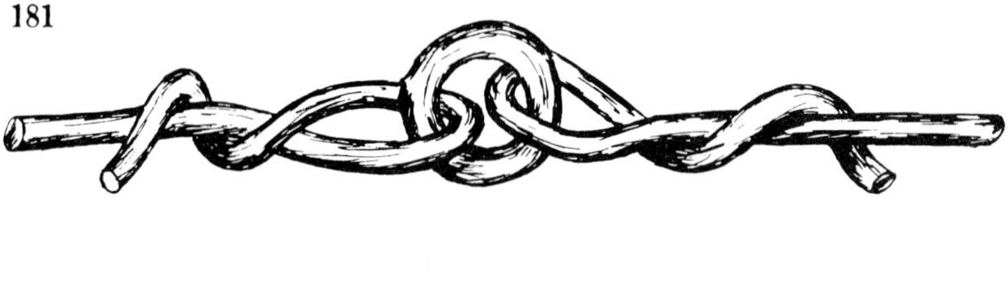

FACTORY SPLICE IN HOG WIRE USING CAST IRON RING

182

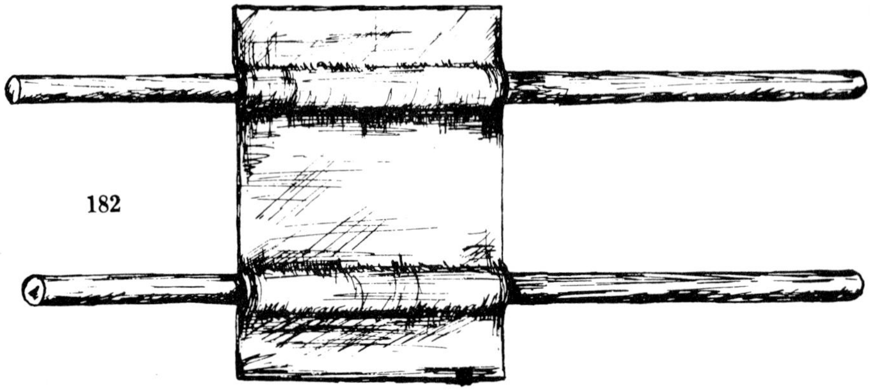

O. P. BRIGGS, PATENTED JANUARY 10, 1882

183

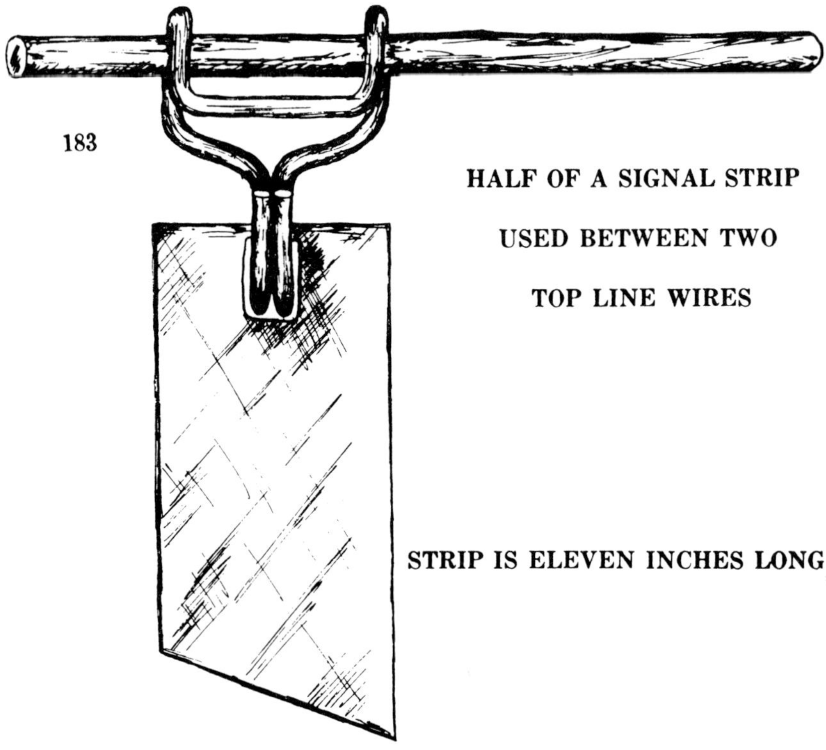

HALF OF A SIGNAL STRIP

USED BETWEEN TWO

TOP LINE WIRES

STRIP IS ELEVEN INCHES LONG

184

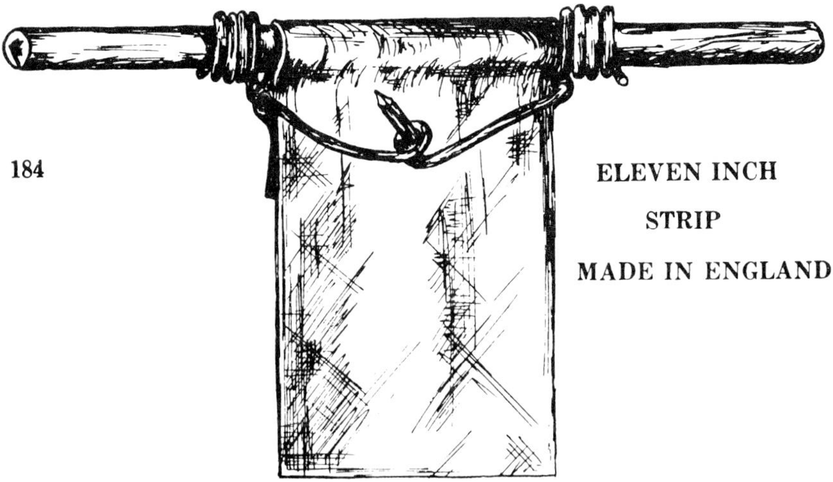

ELEVEN INCH

STRIP

MADE IN ENGLAND

TOP HALF OF SIGNAL STRIP USED BETWEEN TWO TOP LINE WIRES

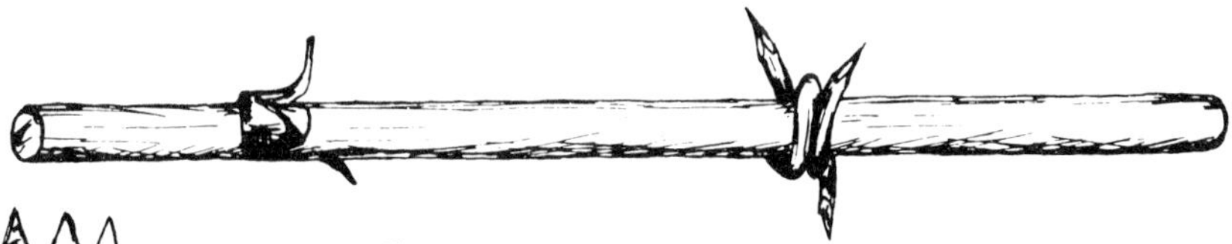

185

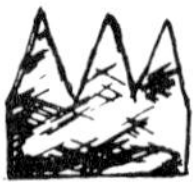

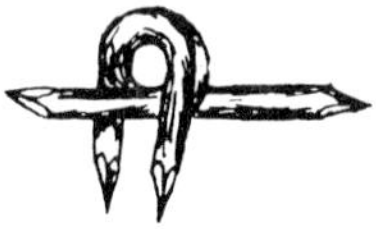

KENNEDY BARB

ON A JAYNE AND HILL

SINGLE LINE WIRE

PAT. APRIL 11, 1876

186

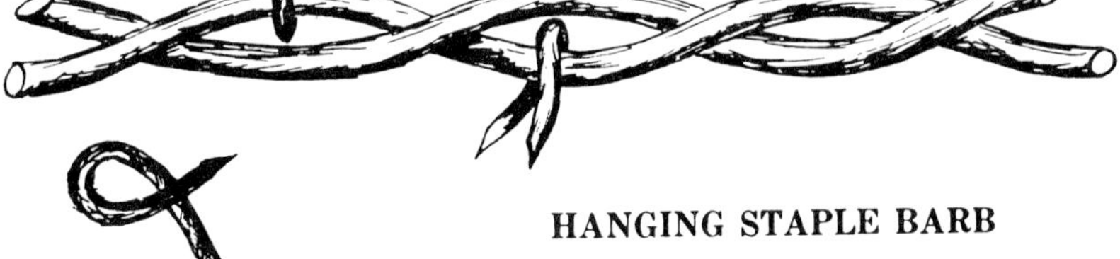

HANGING STAPLE BARB

PAT. A. J. UPHAM, AUG. 29, 1876

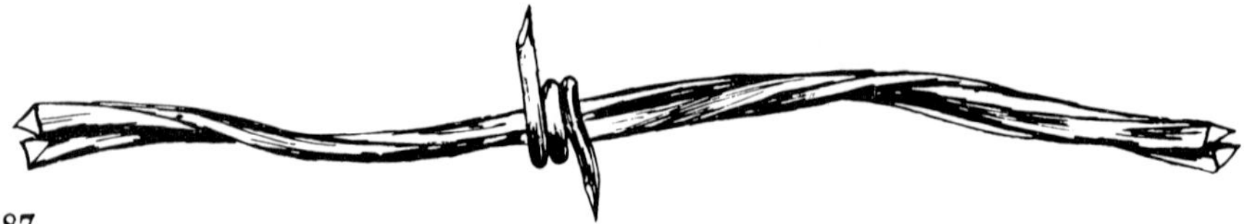

187

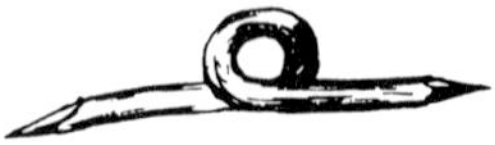

TRIANGLE LINE WIRES WITH ROUND BARB AROUND BOTH . . . FROM MEXICO

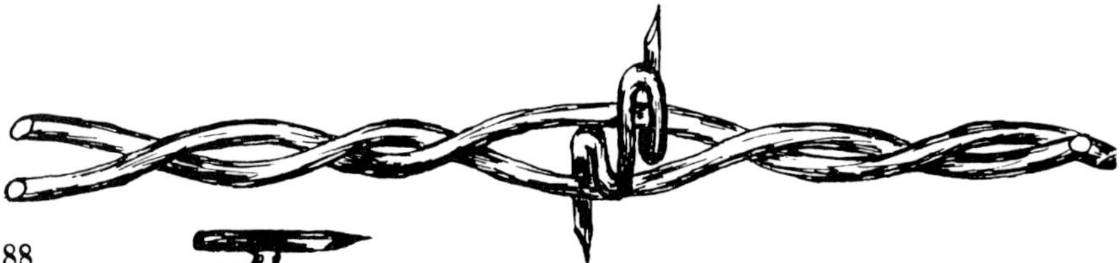

188

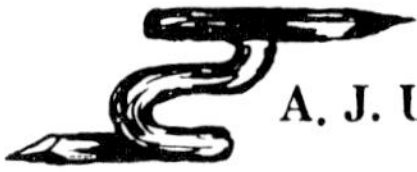

A. J. UPHAM'S "SIDE S", PATENTED JULY 2, 1878

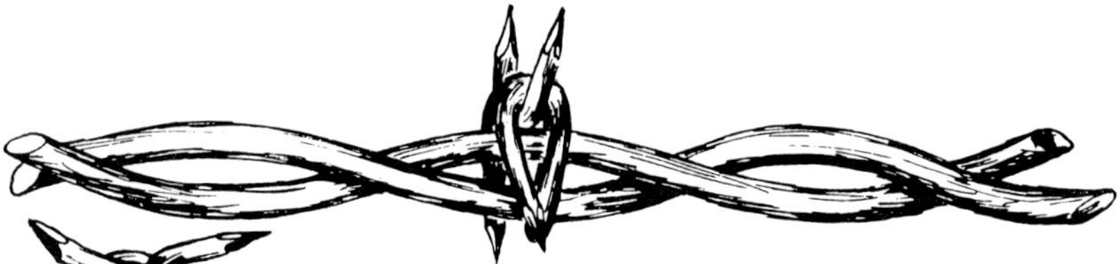

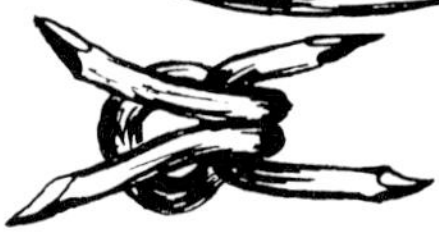

L. T. L. WING TWO LINE WIRE PATENTED FEBRUARY 26, 1878

189

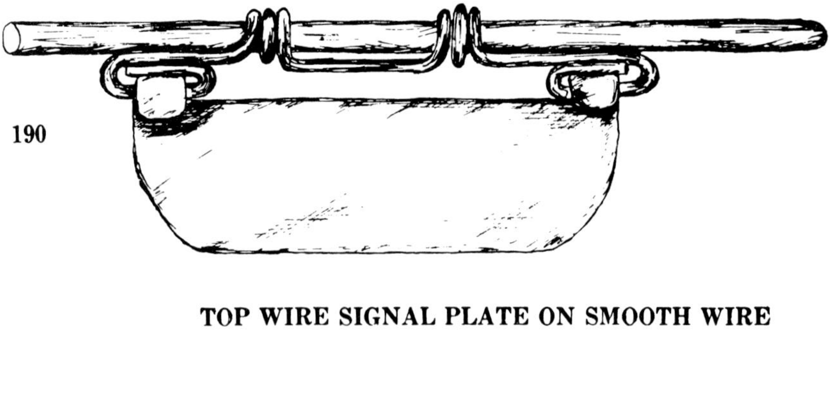

190

TOP WIRE SIGNAL PLATE ON SMOOTH WIRE

191

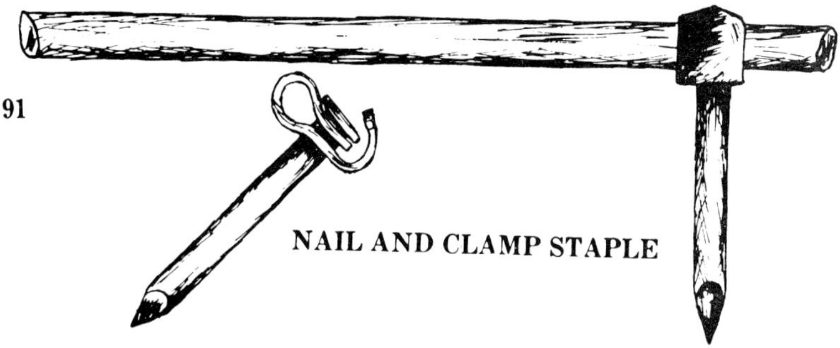

NAIL AND CLAMP STAPLE

192

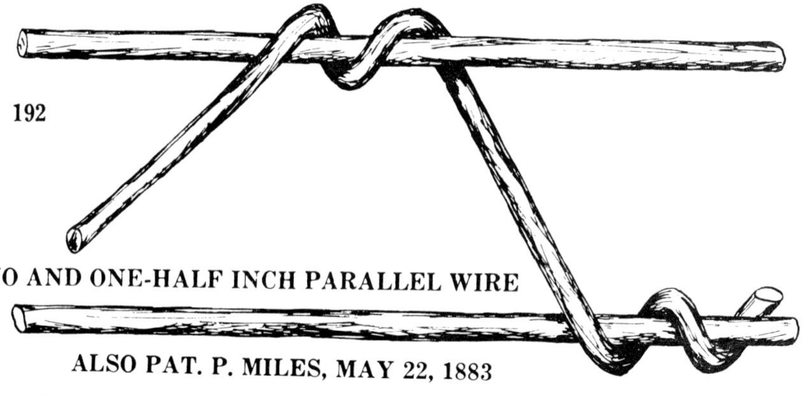

TWO AND ONE-HALF INCH PARALLEL WIRE

ALSO PAT. P. MILES, MAY 22, 1883

MACHINES WERE SOLD TO MAKE THIS WIRE AT HOME

193

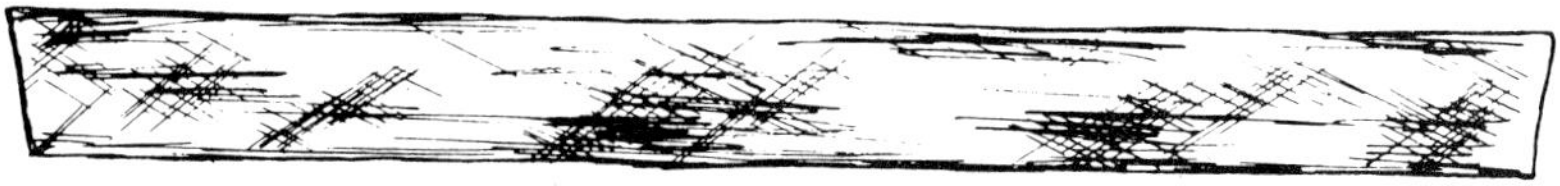

BARBLESS BRINKERHOFF RIBBON WIRE

USED MAINLY FOR SPLICING

PAT. J. BRINKERHOFF, MAY 17, 1881

194
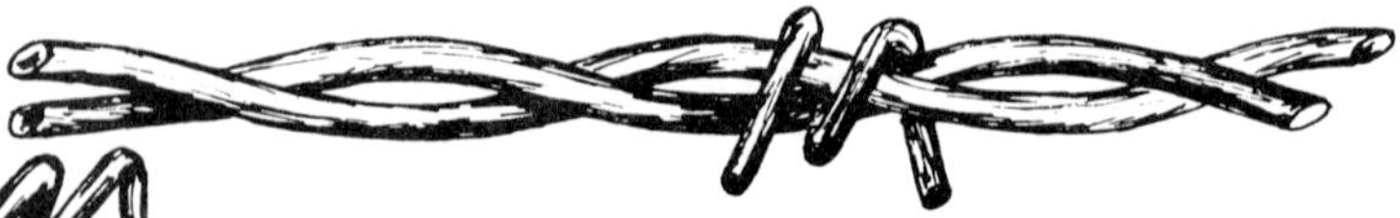
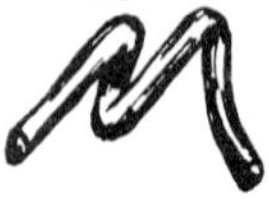
ONE AROUND TWO

NET WIRE

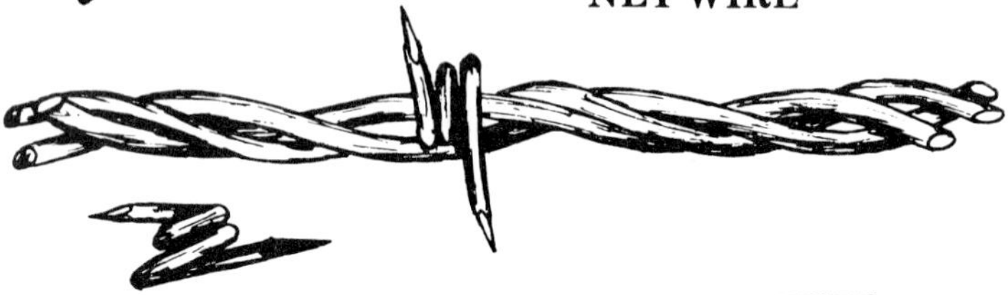
195

THREE LINE BROTHERTON WIRE

PAT. J. BROTHERTON, SEPT. 3, 1887

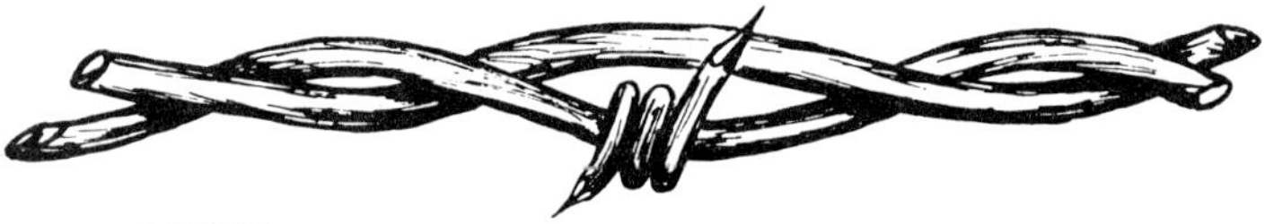
196
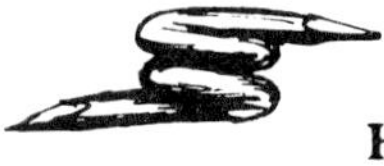

GLIDDEN BARB WITH TWO EXTRA

HEAVY AND SLIGHTLY OVAL LINE WIRES

PAT. J. F. GLIDDEN, NOV. 24, 1874

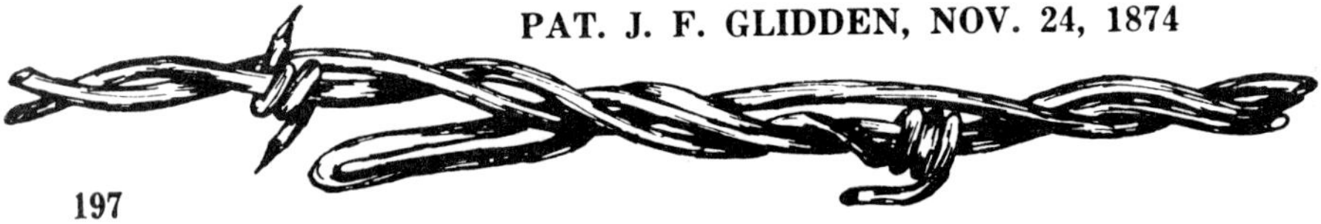
197

THREE LINE SPLICED INTO A TWO LINE WIRE

PAT. J. F. GLIDDEN, NOV. 24, 1874

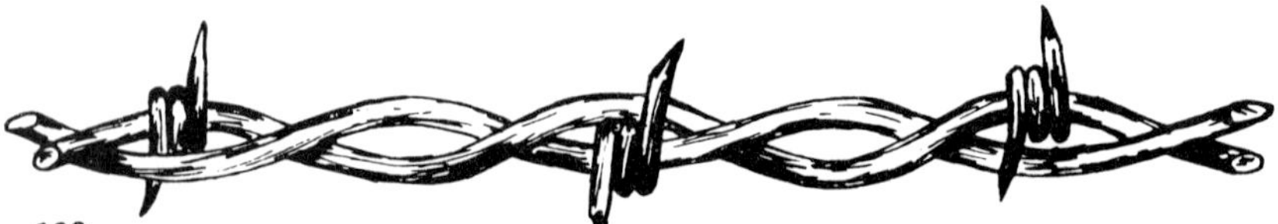
198

GLIDDEN BARB ON BOTH LINES . . .

SPACED ONE INCH APART

PAT. NOV. 24, 1874

199

RIBBON WIRE USED TO LACE SMALL BOIS 'D ARC HEDGES

TOGETHER TO KEEP STOCK FROM GOING BETWEEN THE PLANTS

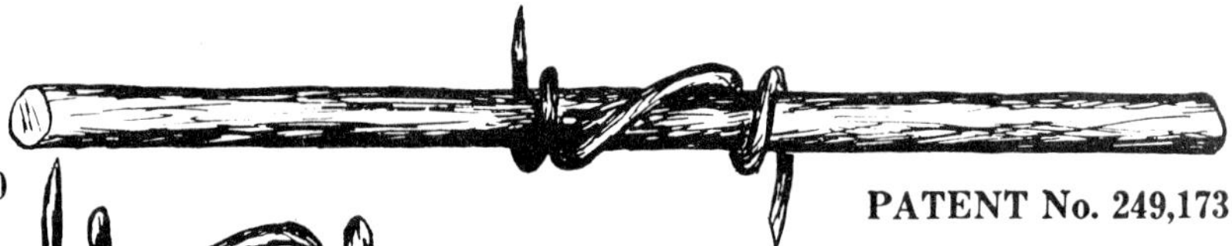

200

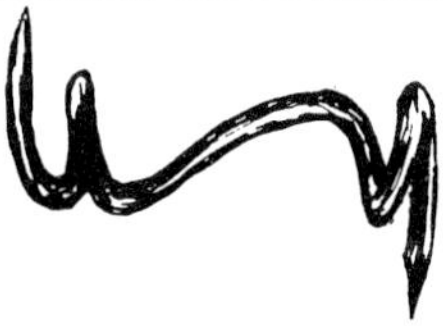

PATENT No. 249,173

GUNDERSON TWO POINT

(DERIVED FROM GUNDERSON FOUR POINT)

PATENTED NOVEMBER 8, 1881

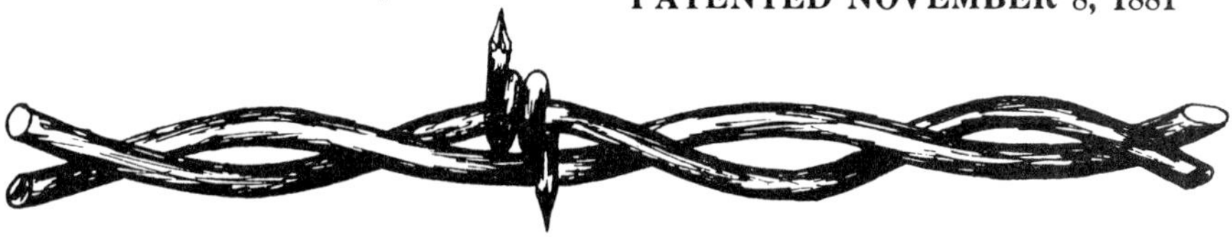

201

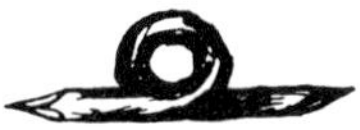

GLIDDEN COPPER LINE WIRE

WITH STEEL BARBS

PAT. J. F. GLIDDEN, NOV. 24, 1874

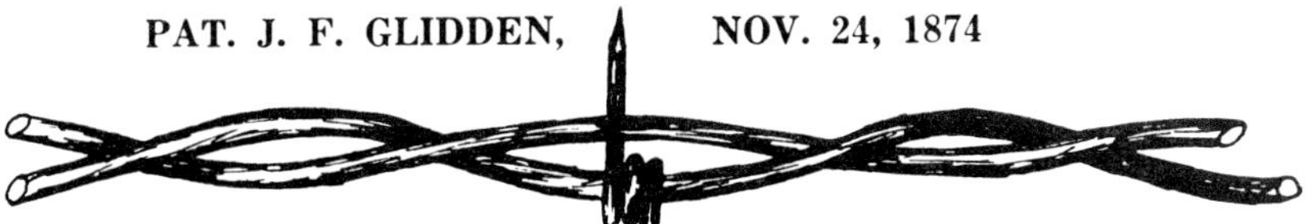

202

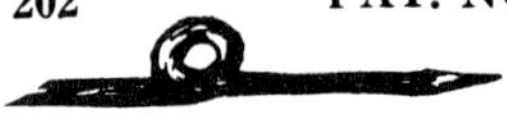

PAT. NOV. 24, 1874

GLIDDEN SMALL LINE WIRE

WITH EXTRA LONG NEEDLE POINT BARB

COMMONLY CALLED "PRICKLY PEAR" WIRE

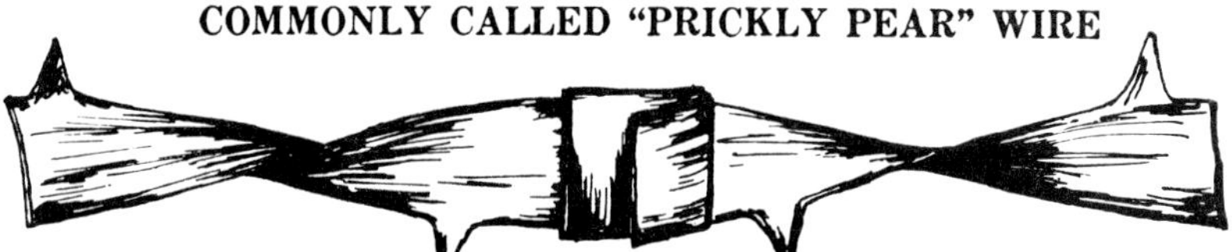

203

FACTORY SPLICE

ON VERY THIN BARBED RIBBON WIRE . . . NOT GALVANIZED

PAT. T. V. ALLIS, JAN. 5, 1892

204 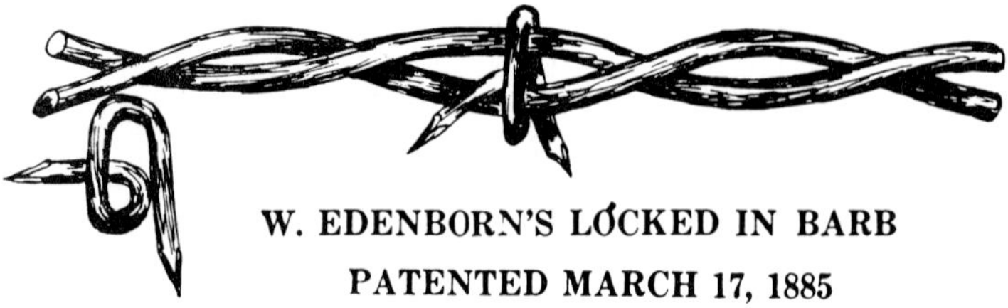

W. EDENBORN'S LOCKED IN BARB
PATENTED MARCH 17, 1885

205 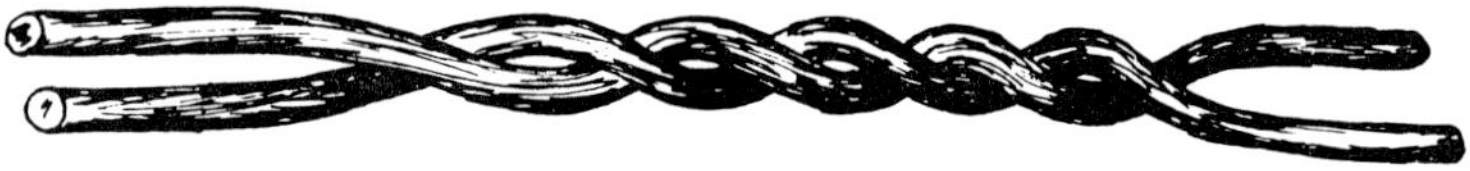

SCUTT'S BARBLESS WOODEN BLOCK
(WITHOUT BLOCKS)
PAT. H. B. SCUTT, FEB. 10, 1880

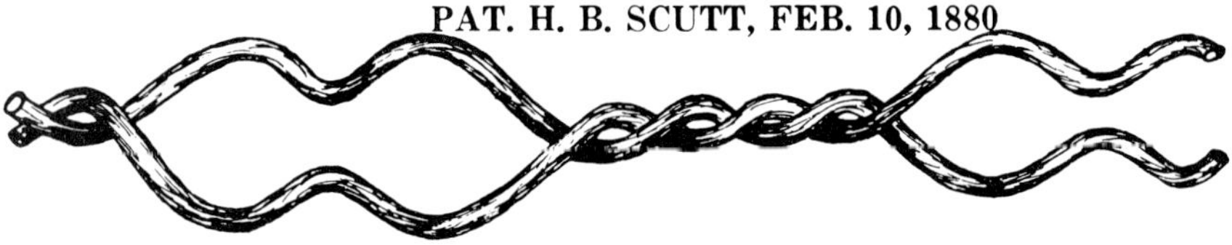

ORNAMENTAL FENCE WIRE

206

207 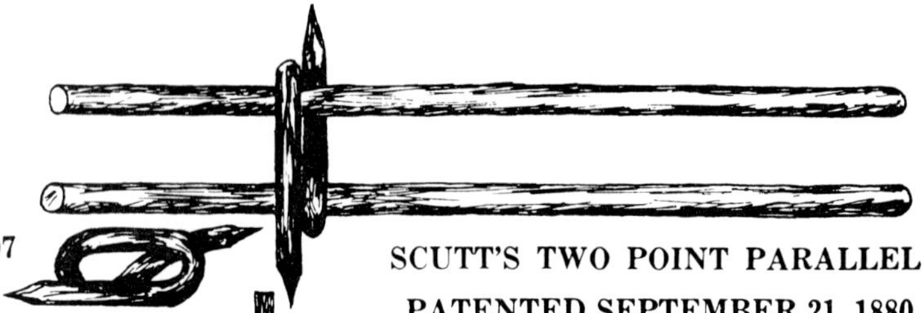

SCUTT'S TWO POINT PARALLEL
PATENTED SEPTEMBER 21, 1880

208 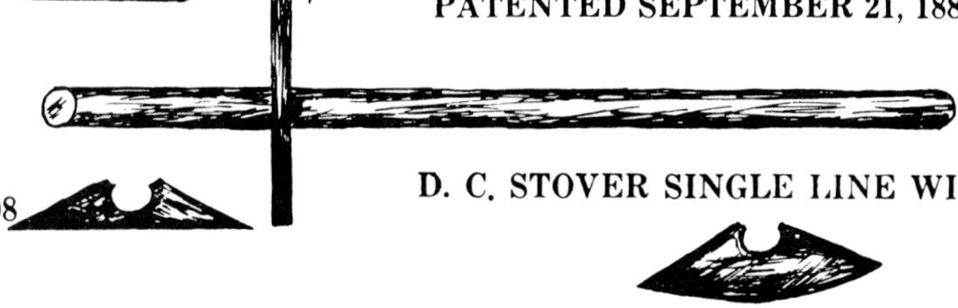

D. C. STOVER SINGLE LINE WIRE

PATENTED JUNE 29, 1875

209

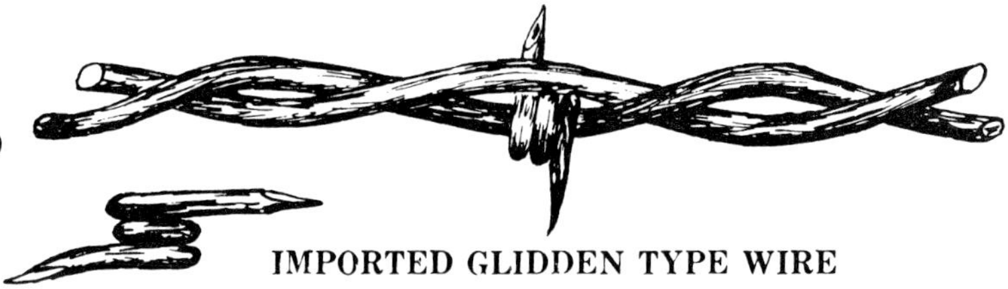

IMPORTED GLIDDEN TYPE WIRE
WITH BARBS PAINTED RED

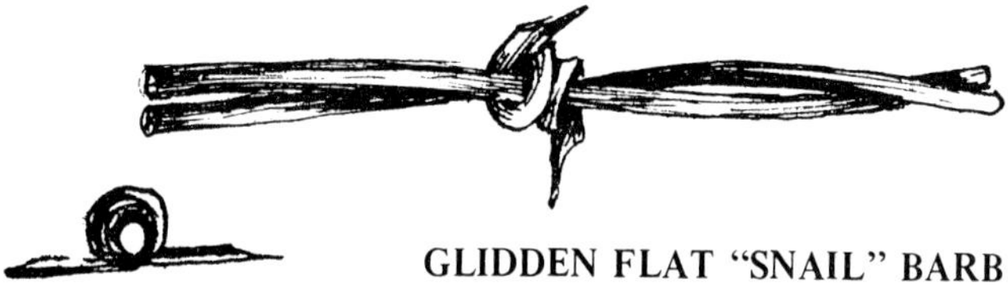

210

GLIDDEN FLAT "SNAIL" BARB

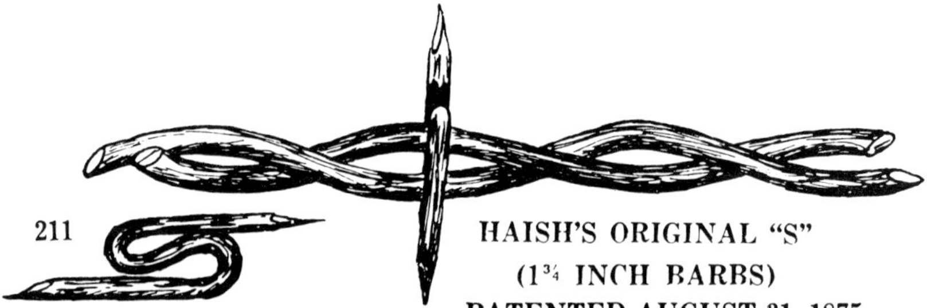

211

HAISH'S ORIGINAL "S"
(1¾ INCH BARBS)
PATENTED AUGUST 31, 1875

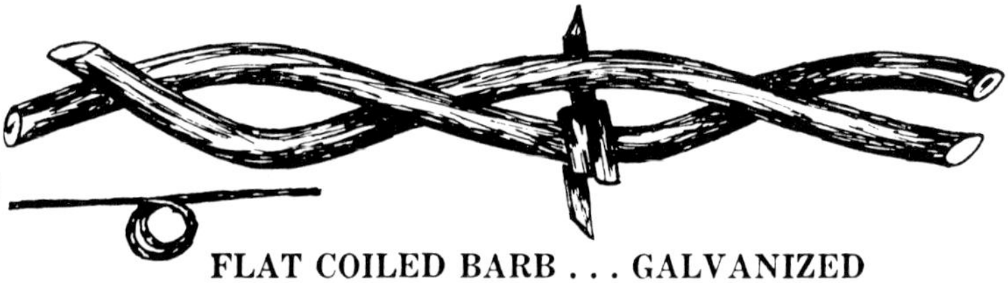

212

FLAT COILED BARB . . . GALVANIZED
PAT. J. F. GLIDDEN, NOV. 24, 1874

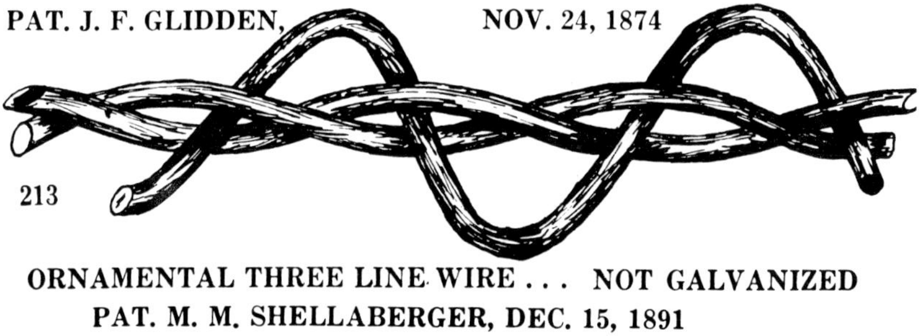

213

ORNAMENTAL THREE LINE WIRE . . . NOT GALVANIZED
PAT. M. M. SHELLABERGER, DEC. 15, 1891

214

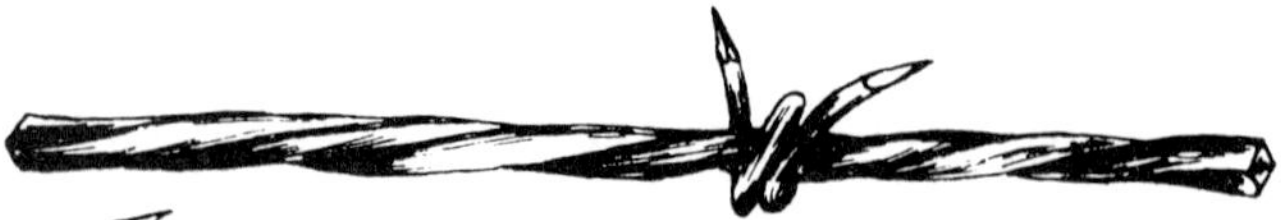

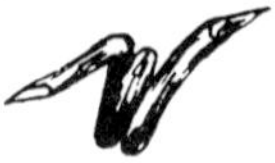

SQUARE TWISTED LINE WIRE
WITH TWO POINT BARBS . . . GALVANIZED
PAT. J. F. GLIDDEN, NOV. 24, 1874

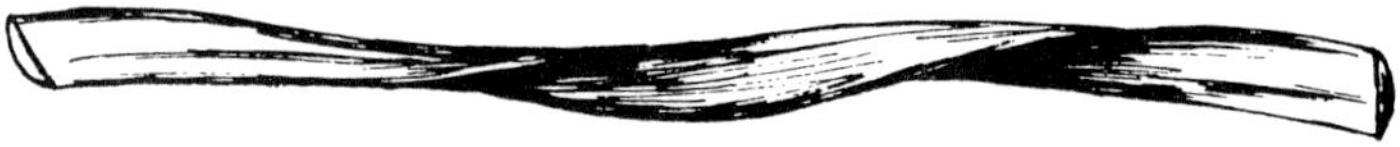

215

TWISTED HALF OVAL RIBBON
COMES IN AT LEAST TWO WIDTHS
COMMONLY CALLED "SCUTT'S CABLE ROD HORSE WIRE"
PAT. JUNE 1, 1875

JAYNE AND HILL
WITH WOOD BLOCK

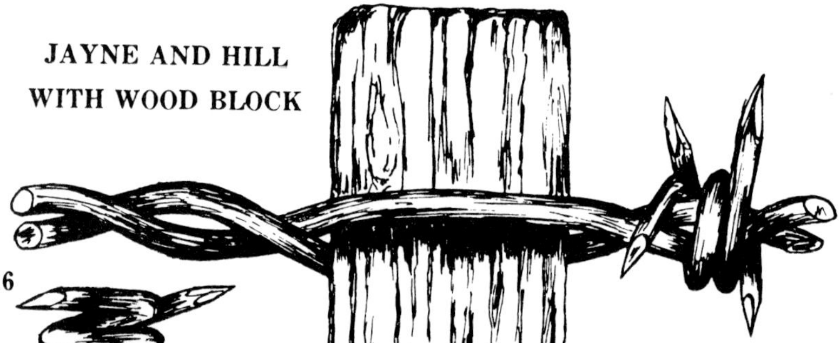

216

PATENTED
APRIL 11, 1876

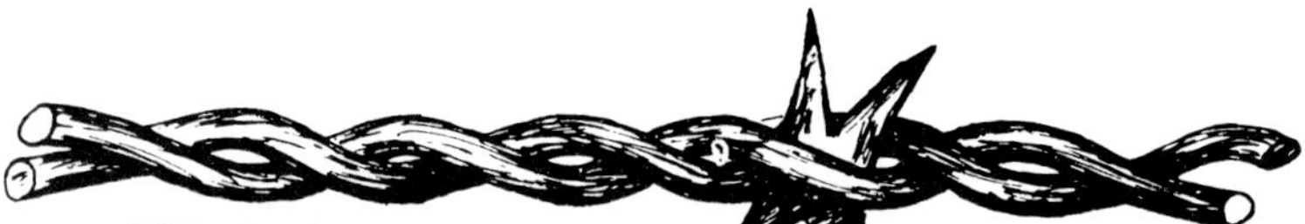

217

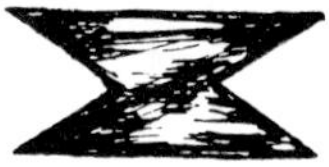

H. B. SCUTT "V" PLATE
BARB IS THICK IN MIDDLE
AND SHARP ON EDGES
PAT. H. B. SCUTT, JUNE 18, 1878

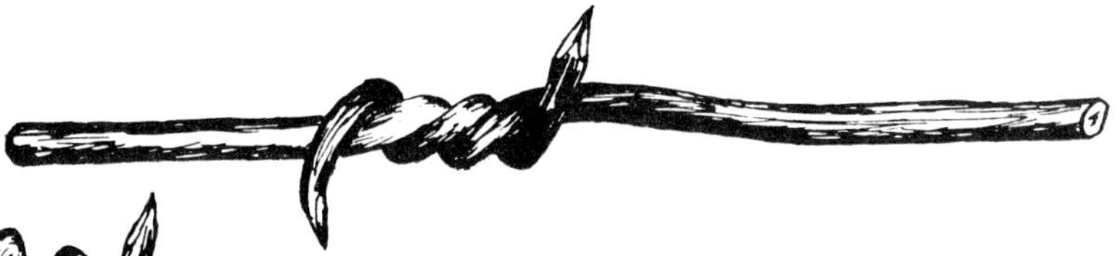

218

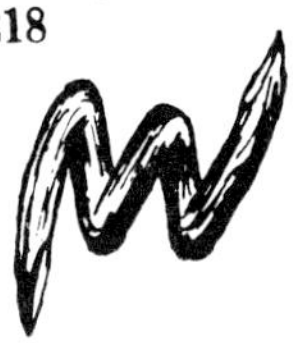

M. M. MACK, PATENTED MAY 4, 1875

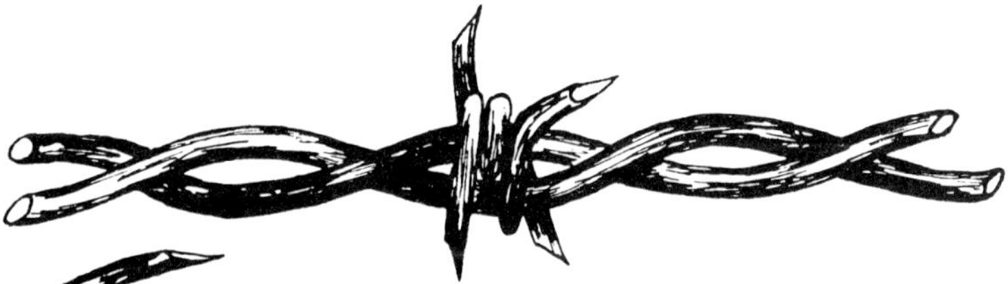

219

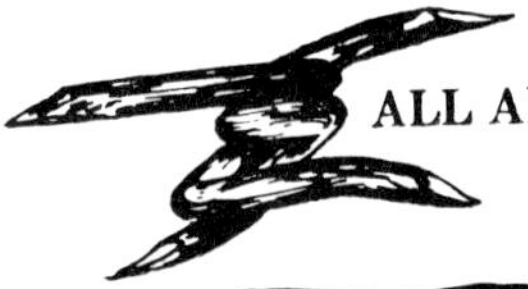

ALL ALUMINUM ROSS TYPE FOUR POINT WIRE

PAT. N. G. ROSS, JUNE 10, 1879

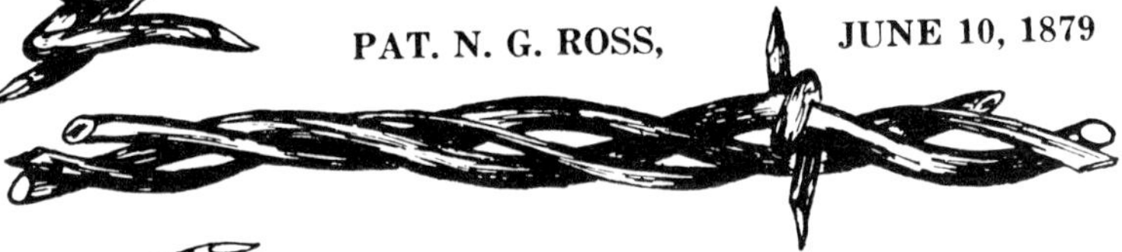

220

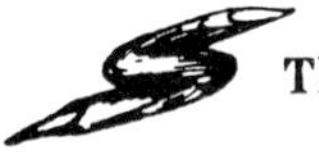

THREE LINE WIRE WITH TWO ROUND AND ONE FLAT

PAT. A. J. UPHAM, AUG. 29, 1876

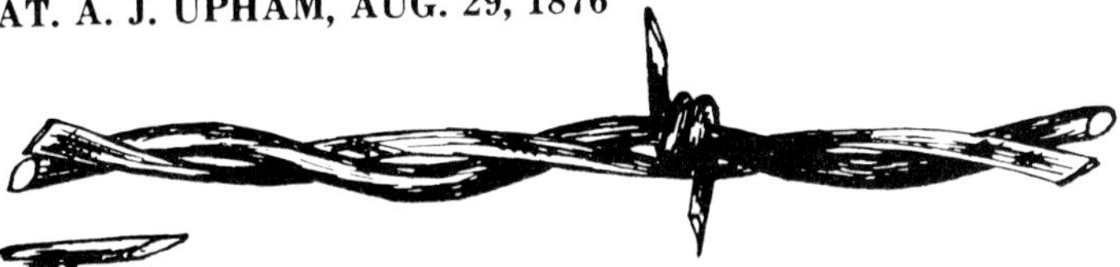

221

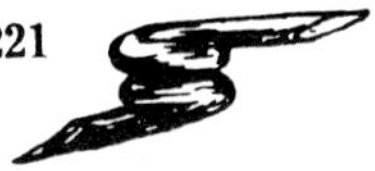

TWO LINE WIRE WITH ONE ROUND AND ONE FLAT

PAT. J. F. GLIDDEN, NOV. 24, 1874

222

HORSE SHOE
CLAMP NET WIRE

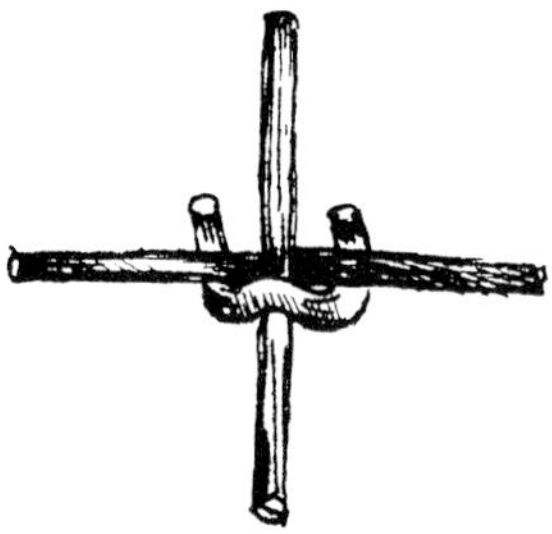

223 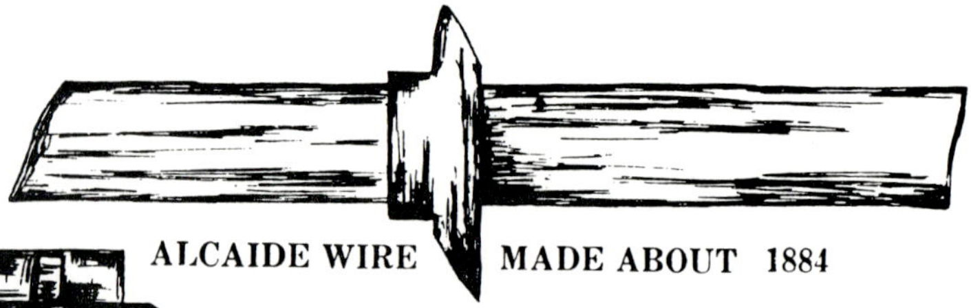

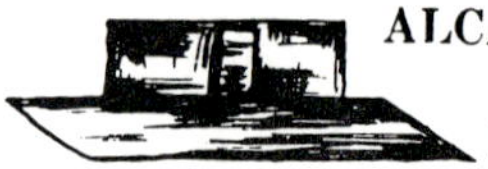

ALCAIDE WIRE MADE ABOUT 1884

MADE OF NATIVE TEXAS IRON ORE

MADE BY PRISONERS WHEN STATE PEN WAS LOCATED AT RUSK

RESEMBLES BRINKERHOFF BUT NOT GALVANIZED

224 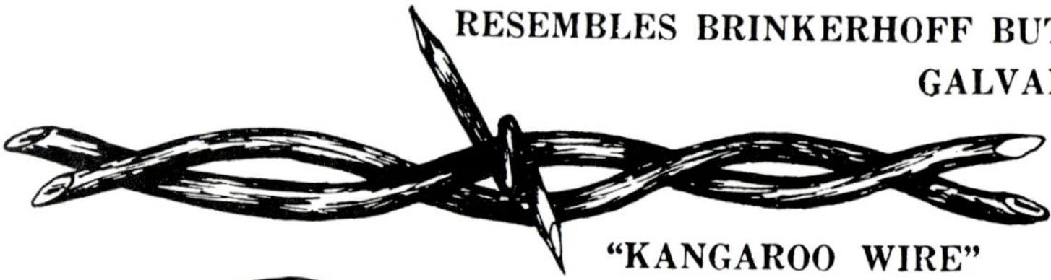

"KANGAROO WIRE"

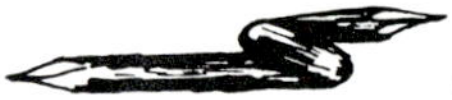

ONE LONG AND ONE SHORT BARB

PAT. A. J. UPHAM, AUG. 29, 1876

225 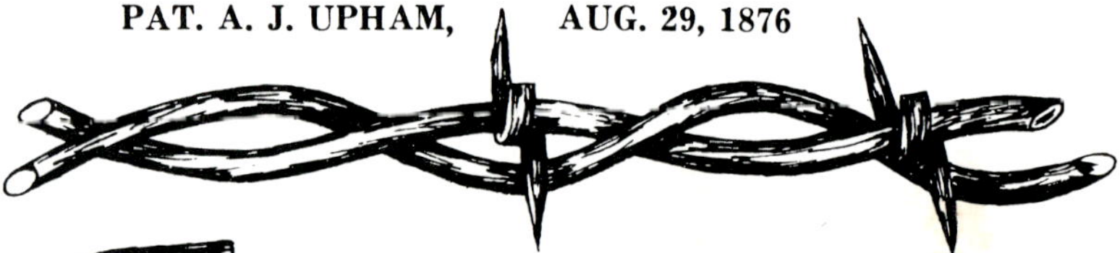

BAKER BARB . . . ONE INCH SPACING

PAT. G. C. BAKER, FEB. 27, 1883

226 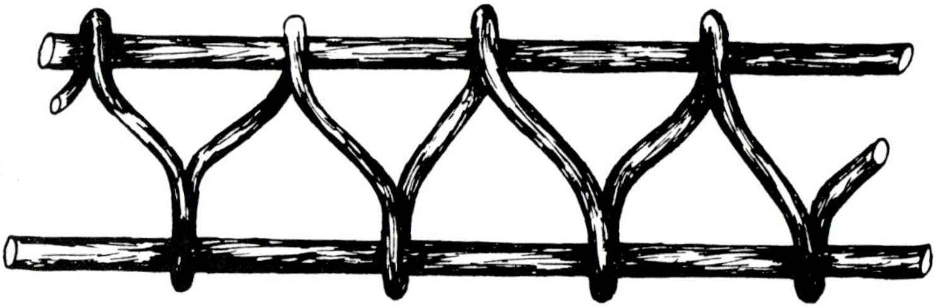

ORNAMENTAL FENCE WIRE

227 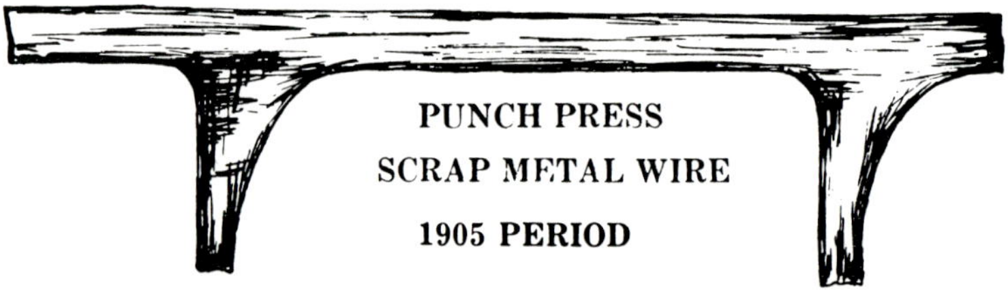

PUNCH PRESS

SCRAP METAL WIRE

1905 PERIOD

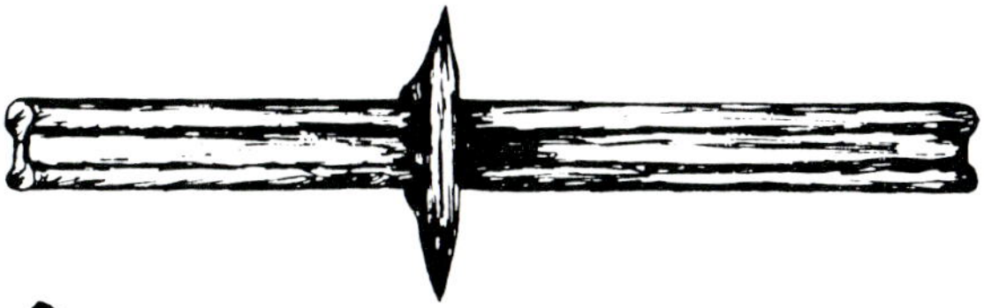

228

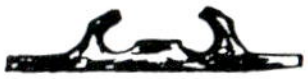 SCUTT'S WEBBED RIBBON . . . H. B. SCUTT

PATENTED OCTOBER 23, 1883

229

SCUTT'S RIDGED RIBBON

HIRAM B. SCUTT

PATENTED OCTOBER 23, 1883

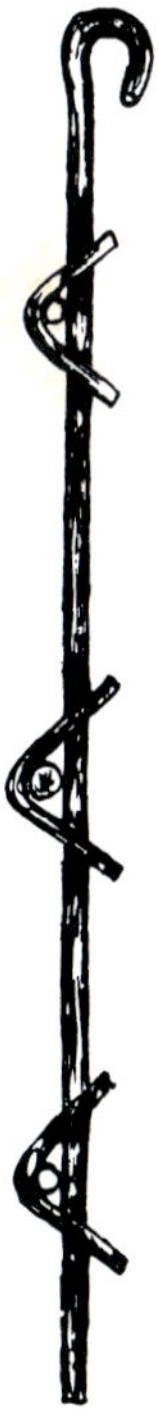

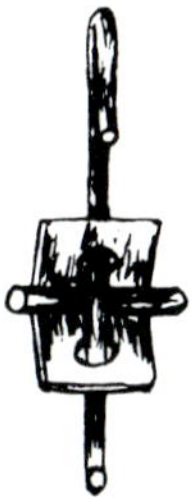

230

"V" PLATE FENCE STAY

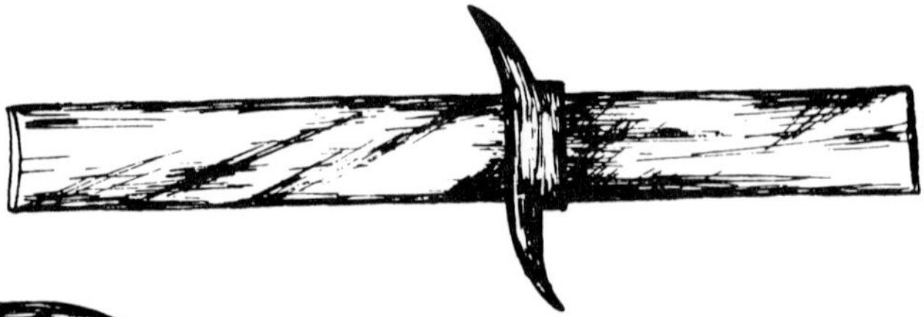

231

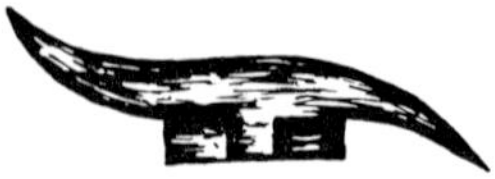

BRINKERHOFF "S" BARB RIBBON
PAT. J. BRINKERHOFF, MAY 17, 1881

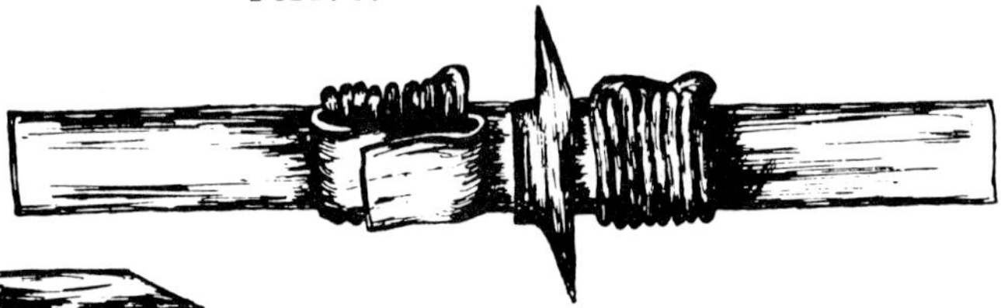

232

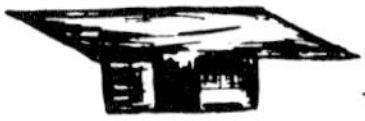

EARLY BRINKERHOFF FACTORY SPLICE
PAT. J. BRINKERHOFF, MAY 17, 1881

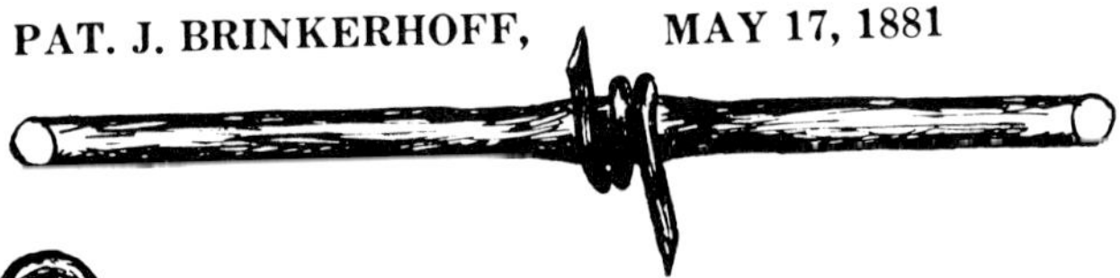

233

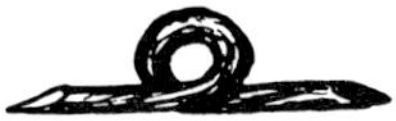

"REVERSE ROGERS" CHARLES D. ROGERS
PATENTED JANUARY 10, 1888

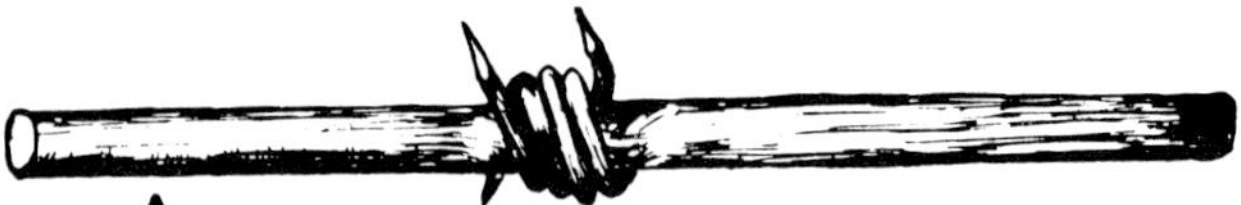

234

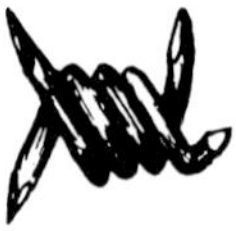

ARGENTINA SINGLE STRAND
"ROUND UNDER BARB", OVAL BETWEEN BARBS

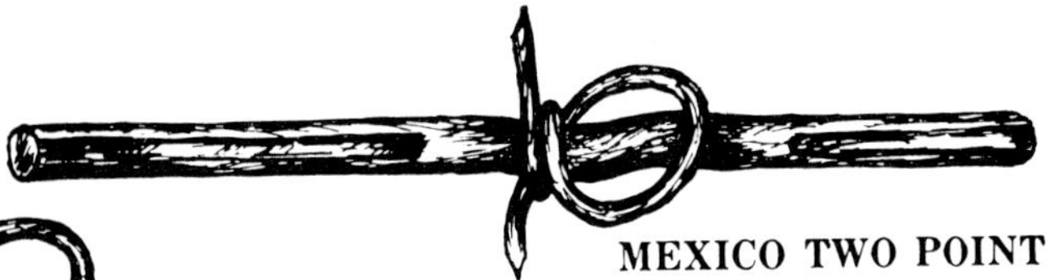

235

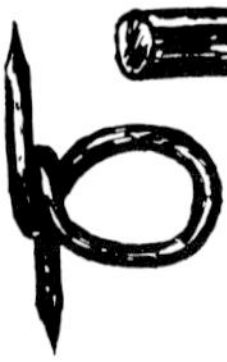

MEXICO TWO POINT
OBVIOUS KINK UNDER BARB

236

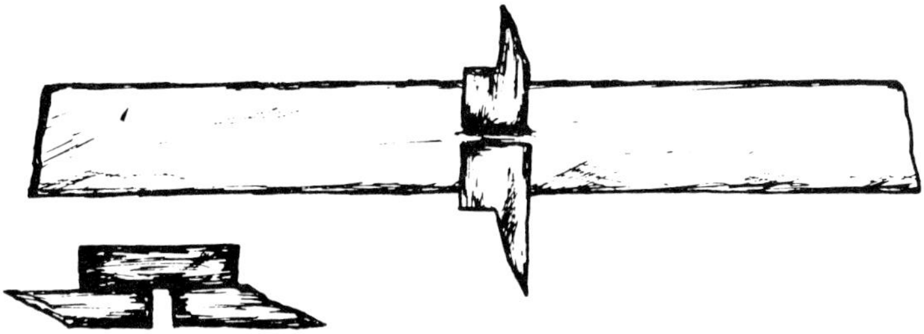

J. AND WM. BRINKERHOFF FRONT CUT BARB
RIBBON WIRE, PATENTED MAY 17, 1881

237

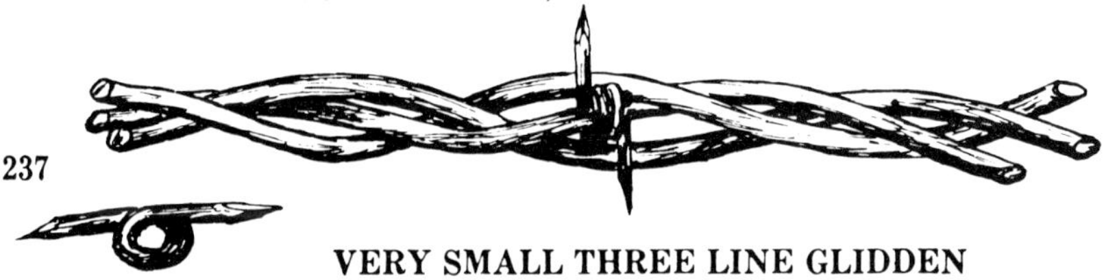

VERY SMALL THREE LINE GLIDDEN
PAT. J. F. GLIDDEN, NOV. 24, 1874

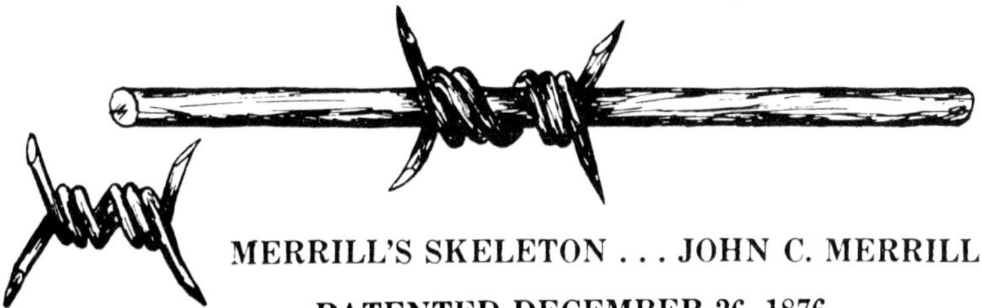

MERRILL'S SKELETON . . . JOHN C. MERRILL
PATENTED DECEMBER 26, 1876

238

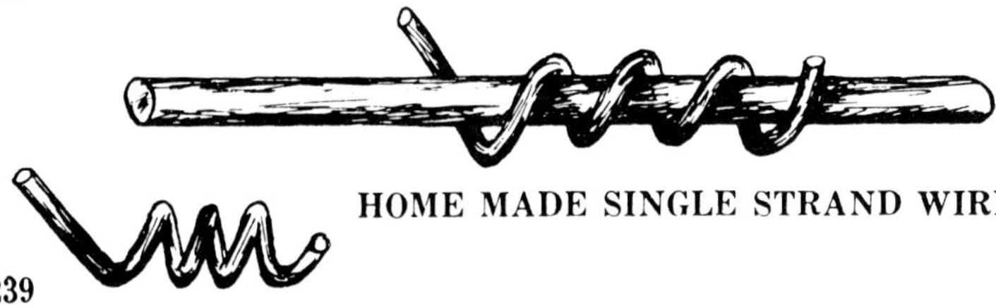

HOME MADE SINGLE STRAND WIRE

239

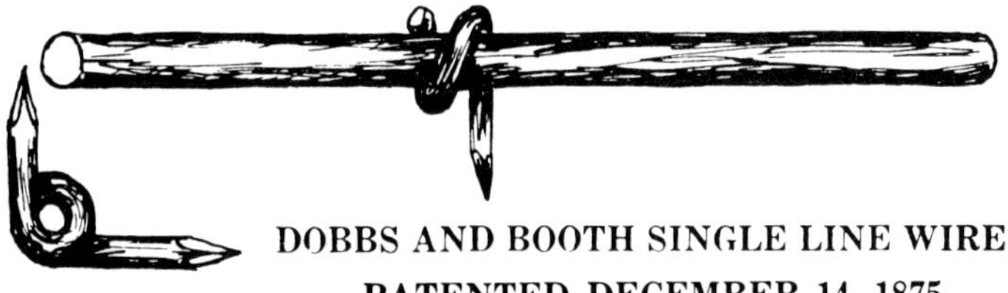

240

DOBBS AND BOOTH SINGLE LINE WIRE
PATENTED DECEMBER 14, 1875

241

SQUARE NAIL STAPLE

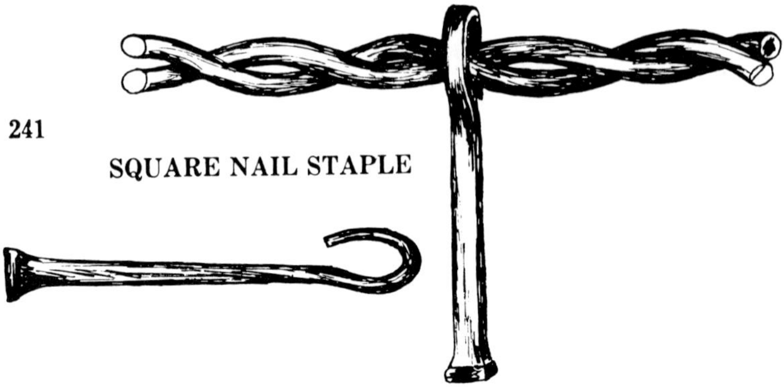

WEDGED IN HOLE DRILLED IN STONE POST BY ANOTHER NAIL

SCUTT'S BARBLESS WOODEN BLOCK WIRE

242

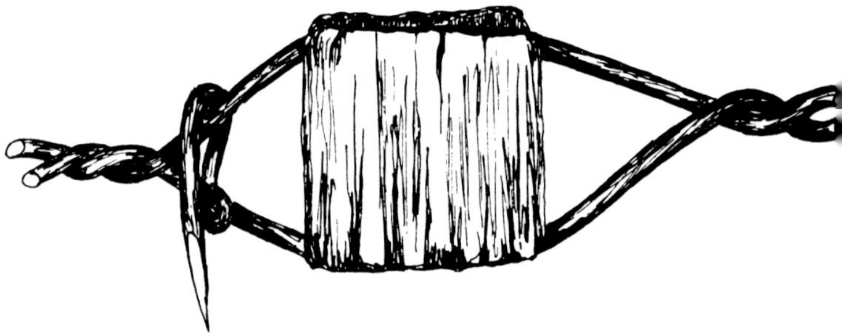

PATENTED BY H. B. SCUTT FEBRUARY 10, 1880
SHOWN WITH ORIGINAL STAPLE

243

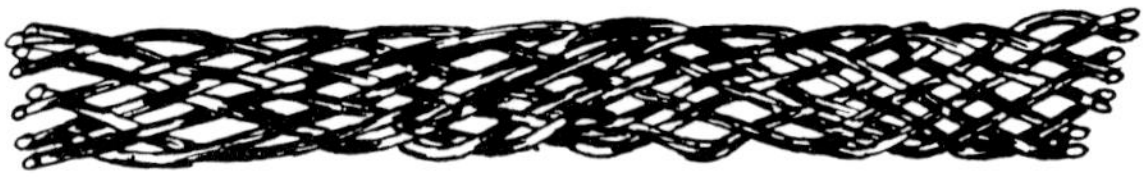

MESH RIBBON WIRE . . . SEVENTEEN STRAND BRAID
USED AS FENCING

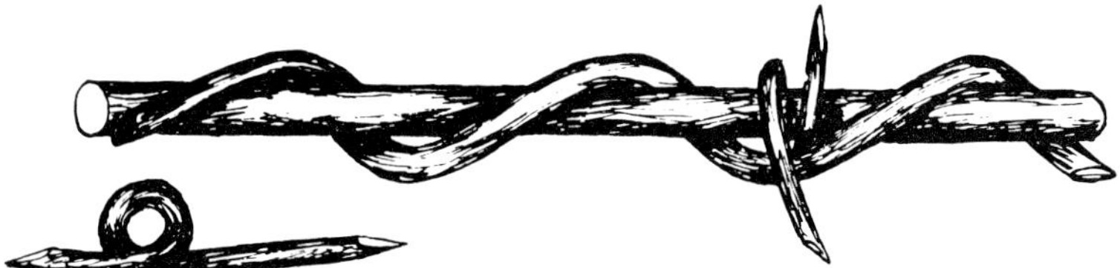

STAPLE BARB ON STRAIGHT LINE

244 WITH SECOND LINE WRAPPED AROUND

PAT. A. J. UPHAM, AUG. 29, 1876

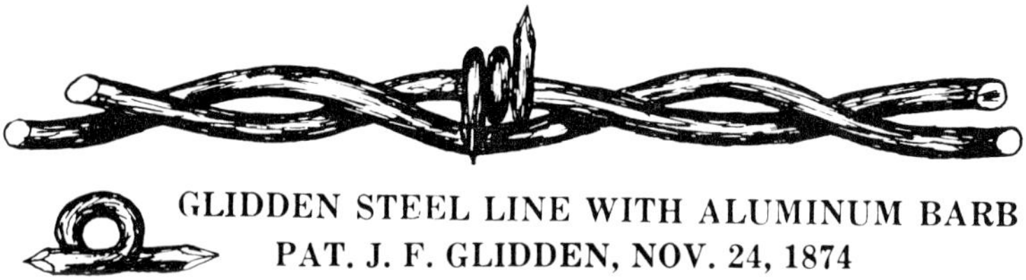

GLIDDEN STEEL LINE WITH ALUMINUM BARB

PAT. J. F. GLIDDEN, NOV. 24, 1874

245

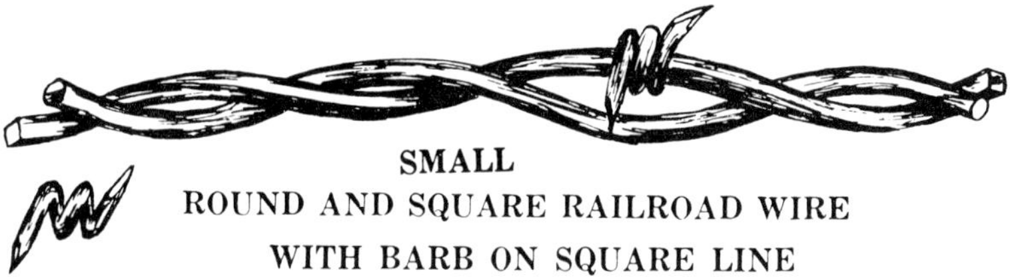

SMALL

ROUND AND SQUARE RAILROAD WIRE

WITH BARB ON SQUARE LINE

246 PAT. J. F. GLIDDEN, NOV. 24, 1874

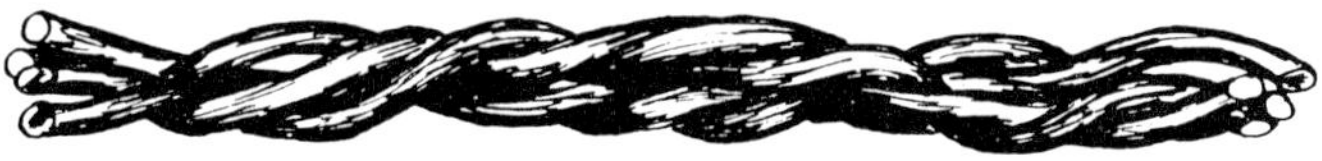

247 TWO TWISTED TWO LINES

TWISTED TOGETHER RESEMBLING FOUR STRAND BRAID

PAT. J. B. CLEAVELAND NOV. 22, 1892

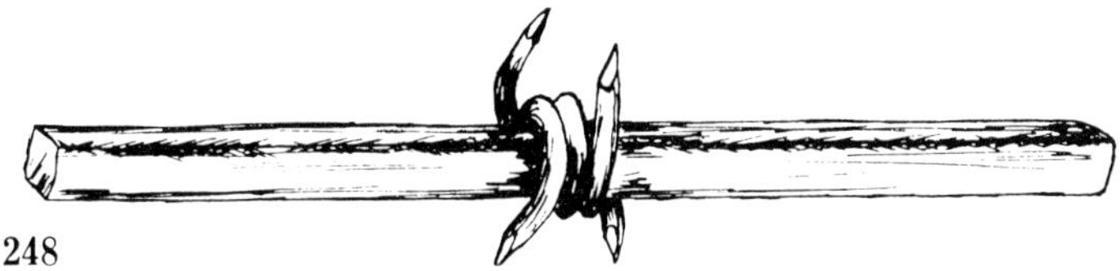

248

FLATTENED OVAL WIRE WITH GLIDDEN FOUR POINT BARB

PAT. J. F. GLIDDEN, FEB. 8, 1876

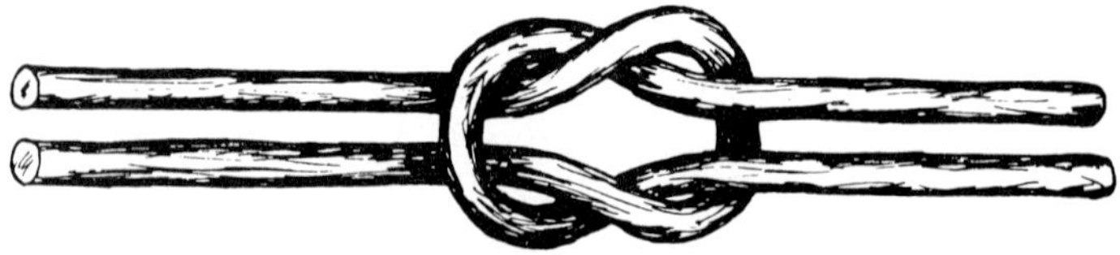

249 DOUBLE HALF HITCH HORSE WIRE
WITH HITCH EVERY TWENTY-FOUR INCHES

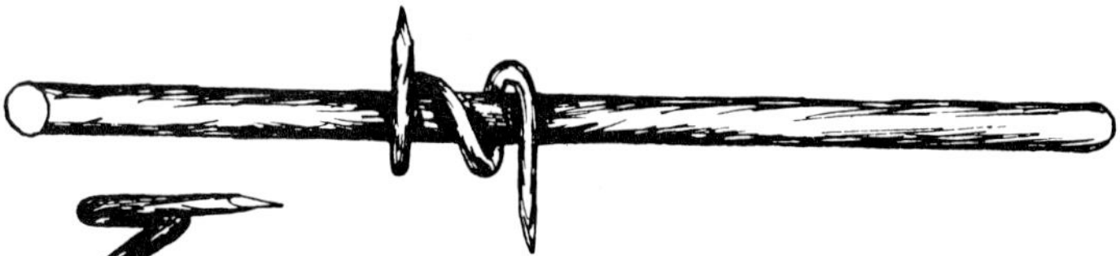

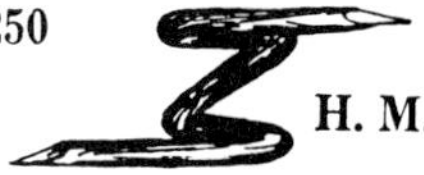

250 H. M. ROSE ONE-QUARTER INCH WRAPPED BARB
PATENTED DECEMBER 25, 1877

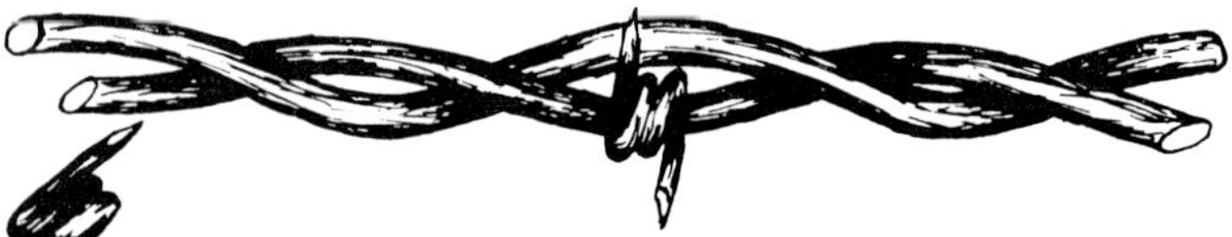

251 GLIDDEN WIRE WITH ONE LINE OF SOLID COPPER
PAT. J. F. GLIDDEN, NOV. 24, 1874

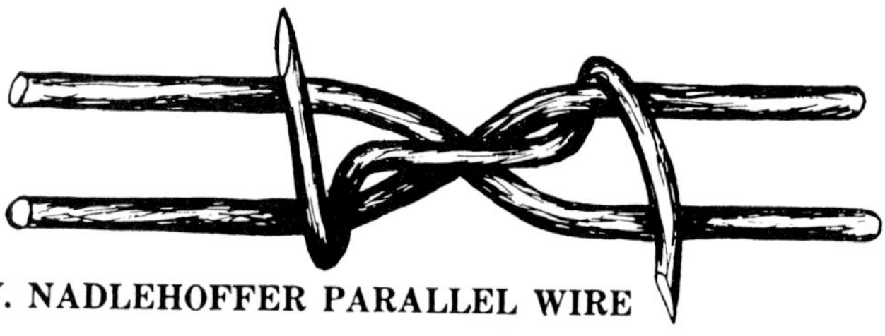

252 J. W. NADLEHOFFER PARALLEL WIRE
PATENTED JANUARY 2, 1883

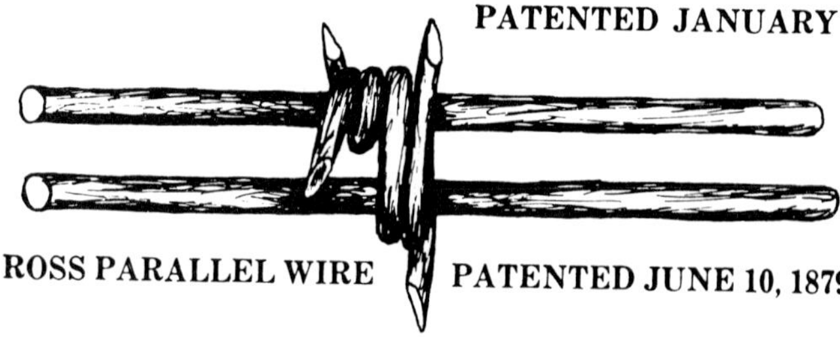

253 ROSS PARALLEL WIRE PATENTED JUNE 10, 1879

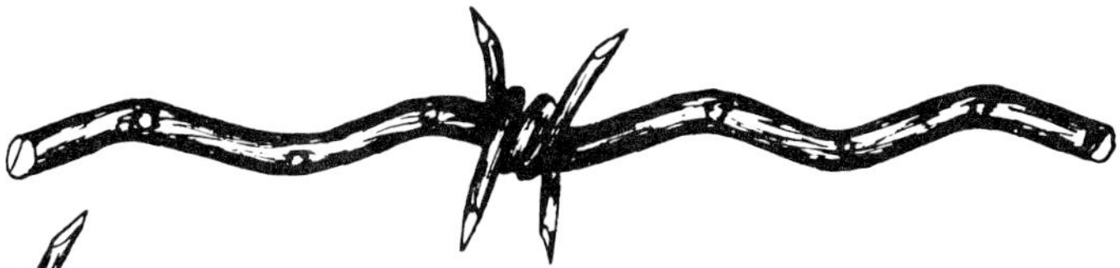

254

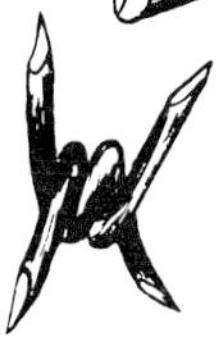

KINKED SINGLE LINE WIRE
COMMONLY CALLED "SOLO" WIRE

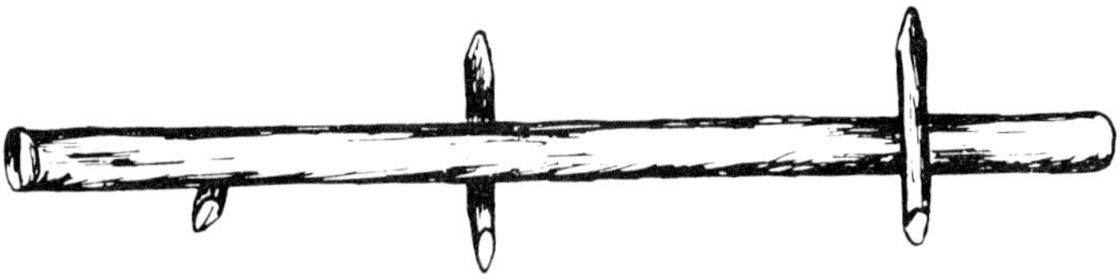

255

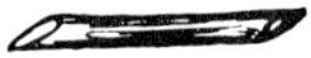

WELDED PIN WIRE FROM AUSTRALIA
PAT. J. C. PERRY, AUG. 24, 1897

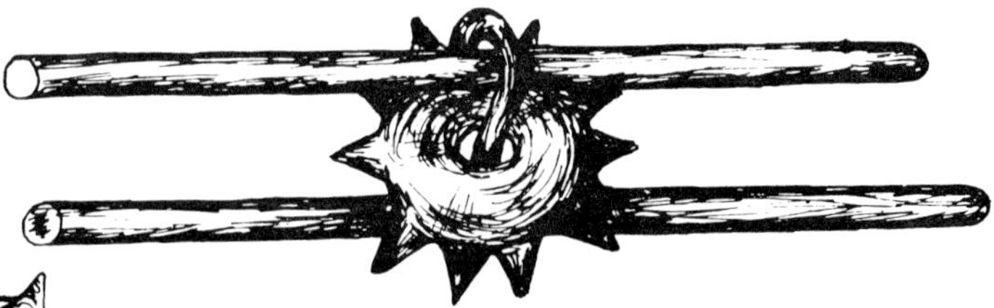

256

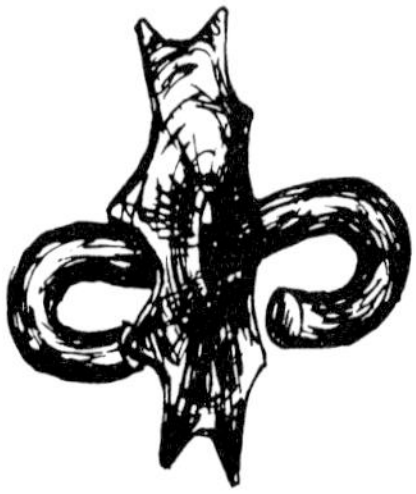

HODGE'S PARALLEL ROWELL
PATENTED AUGUST 2, 1887

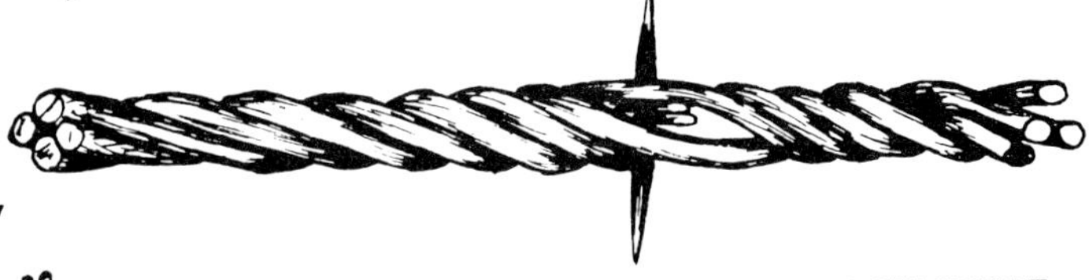

257

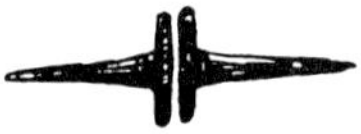

H. M. UNDERWOOD'S DOUBLE TACK WIRE
PATENTED AUGUST 6, 1878

258 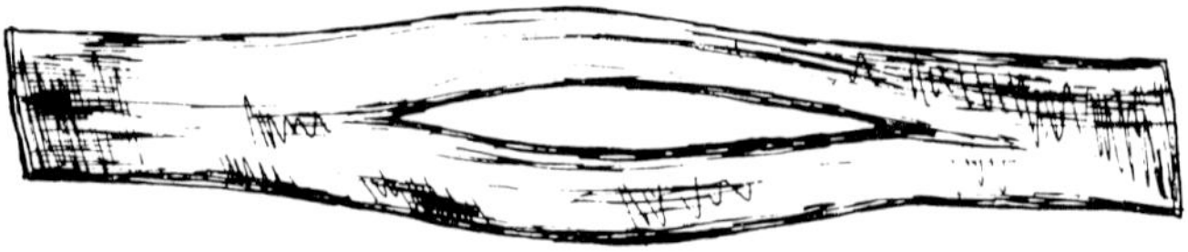

M. KELLEY SPLIT RIBBON WIRE
DOUBLE POINTED PINS USED FOR BARBS
PATENTED NOVEMBER 17, 1868

259 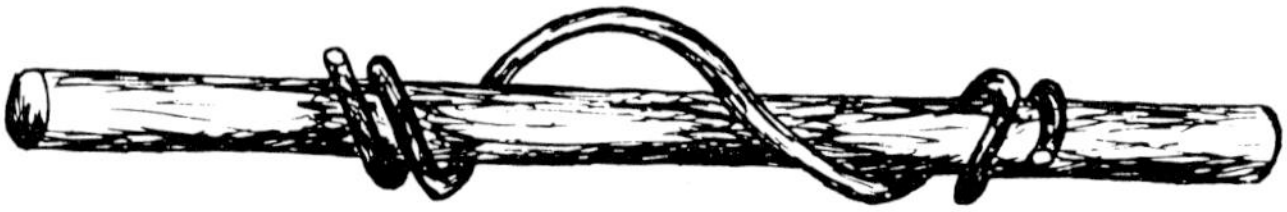

TOP RUNNER WIRE

LOOPS SPACED THREE AND ONE-HALF FEET APART
PAT. M. B. CHAPPELL, APRIL 9, 1889

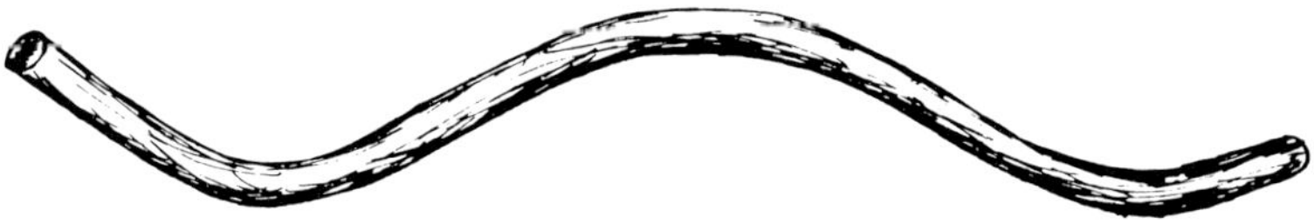

260 HIGH TENSILE SPIRAL NON-SAG WIRE
PAT. J. L. RITER, OCT. 10, 1893

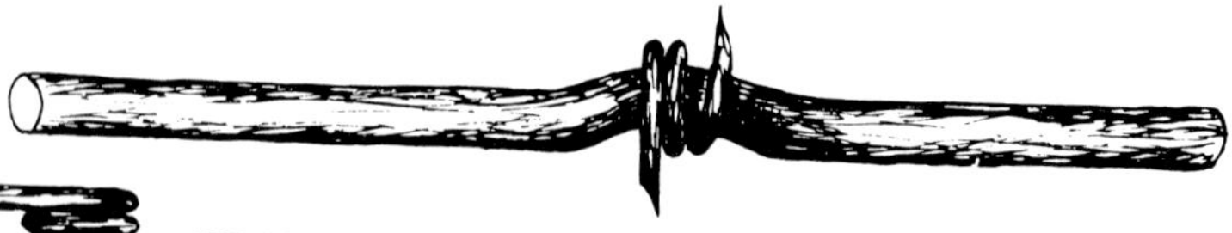

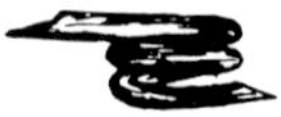

PROBABLY A VARIATION OF H. W. PUTNAM
261 PATENT No. 187,172 PATENTED FEBRUARY 6, 1877

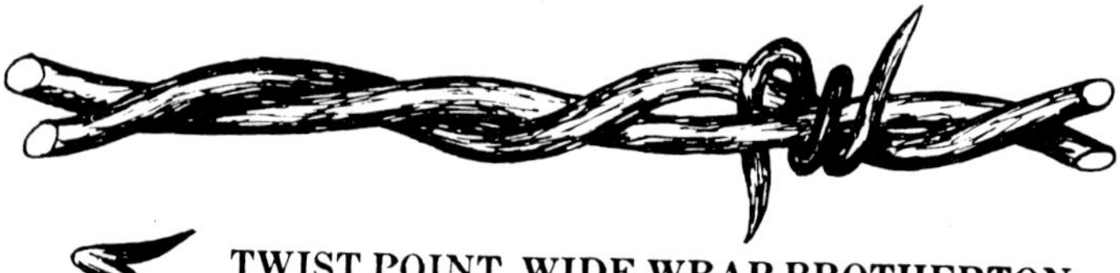

262 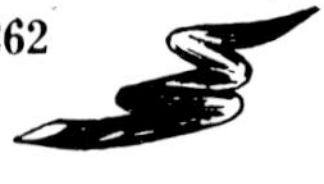 TWIST POINT, WIDE WRAP BROTHERTON
PATENTED SEPTEMBER 3, 1878

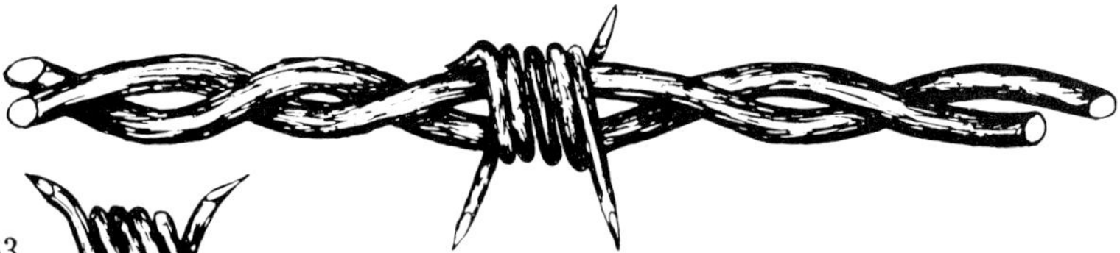

263 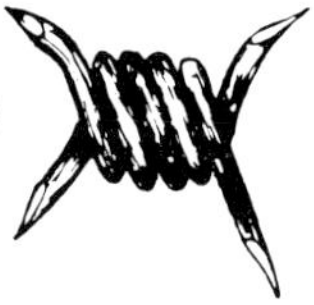

MODERN RIGHT AND LEFT TWIST
LIGHT GAUGE, HIGH TENSILE LINE WIRE

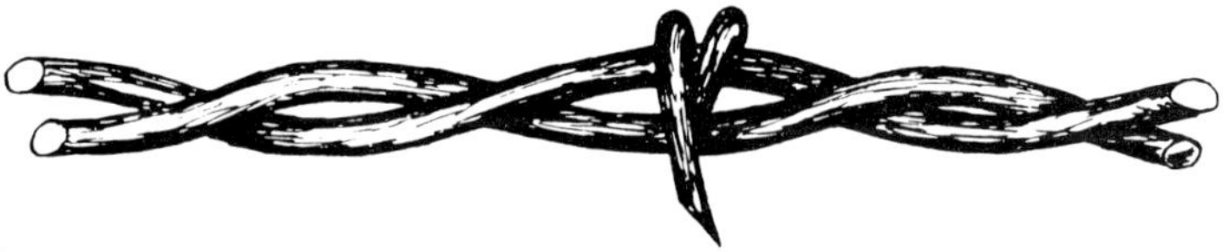

264 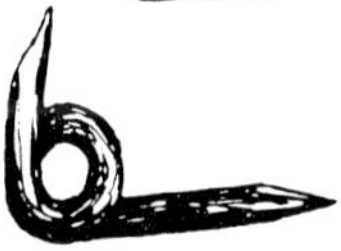

EDENBORN TWO LINE WIRE
PATENTED MARCH 17, 1885

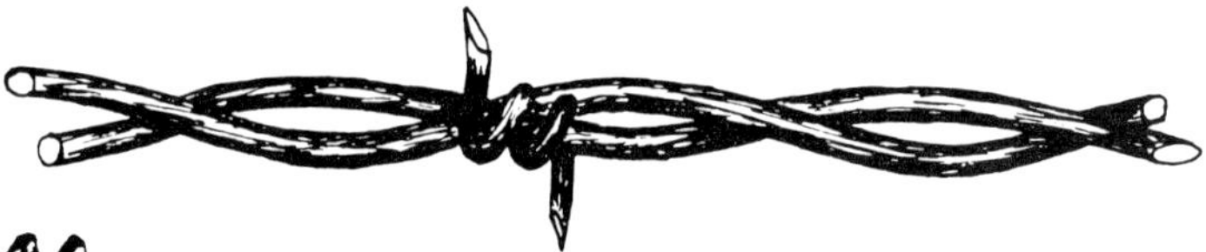

265 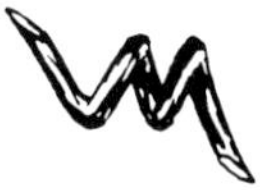

MODERN ROSE TWO LINE WIRE

PAT. H. M. ROSE, DEC. 25, 1877

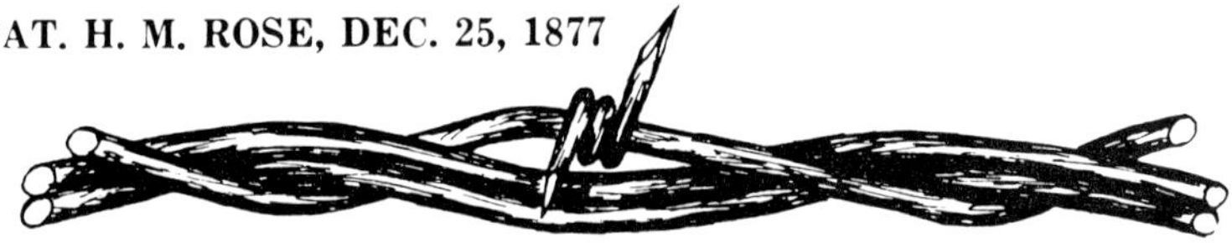

266 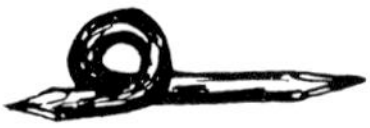

GLIDDEN THREE LINE
TWO LINE RUNNING PARALLEL
PAT. J. F. GLIDDEN, NOV. 24, 1874

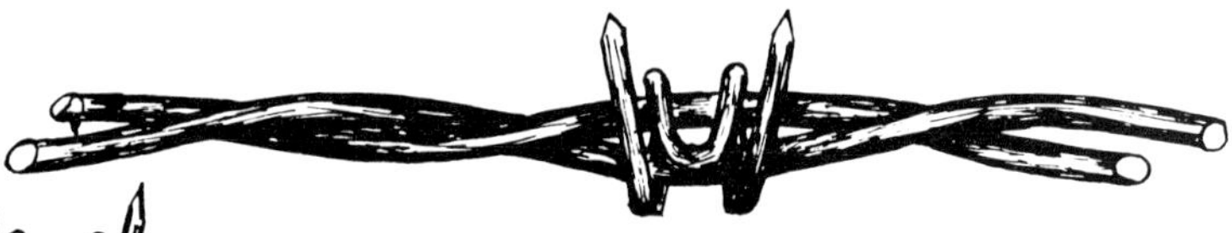

267 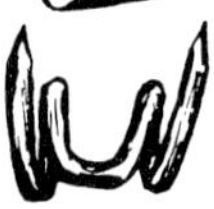

FACTORY FREAK ON MERRILL'S TWIRL

268

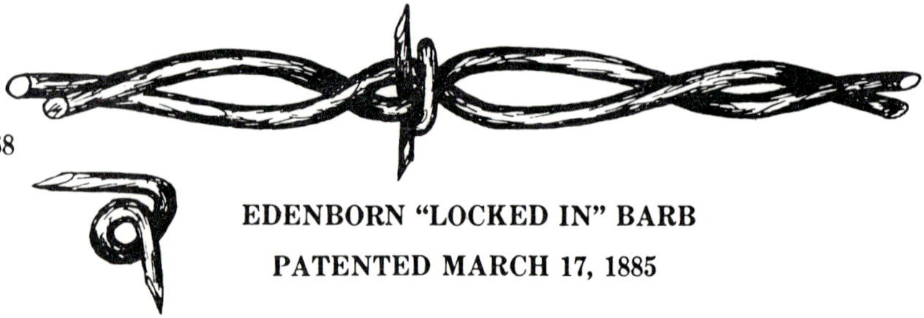

EDENBORN "LOCKED IN" BARB

PATENTED MARCH 17, 1885

269

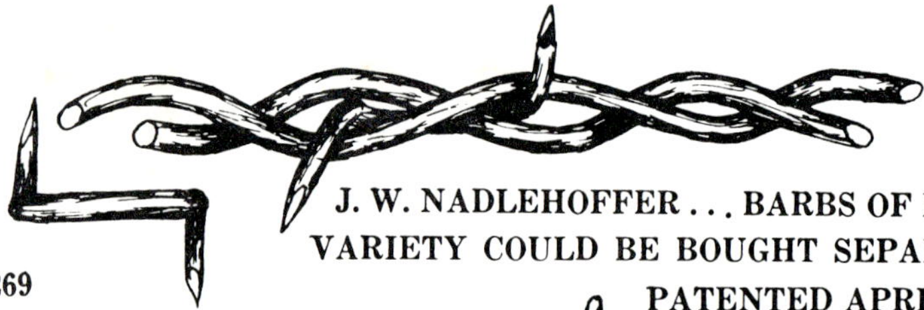

J. W. NADLEHOFFER . . . BARBS OF EARLY VARIETY COULD BE BOUGHT SEPARATELY

PATENTED APRIL 2, 1878

270

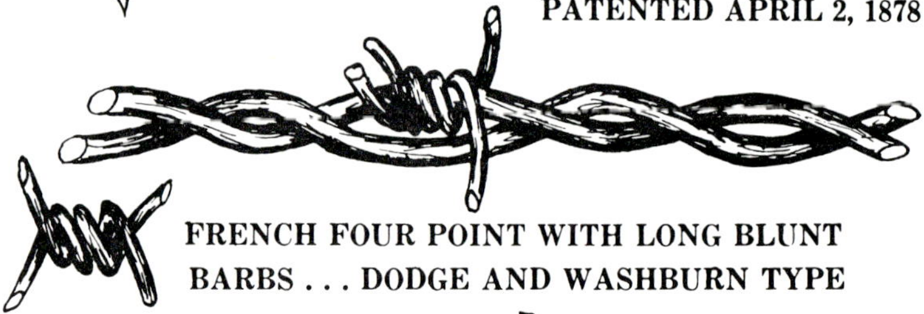

FRENCH FOUR POINT WITH LONG BLUNT BARBS . . . DODGE AND WASHBURN TYPE

271

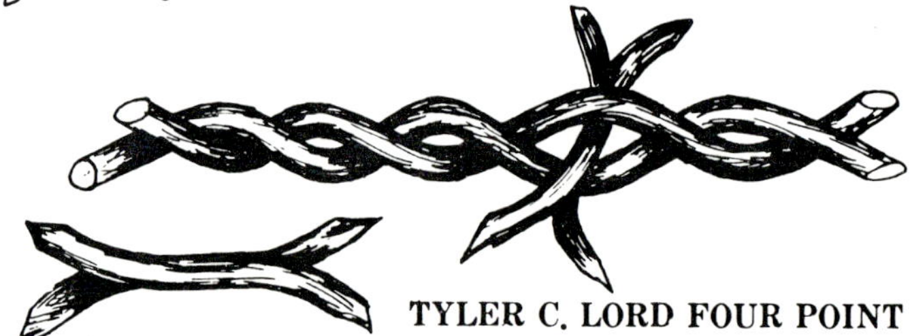

TYLER C. LORD FOUR POINT WIRE

PATENTED FEBRUARY 12, 1884

272

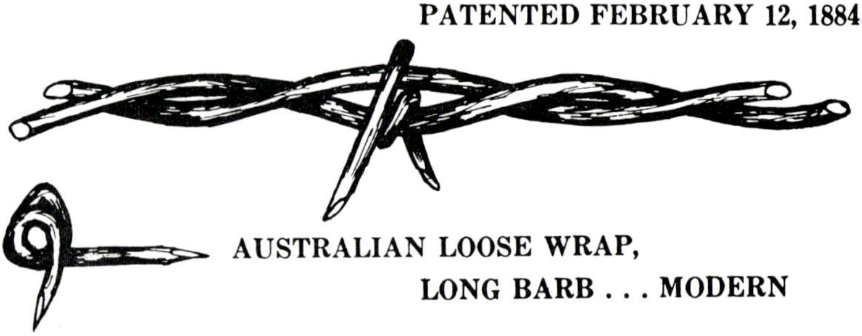

AUSTRALIAN LOOSE WRAP, LONG BARB . . . MODERN

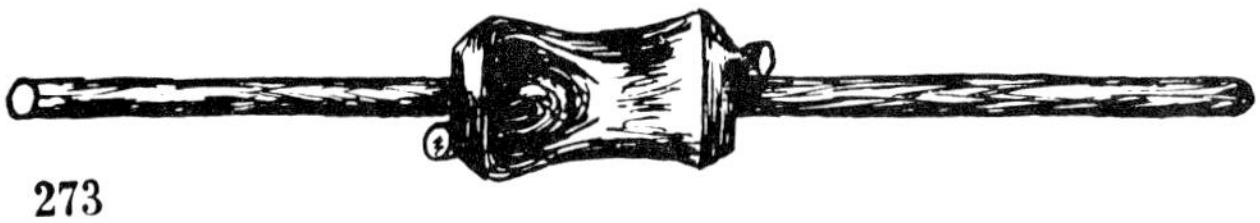

273

PLANTER WIRE

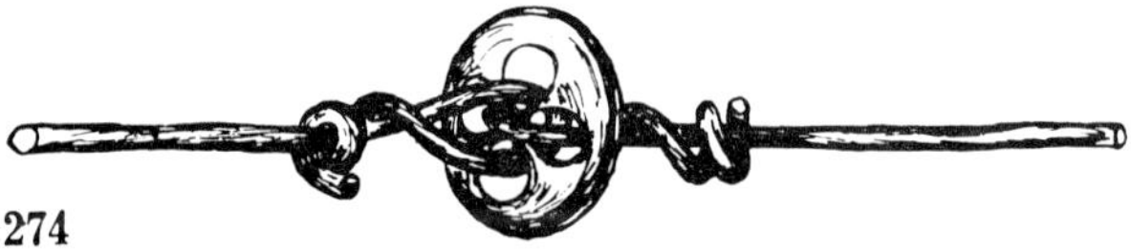

274

PLANTER WIRE

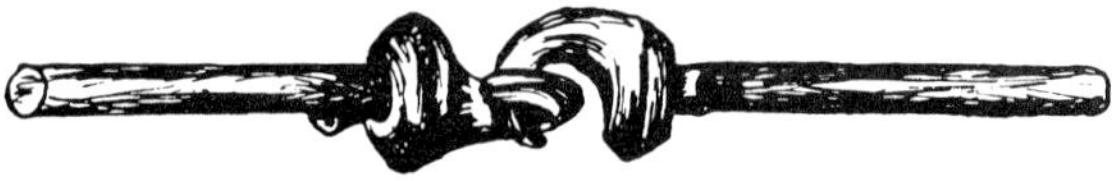

275

PLANTER WIRE

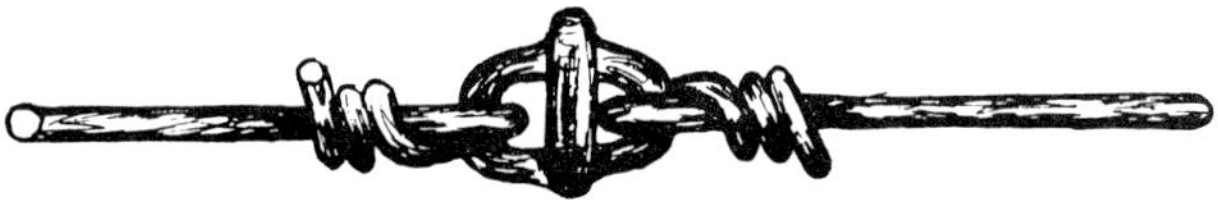

276

PLANTER WIRE

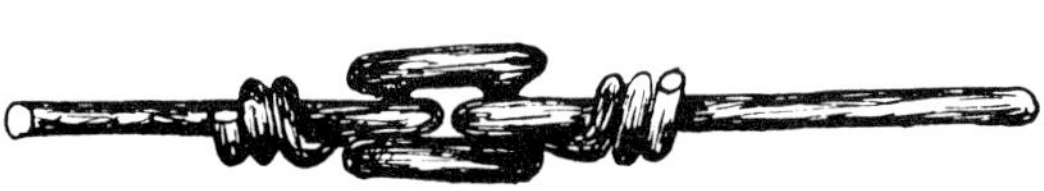

277

PLANTER WIRE

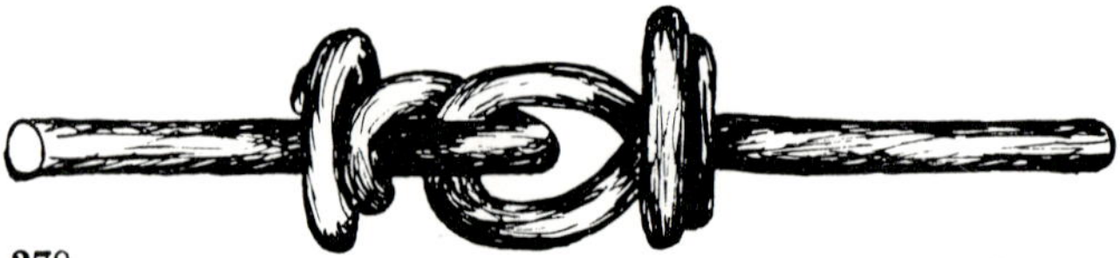

278

PLANTER WIRE

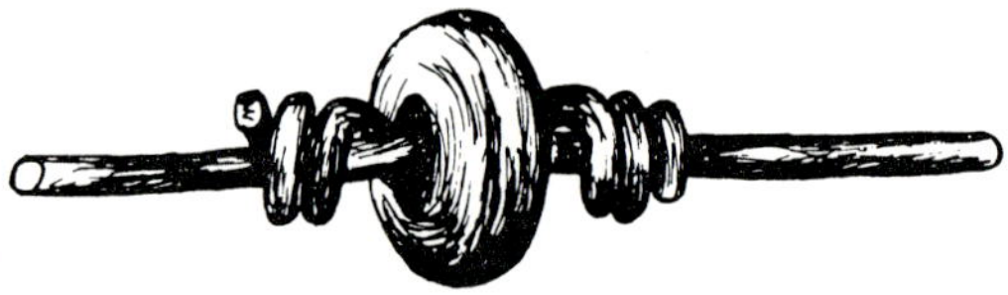

279

PLANTER WIRE

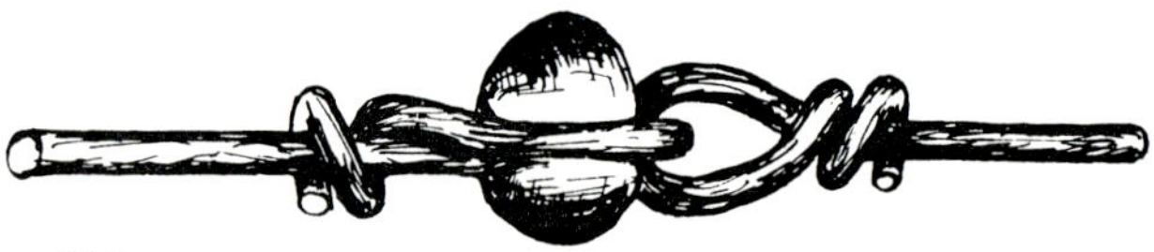

280

PLANTER WIRE

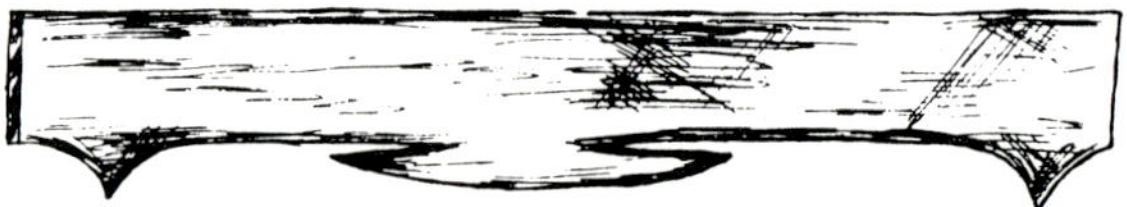

SIDE

281

T. V. ALLIS' RIBBON WIRE

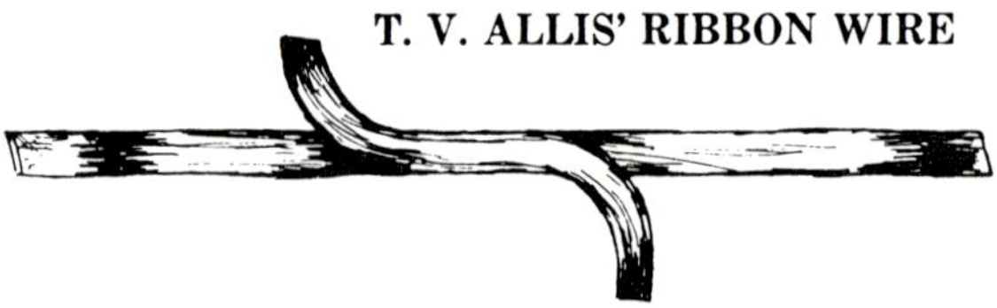

TOP

PATENTED FEBRUARY 17, 1891

282

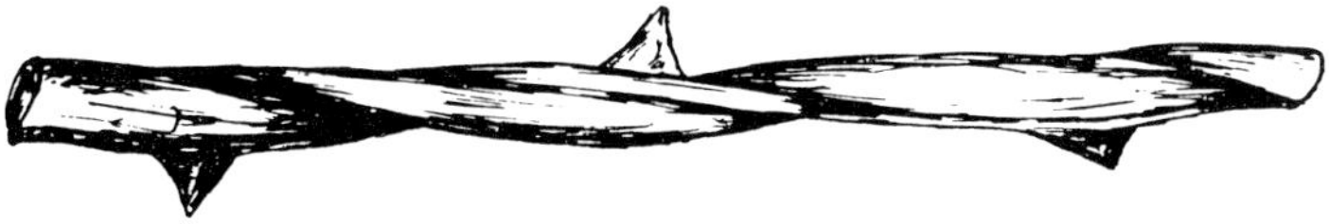

A ROUND ROD "BUCKTHORN"

PAT. T. V. ALLIS, JULY 26, 1881

283

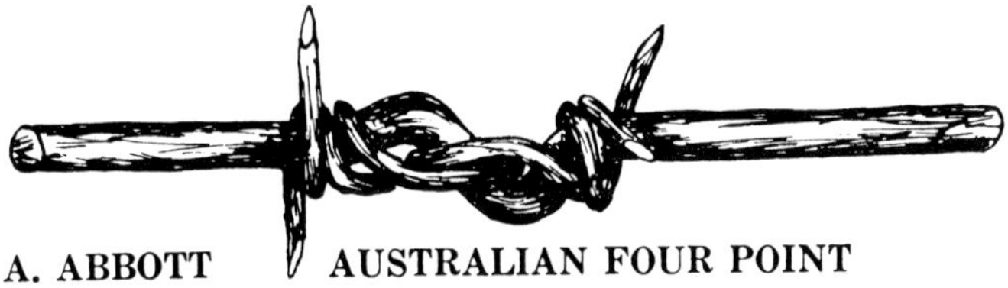

H. A. ABBOTT

FEB. 13, 1939

AUSTRALIAN FOUR POINT

FROM NEW SOUTH WALES AUSTRALIA

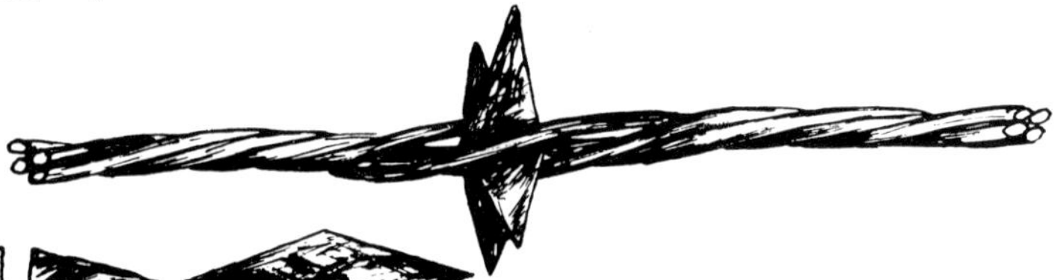

284

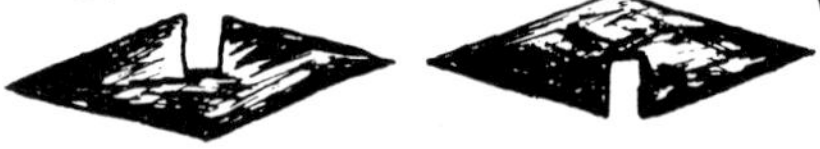

W. T. BURROWS PATENTED JULY 3, 1877

285

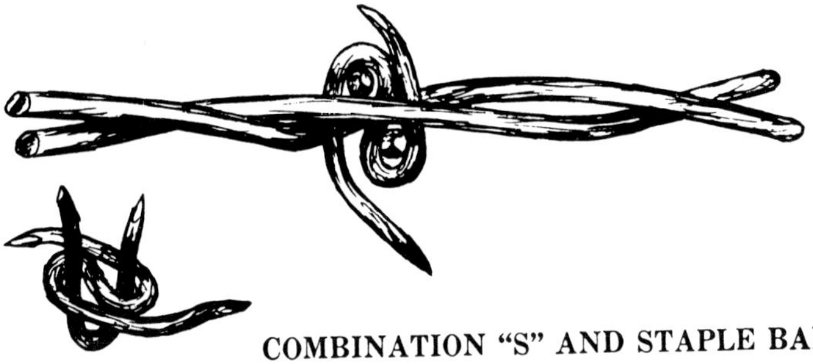

COMBINATION "S" AND STAPLE BARB

SPENCER H. ST. JOHN
PATENTED JANUARY 15, 1878

286

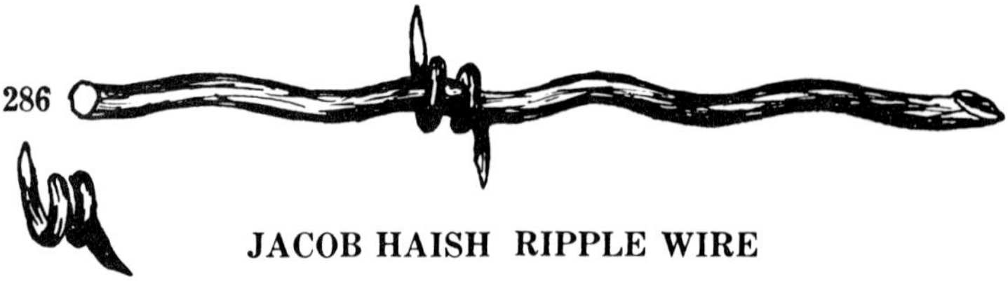

JACOB HAISH RIPPLE WIRE

PATENTED FEBRUARY 1, 1887

287

TEXAS AND PACIFIC RAILROAD WIRE
PAT. A. J. UPHAM, SEPT. 4, 1883

288

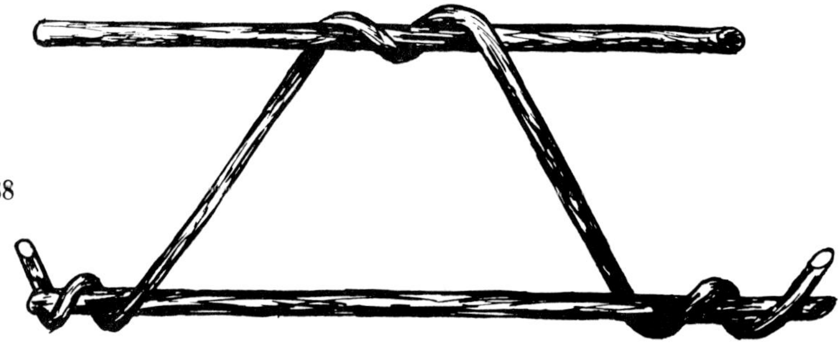

HORSE PEN FINISH WIRE

289

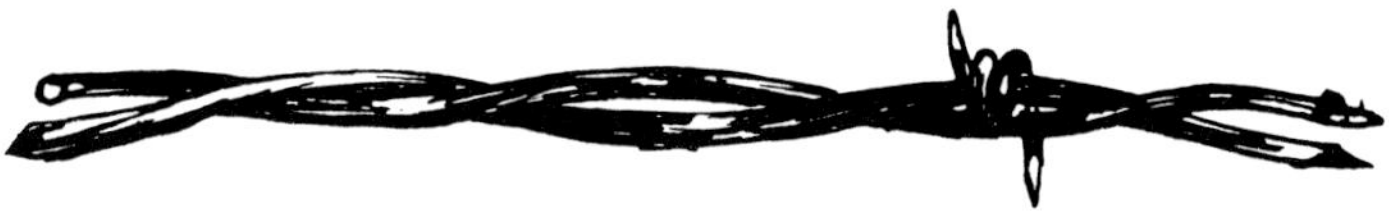

GLIDDEN UNION-PACIFIC RAILROAD WIRE
(ONE ROUND, ONE TRIANGLE)
PAT. J. F. GLIDDEN, NOV. 24, 1874

290

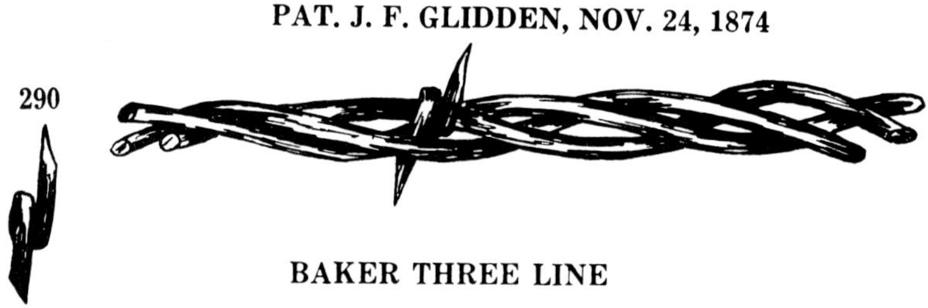

BAKER THREE LINE

PAT. G. C. BAKER, FEB. 27, 1883

291

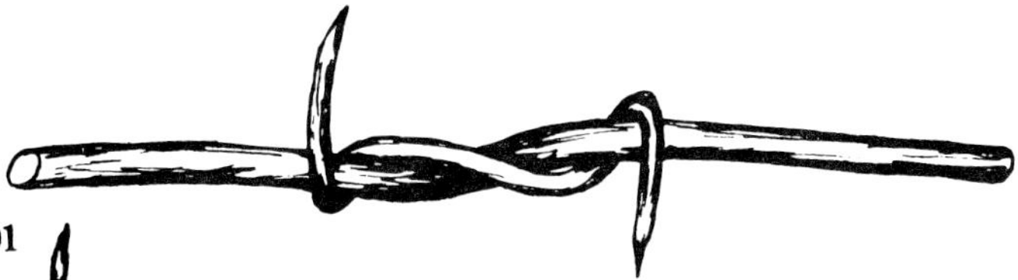

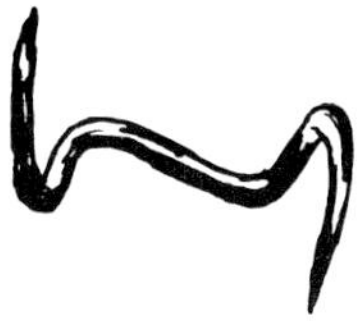

TWO POINT SINGLE LINE
MARTIN M. MACK
PATENTED MAY 4, 1875

292

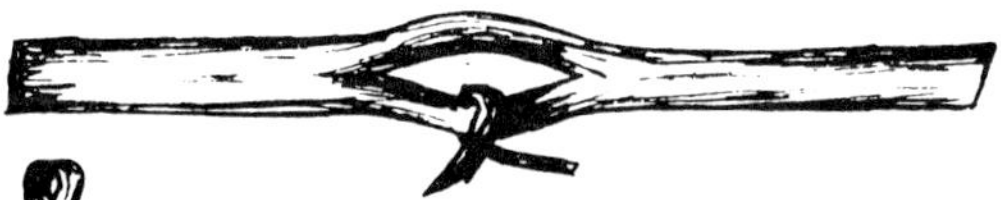

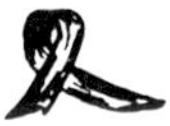

VERY LIGHT AND NARROW RIBBON
WITH LOOSE BARBS
PAT. M. KELLEY, NOV. 17, 1868

293

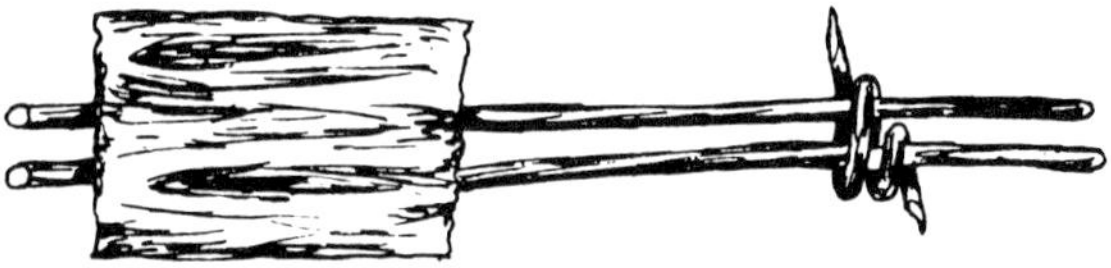

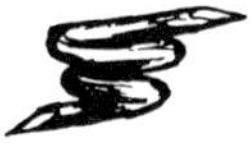

J. BROTHERTON PARALLEL
WITH WOODEN BLOCKS
PATENTED SEPTEMBER 3, 1887

294

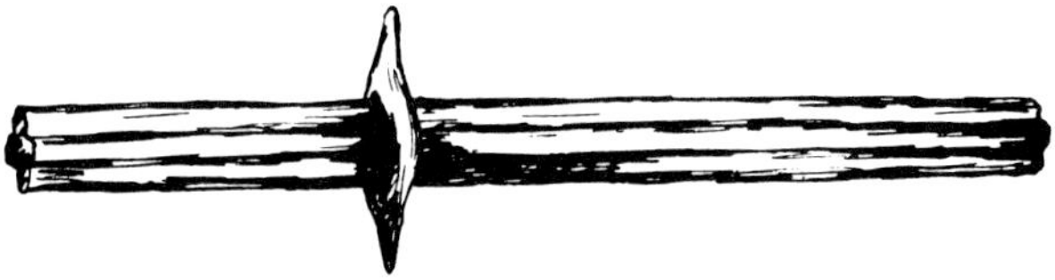

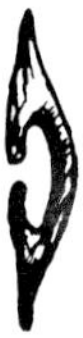

SCUTT'S ROD AND RAIL WIRE
PATENTED OCTOBER 23, 1883
BY HIRAM B. SCUTT

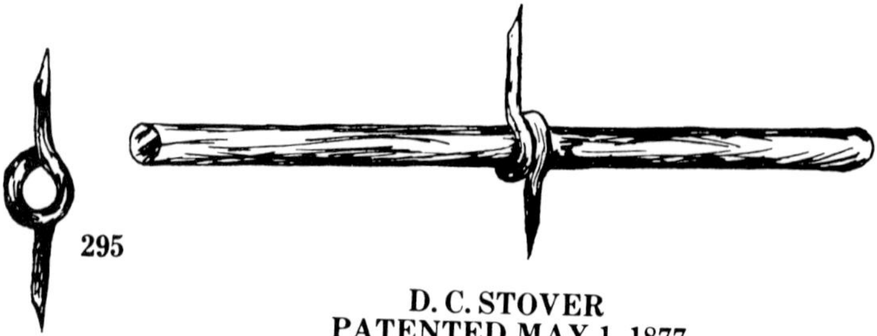

295

D. C. STOVER
PATENTED MAY 1, 1877

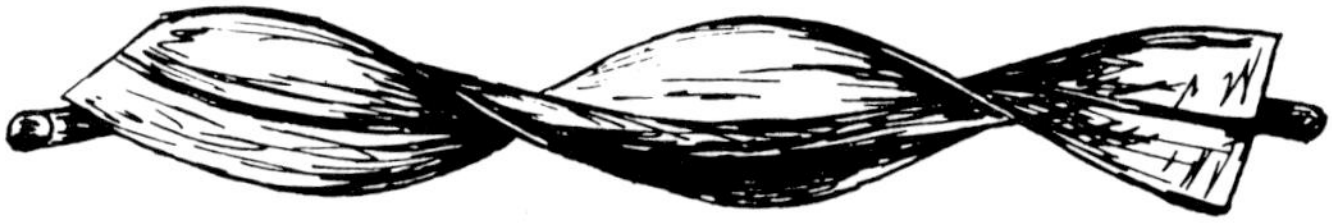

296

T. V. ALLIS
BARBLESS RIBBON AND SINGLE WIRE
PATENTED JULY 26, 1881

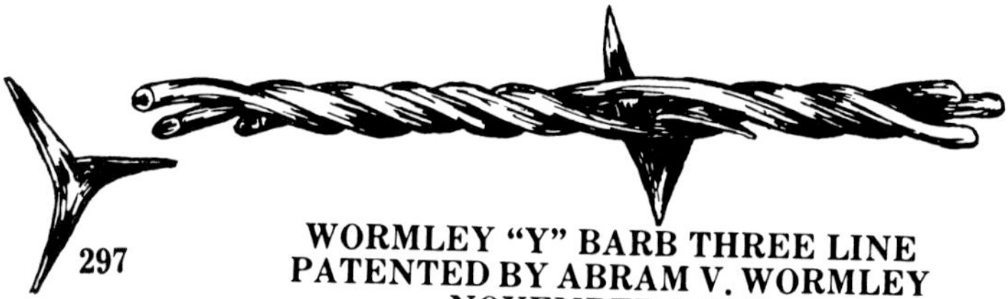

297

WORMLEY "Y" BARB THREE LINE
PATENTED BY ABRAM V. WORMLEY
NOVEMBER 2, 1875

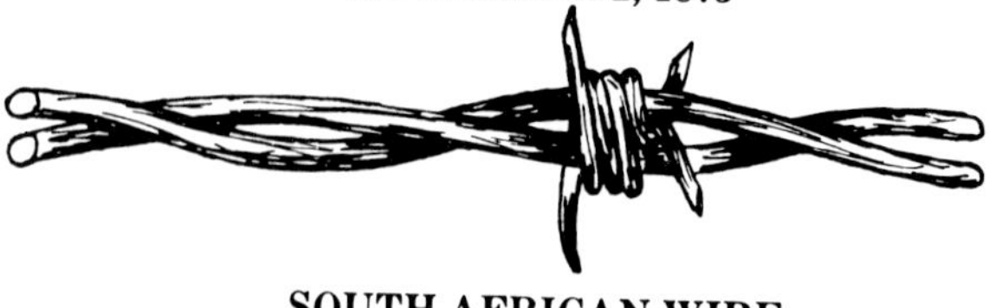

298

SOUTH AFRICAN WIRE
(MODERN MANUFACTURE)

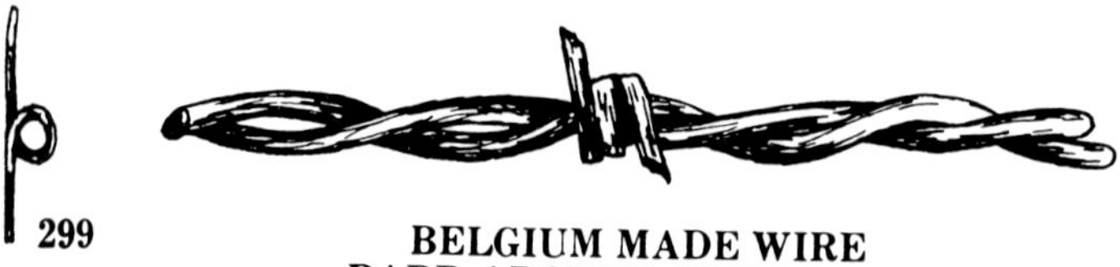

299

BELGIUM MADE WIRE
BARB AROUND BOTH LINES

300

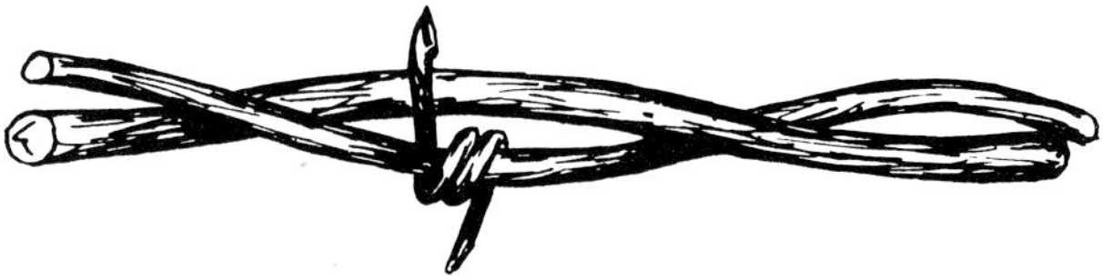

GLIDDEN SMALL AND LARGE
TWO LINE — A VARIATION

PAT. J. F. GLIDDEN, NOV. 24, 1874

301

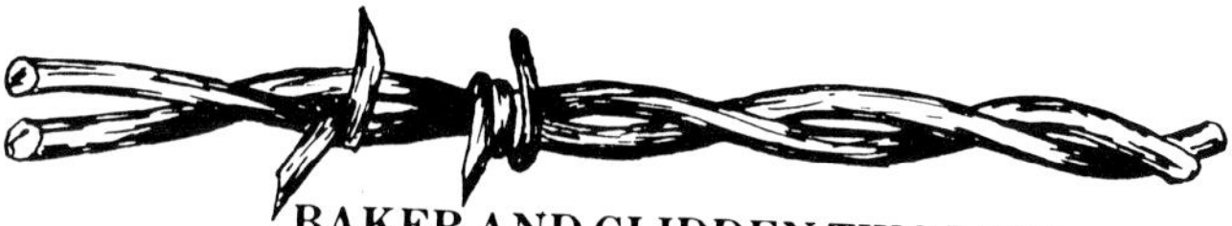

BAKER AND GLIDDEN TWO LINE
BAKER BARB ON ONE LINE, GLIDDEN
BARB ON OTHER LINE

PAT. G. C. BAKER, FEB. 27, 1883

302

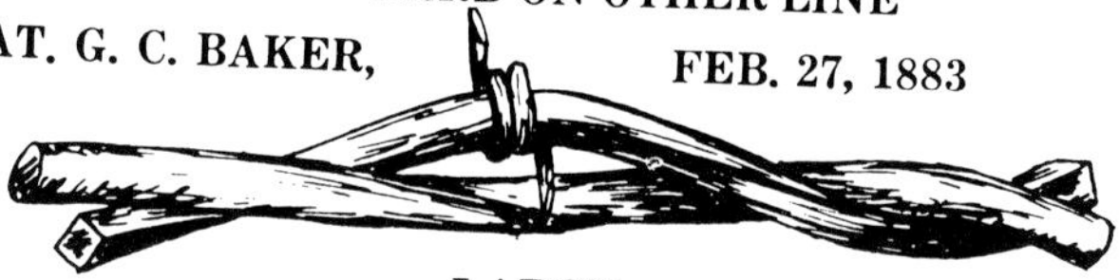

LARGE
ROUND AND SQUARE GLIDDEN
RAILROAD WIRE
(BARB ON SQUARE LINE)
PAT. J. F. GLIDDEN, NOV. 24, 1874

303

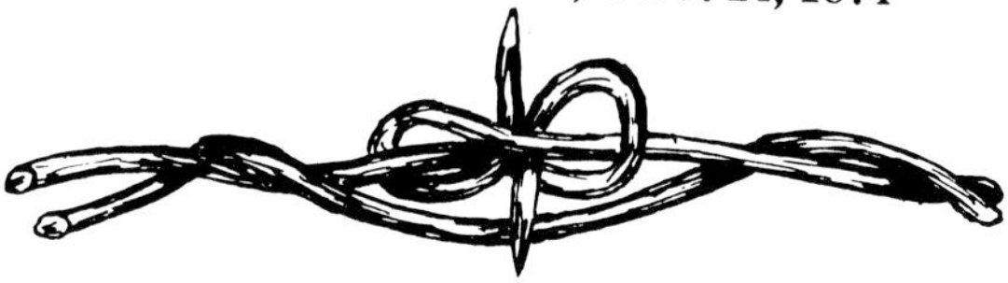

BRONSON LINK WIRE
PATENTED BY A. E. BRONSON, APRIL 24, 1877

304

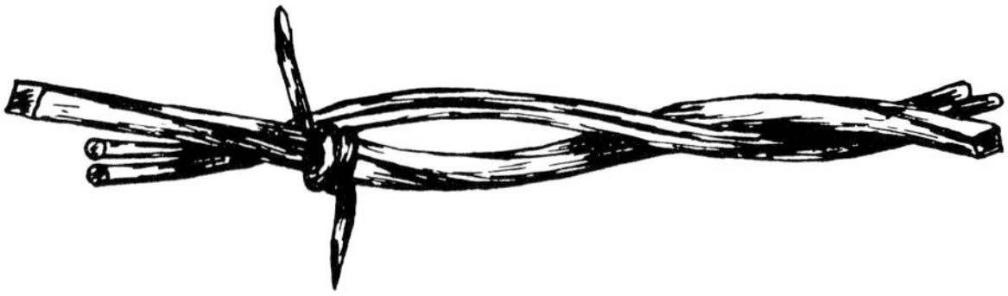

GLIDDEN THREE LINE RAILROAD WIRE
(BARB ON SQUARE LINE)

PAT. J. F. GLIDDEN, NOV. 24, 1874

305 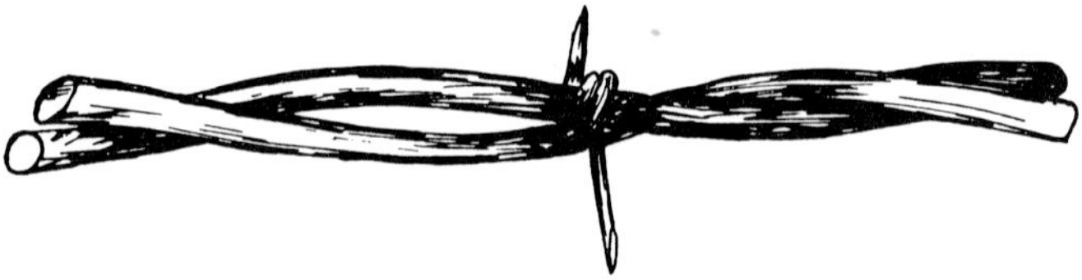

GLIDDEN OVAL AND ROUND TWO LINE
RAILROAD WIRE

PAT. NOV. 24, 1874

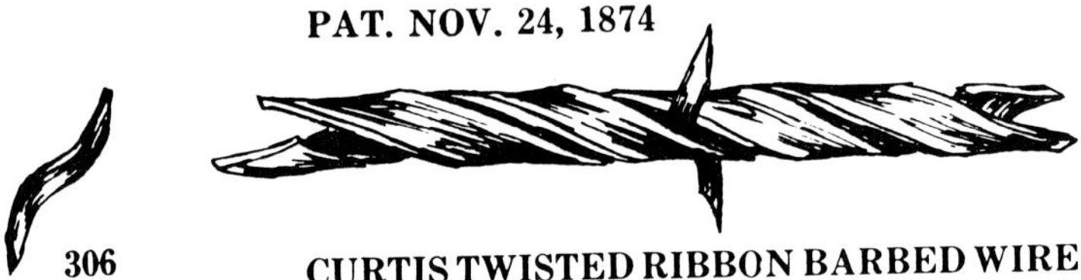

306

CURTIS TWISTED RIBBON BARBED WIRE
(OFFSET BARB) PAT. FEBRUARY 13, 1894
BY J. D. CURTIS

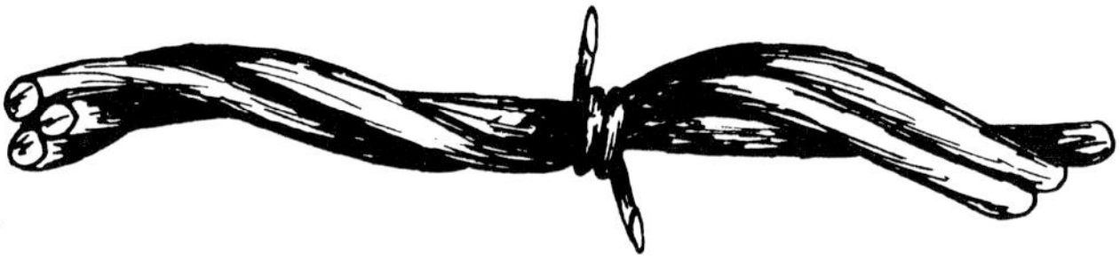

307

GLIDDEN THREE LINE
(TWO LINES TWISTED AROUND
HORIZONTAL LINE)

PAT. NOV. 24, 1874

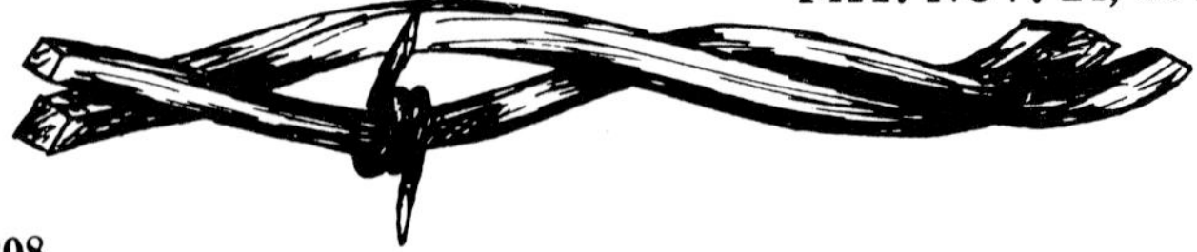

308

GLIDDEN LARGE AND SMALL SQUARE
TWO LINE RAILROAD WIRE

PAT. NOV. 24, 1874

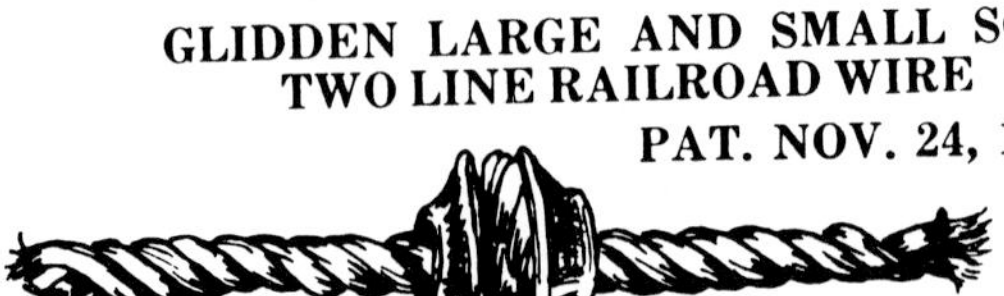

309

ROPE PLANTER WIRE, FORERUNNER
OF METAL PLANTER WIRE

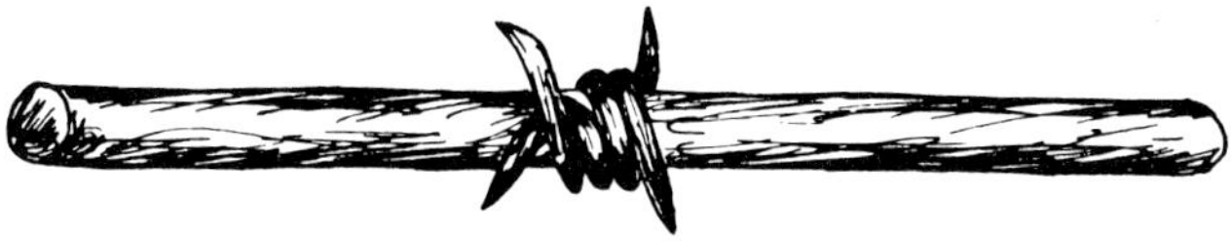

310

SHINN ROUND LINE LOCK BARB
PATENTED BY M. C. SHINN, MARCH 1, 1881

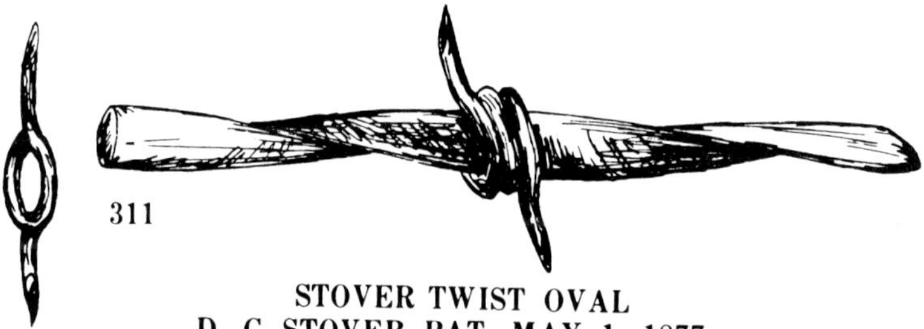

311

STOVER TWIST OVAL
D. C. STOVER PAT. MAY 1, 1877

312

ALLIS FLAT RIBBON BARBED WIRE
PATENTED T. V. ALLIS, JANUARY 5, 1892

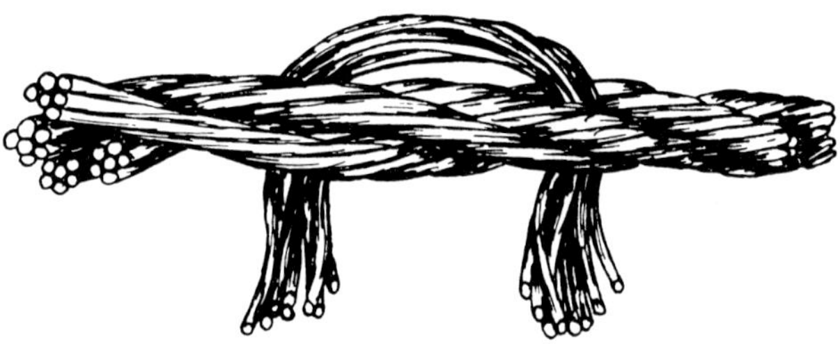

313

MINE CABLE USED FOR FENCING
AFTER BEING CUT IN HALF
(NEVADA AND COLORADO TERRITORY)

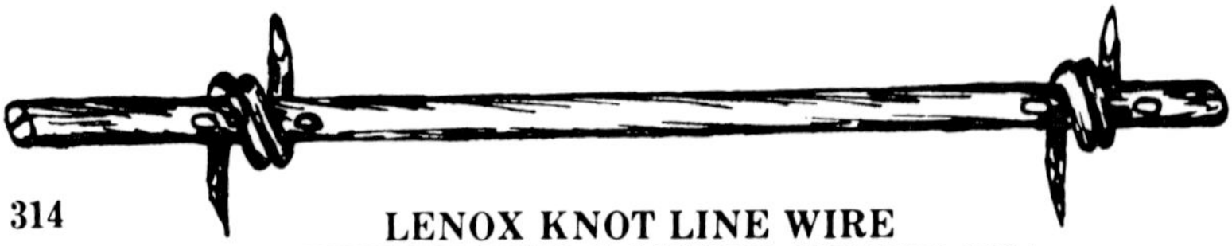

314

LENOX KNOT LINE WIRE
PATENTED BY E. S. LENOX, JUNE 24, 1884

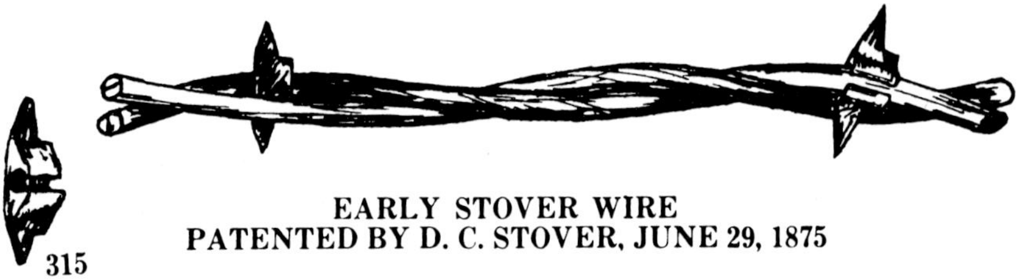

315

EARLY STOVER WIRE
PATENTED BY D. C. STOVER, JUNE 29, 1875

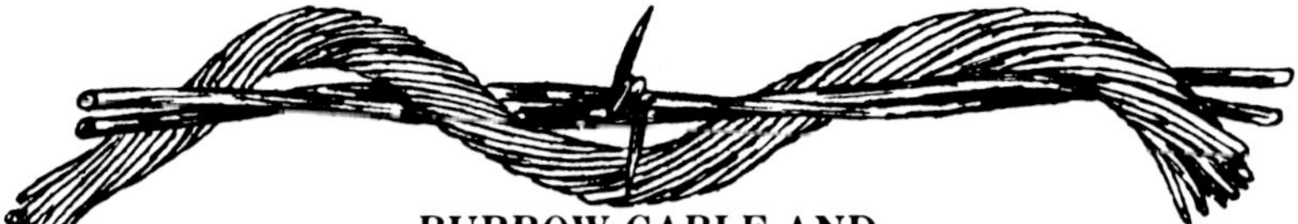

316

BURROW CABLE AND
GLIDDEN FENCE WIRE
(STRAWN COAL MINES IN TEXAS)

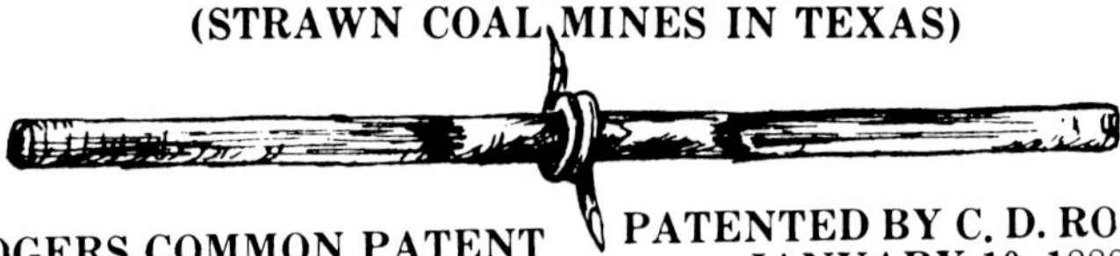

317

ROGERS COMMON PATENT

PATENTED BY C. D. ROGERS,
JANUARY 10, 1888

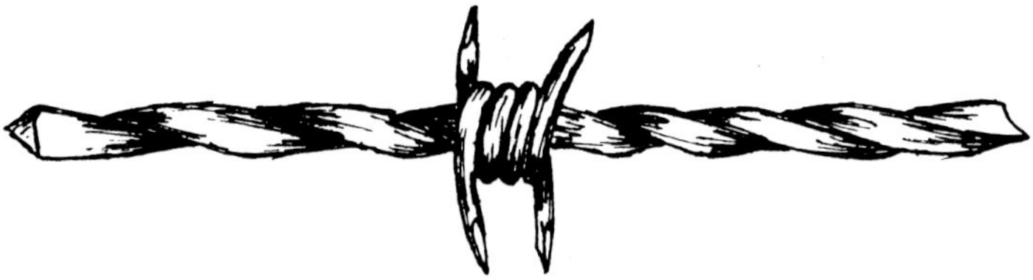

318

TRIANGLE FOUR POINT WIRE
(SOUTH AFRICA)

319 HALF ROUND, RAISED "V" IN CENTER
RAILROAD WIRE (TEXAS AND PACIFIC- 1880's)

PAT. JACOB HAISH, DEC. 15, 1885

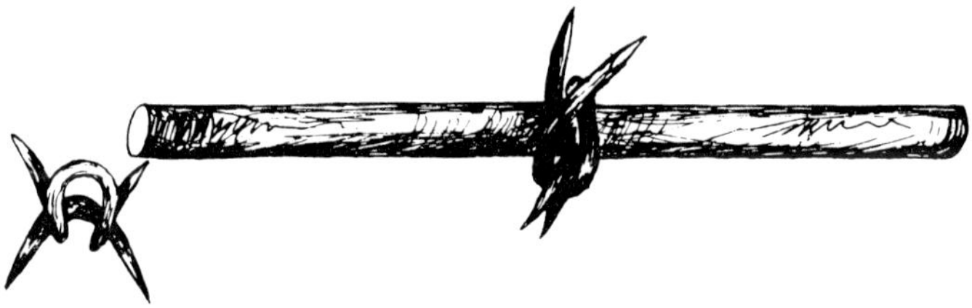

320 WING'S OVAL LINE
PATENTED BY L. T. WING, FEBRUARY 26, 1878

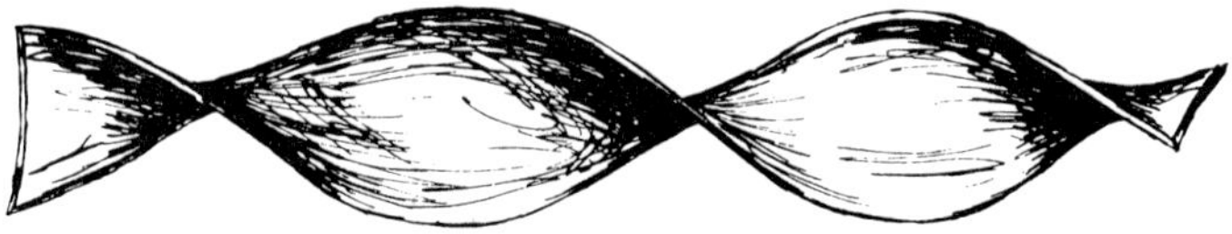

321 THREE-QUARTER INCH RIBBON WIRE
FROM FIRST CATTLE HOLDING PENS
IN SOUTH TEXAS TO GO UP TRAILS

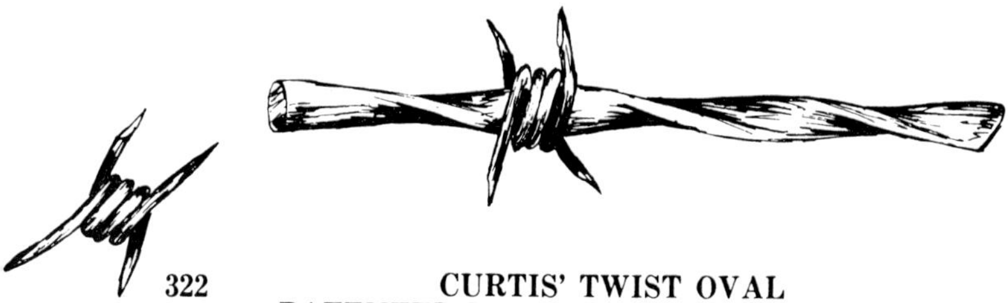

322 CURTIS' TWIST OVAL
PATENTED BY J. D. CURTIS, MARCH 28, 1893

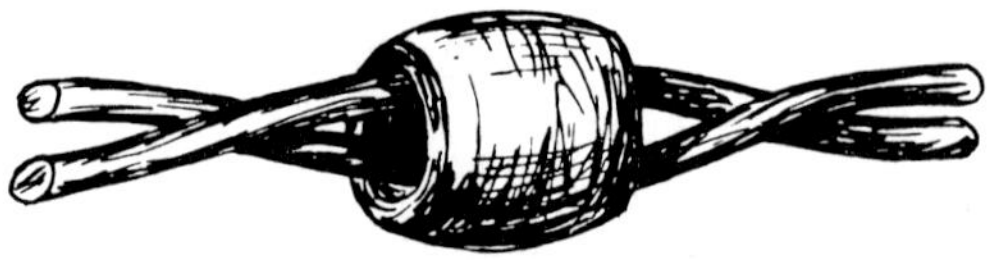

323

BARREL FENCE SPLICER

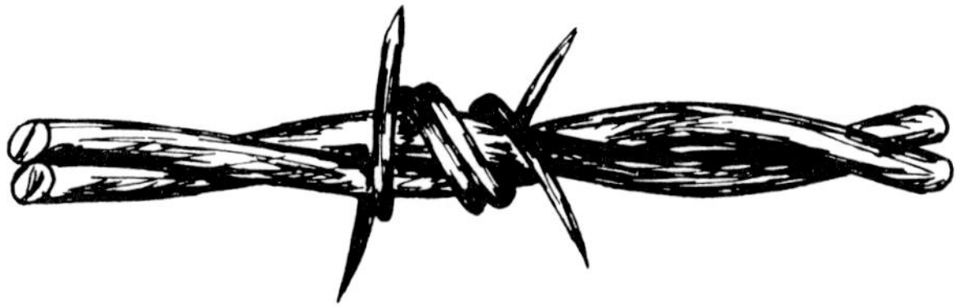

324

W. H. RODDEN
PATENTED BY MARCH 20, 1888

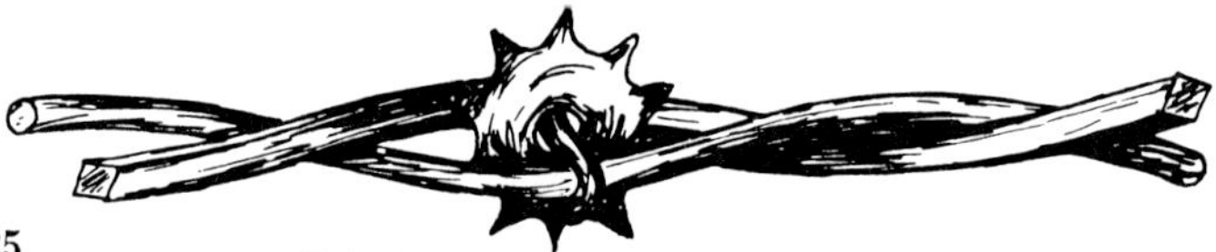

325

RAILROAD SPUR ROWELL
(ONE SQUARE LINE)
PATENTED C. A. HODGE, AUGUST 2, 1887

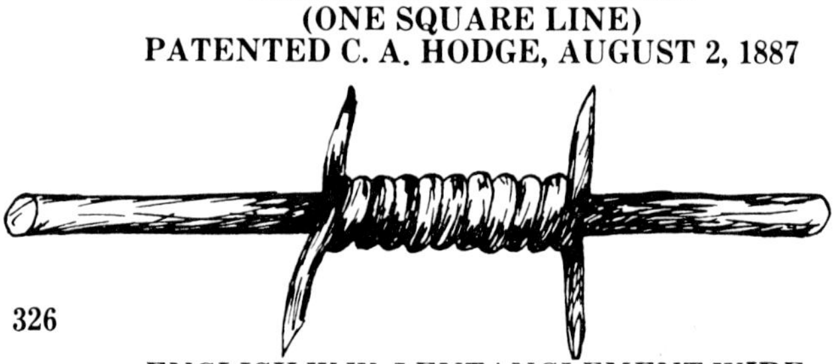

326

ENGLISH W.W. I ENTANGLEMENT WIRE
APPROXIMATELY 1916

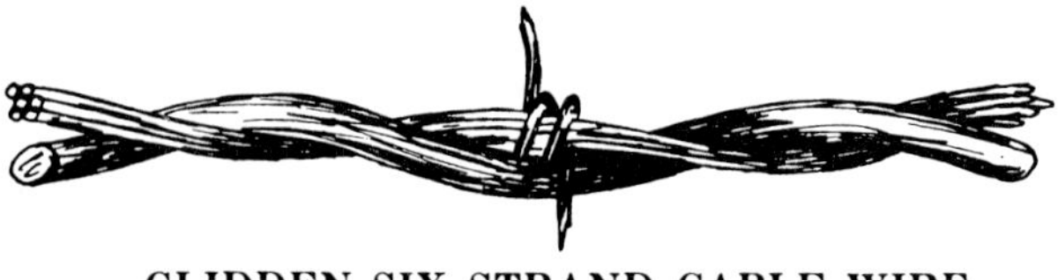

327

GLIDDEN SIX STRAND CABLE WIRE
PATENTED APRIL 16, 1886

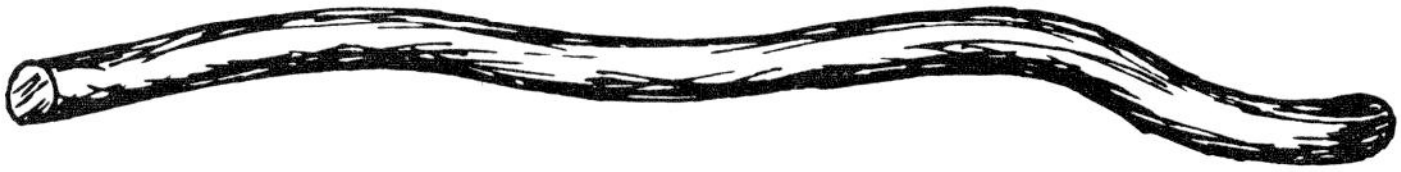

328

NEVADA TELEGRAPH LINE USED AS FENCING, STEEL CENTER, LEAD COATED (APPROXIMATELY 1868)

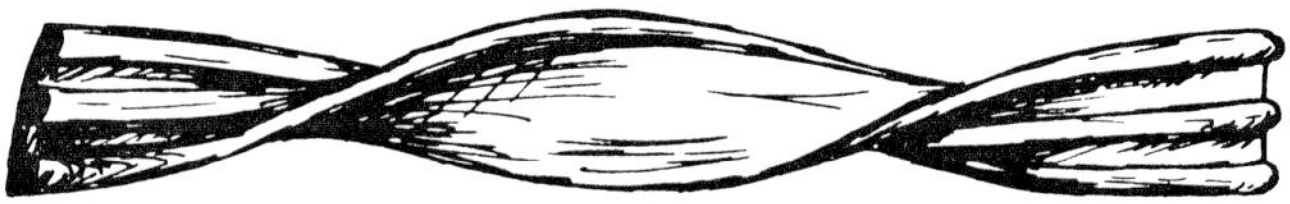

329

HARBAUGH'S TWISTED RIBBON
PATENTED BY J. W. HARBAUGH, JUNE 7, 1881

330

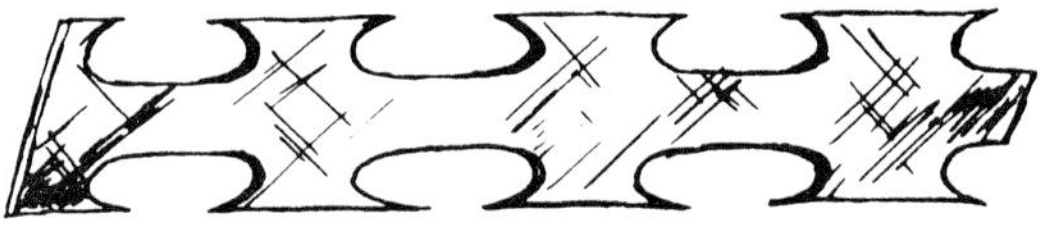

GERMAN MADE MILITARY ENTANGLEMENT (MODERN MANUFACTURE)

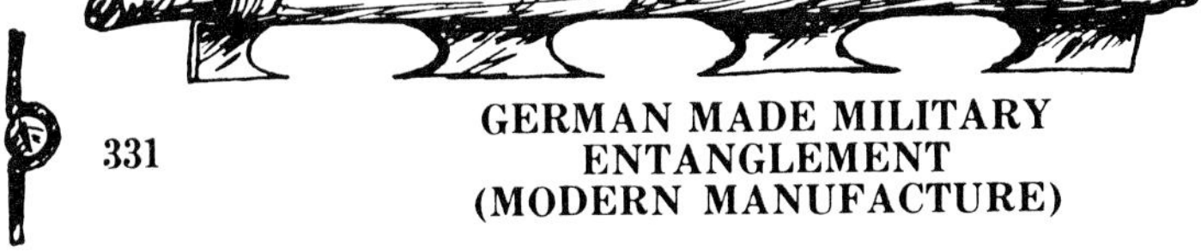

331

GERMAN MADE MILITARY ENTANGLEMENT (MODERN MANUFACTURE)

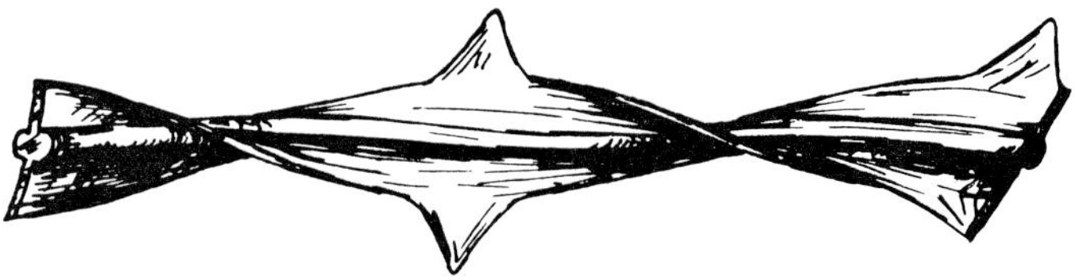

332

T. V. ALLIS' PARALLEL BARB
PATENTED NOVEMBER 30, 1880

ARMSTRONG AND DOOLITTLE
NOTCHED BARB

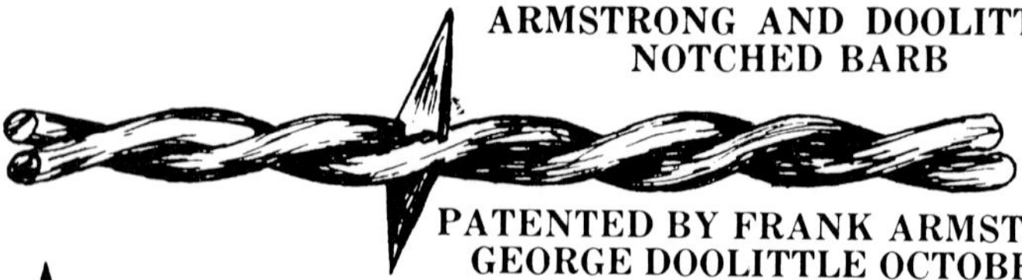

PATENTED BY FRANK ARMSTRONG AND
GEORGE DOOLITTLE OCTOBER 11, 1875

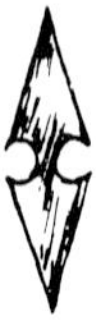

333

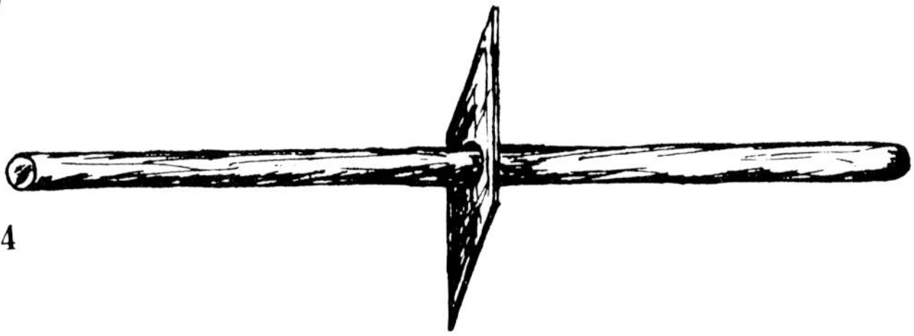

334

KELLY SINGLE LINE
PATENTED BY M. KELLY, FEBRUARY 11, 1868

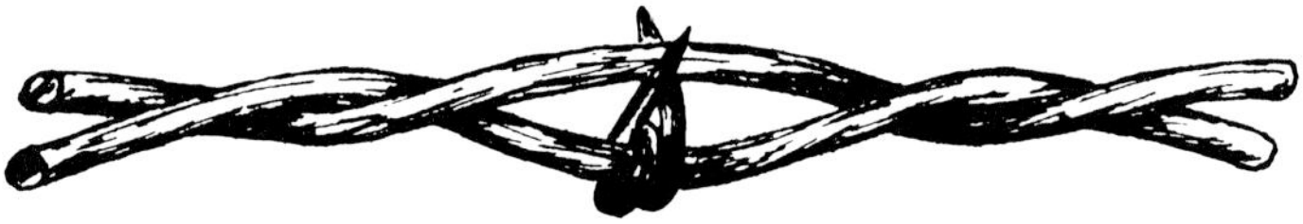

335

GLIDDEN TWO LINE
(ONE LINE SPLITS AND HOLDS BARB)

PAT. NOV. 24, 1874

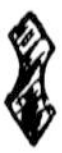

336

PECOS-HERRING GROOVED
SPIRAL RIBBON, PATENTED 1869

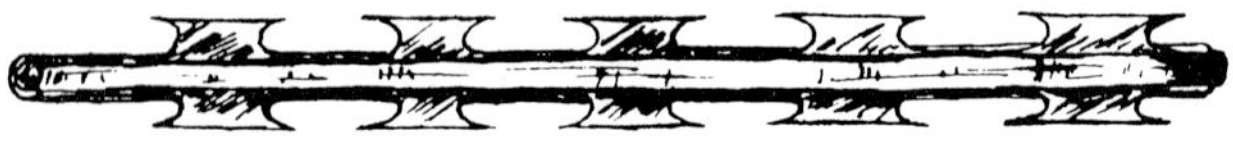

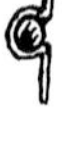

337

GERMAN MADE MILITARY
ENTANGLEMENT
(MODERN MANUFACTURE)

338 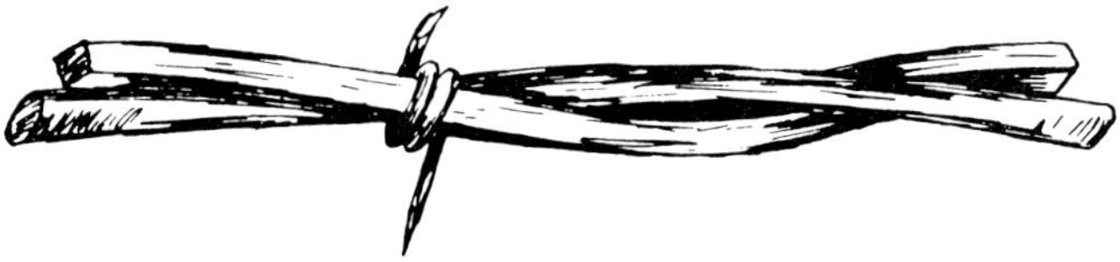

GLIDDEN SQUARE AND OVAL R. R. WIRE
PATENTED 1880's

PAT. NOV. 24, 1874

339 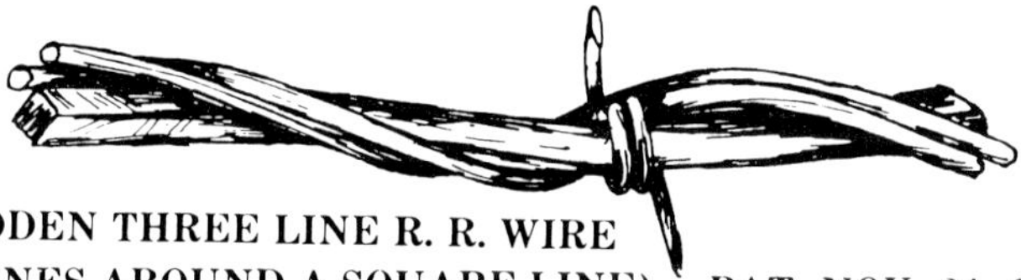

GLIDDEN THREE LINE R. R. WIRE
(TWO LINES AROUND A SQUARE LINE) PAT. NOV. 24, 1874

340

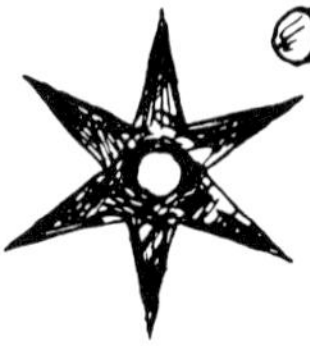

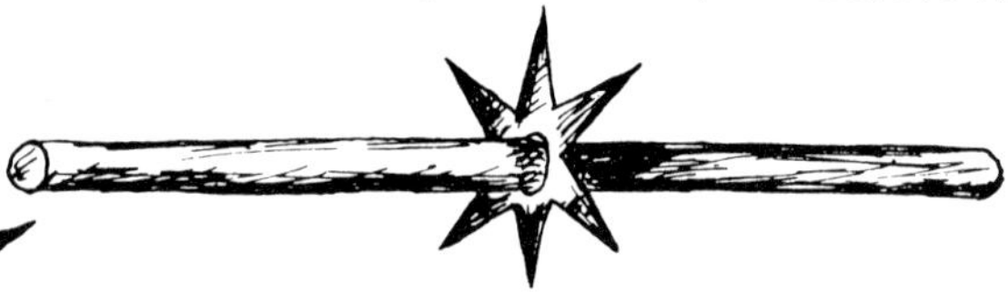

DODGE ROWELL WIRE
PATENTED BY F. H. DODGE, NOVEMBER 29, 1881

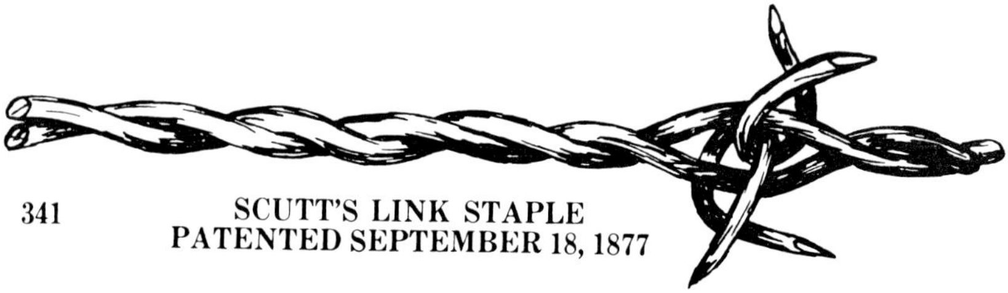

341 SCUTT'S LINK STAPLE
PATENTED SEPTEMBER 18, 1877

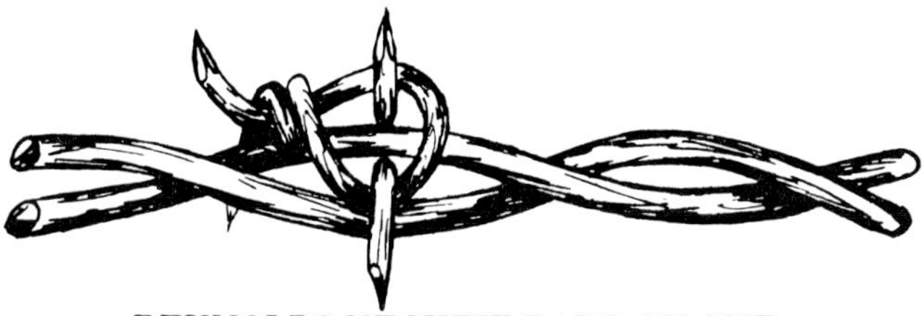

342

REYNOLDS NECKTIE BARB ON ONE
LINE, TWISTED INTO THE SECOND LINE
PATENTED BY H. REYNOLDS, MAY 14, 1878

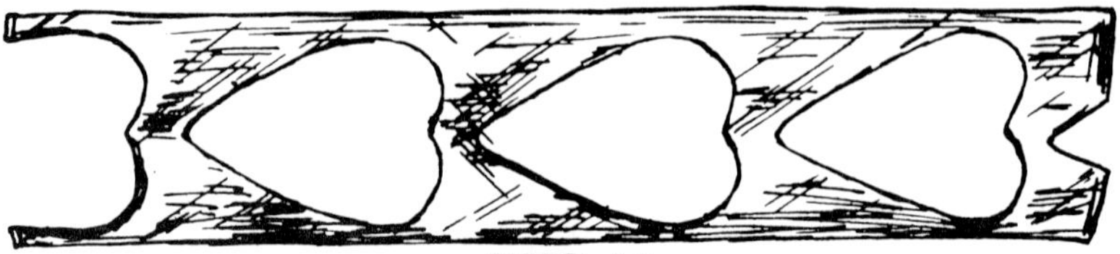

USED AS

343 ORNAMENTAL FENCE WIRE
(WAS APPLIED RUNNING HORIZONTAL
OR VERTICAL)

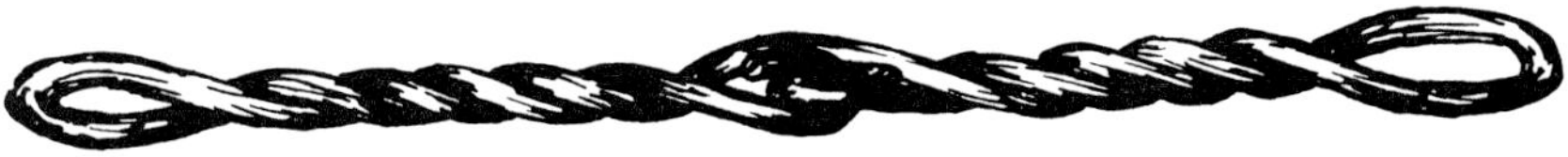

344 KANSAS GATE LINK WIRE
(MACHINE MADE) EACH LINK NINE
INCHES LONG, EARLY 1880's

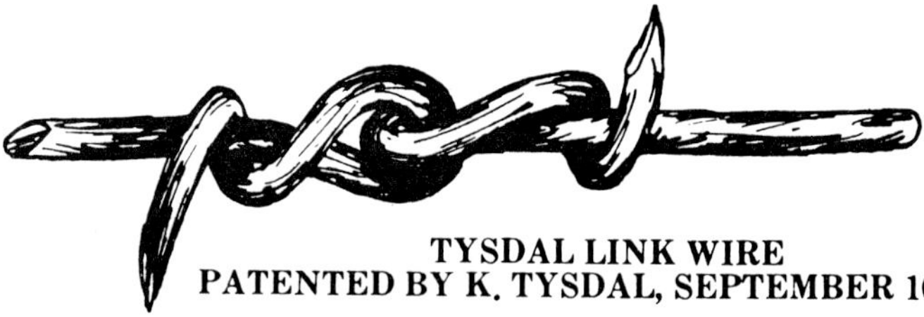

345 TYSDAL LINK WIRE
PATENTED BY K. TYSDAL, SEPTEMBER 10, 1878

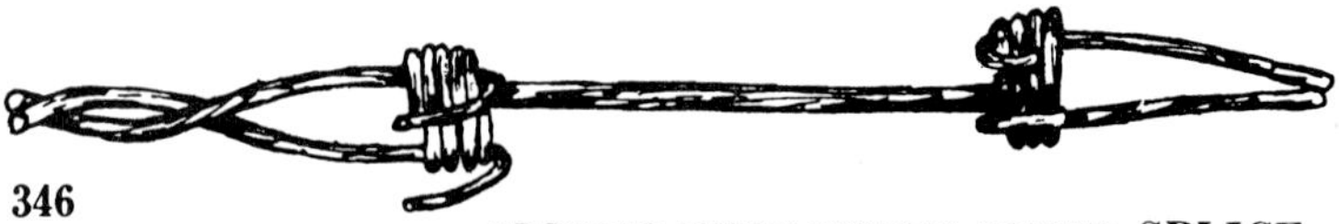

346 SPRING EXPANSION JOINT SPLICE
(WILL GIVE OR TAKE UP 3-4 INCHES OF SLACK)
FOUND IN MERRILL'S HOLD-FAST FENCE

347 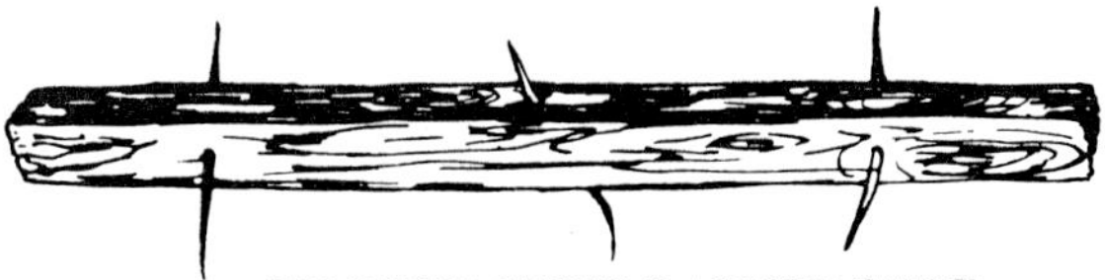

WOODEN STRIP BARBED WIRE
(IN A DEKALB, ILLINOIS MUSEUM)
H. M. ROSE PAT. MAY 13, 1873

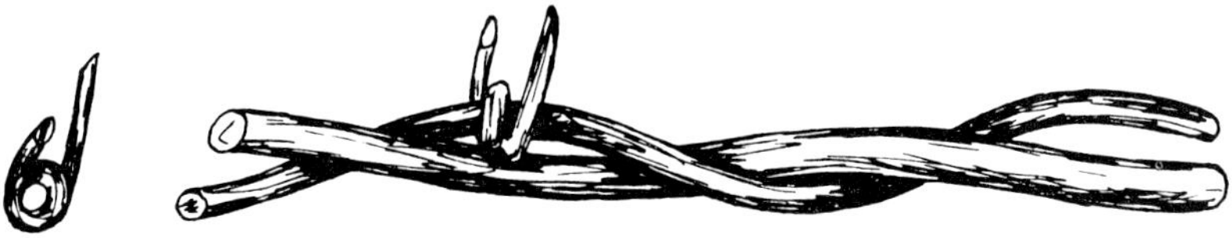

348 GLIDDEN TWO LINE GOAT BARB
(LINES ARE TWO DIFFERENT SIZES)

PAT. NOV. 24, 1874

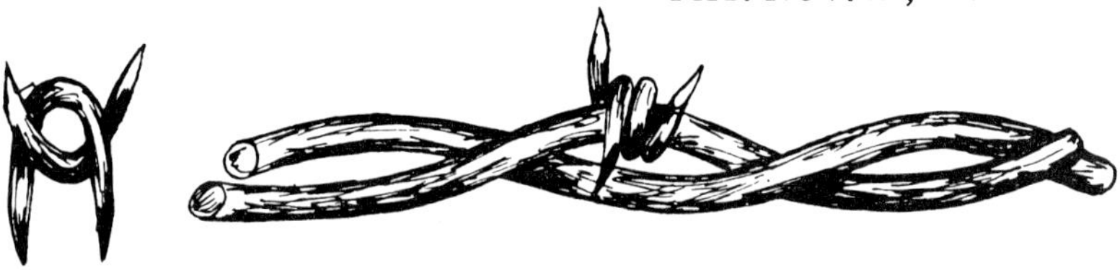

349 JAYNE AND HILL TWO LINE ON ONE LINE
PATENTED APRIL 11, 1876

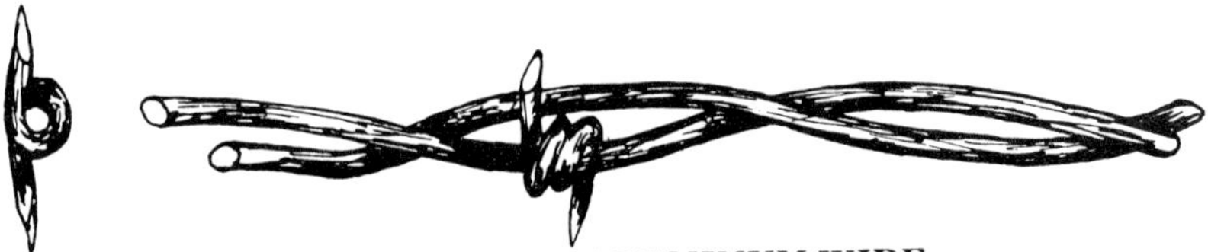

350 GLIDDEN ALL ALUMINUM WIRE

PAT. NOV. 24, 1874

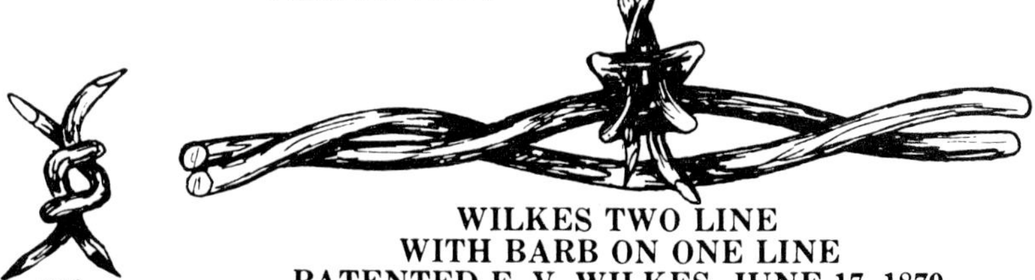

351 WILKES TWO LINE
WITH BARB ON ONE LINE
PATENTED E. V. WILKES, JUNE 17, 1879

352

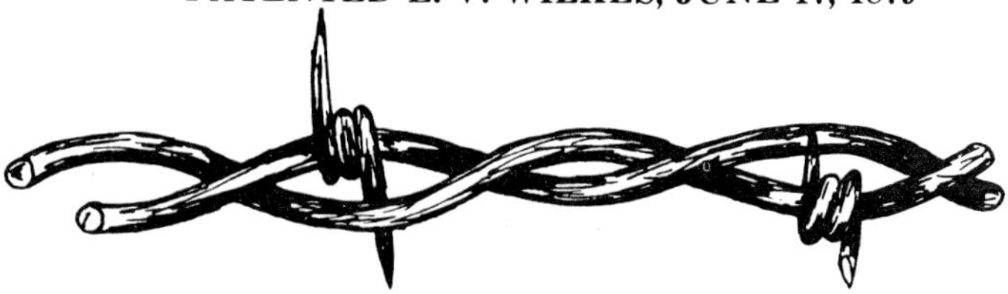

GLIDDEN BARB ON BOTH LINES
(ALTERNATE LONG AND SHORT BARBS)

PAT. NOV. 24, 1874

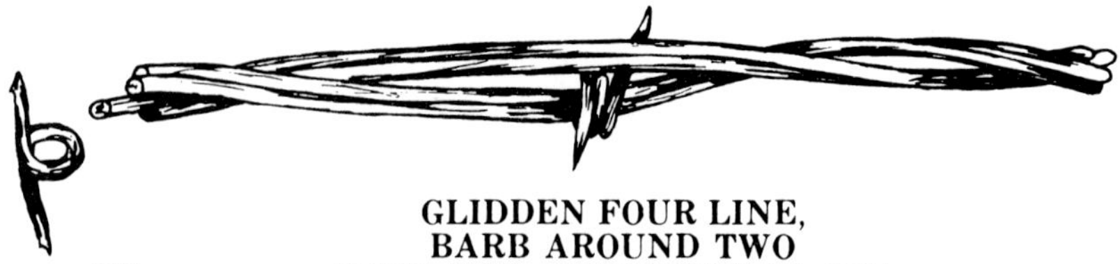

353 GLIDDEN FOUR LINE, BARB AROUND TWO
PATENTED NOVEMBER 24, 1874

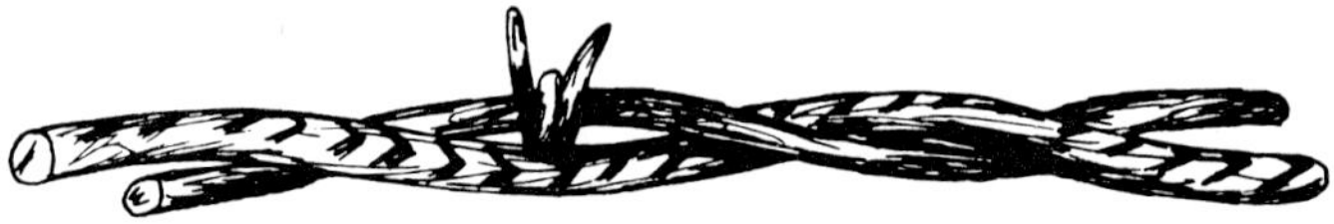

354

GLIDDEN HERRINGBONE
PATENTED BY J. F. GLIDDEN, AUGUST 22, 1876

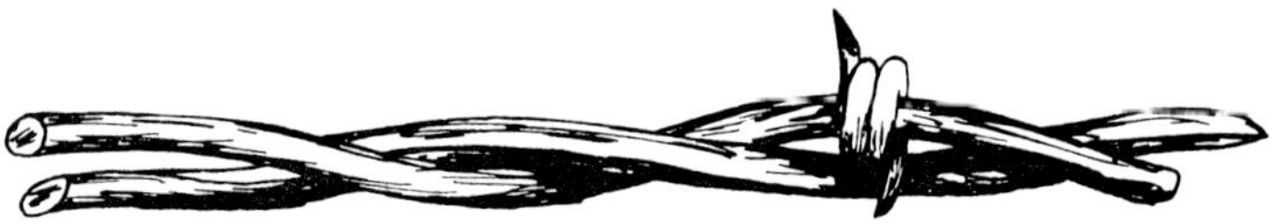

355

GLIDDEN BARELY BARB
(SHARP POINT OF BARB ALL THAT IS OUT)
PAT. NOV. 24, 1874

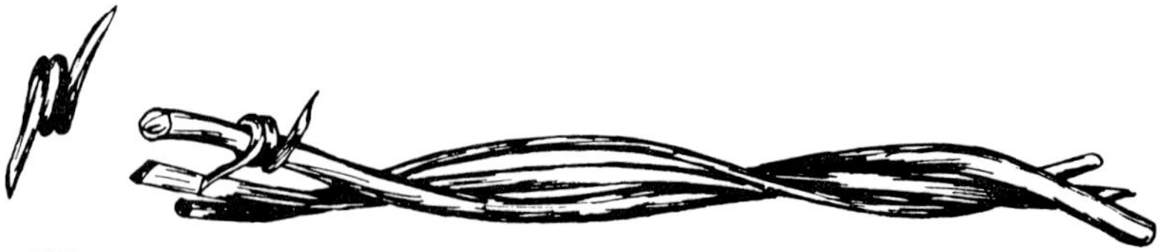

356 GLIDDEN THREE LINE BAR WIRE
(TWO LINES DIFFERENT GAUGE WITH SMALL HIGH TINSEL BAR TWISTED IN TOGETHER)
PAT. NOV. 24, 1874

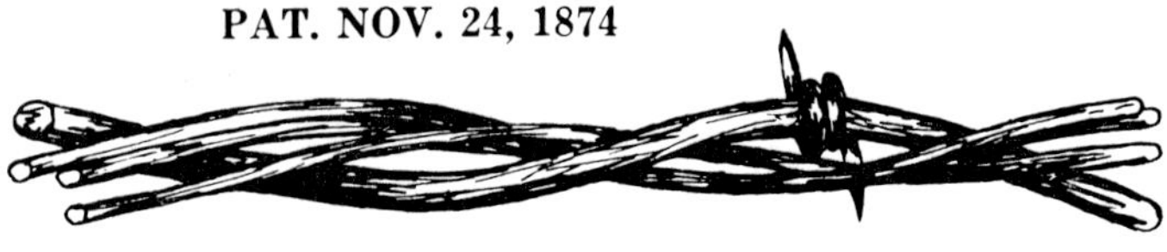

357

GLIDDEN FOUR LINE
(THREE SMALL, ONE LARGE WITH BARB)
PATENTED NOVEMBER 24, 1874

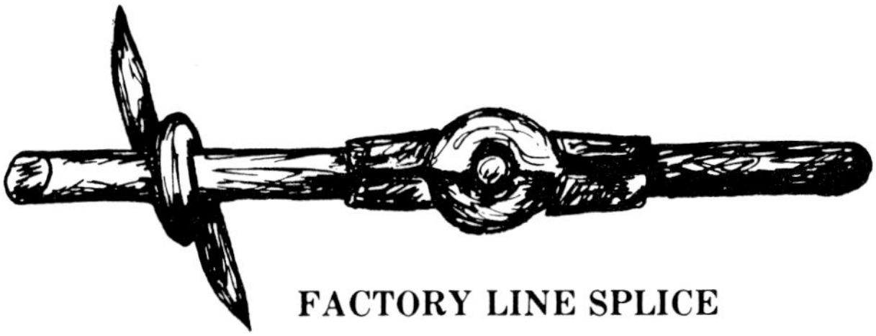

358

FACTORY LINE SPLICE

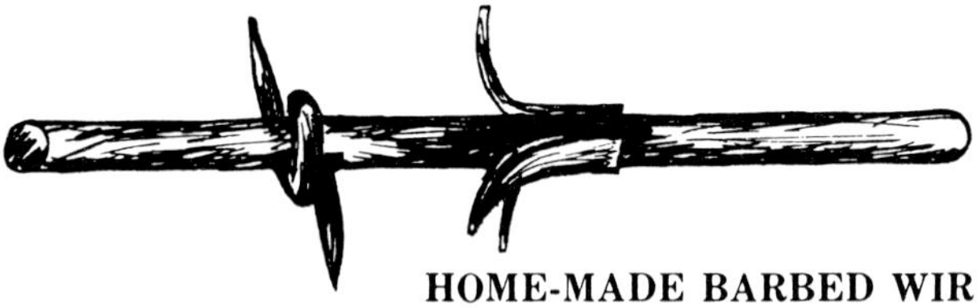

359

HOME-MADE BARBED WIRE
(PLAIN WIRE WITH STAPLE AND
KENNEDY BARB)
PATENT AUGUST 11, 1878 ON KENNEDY BARB

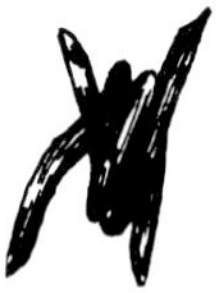

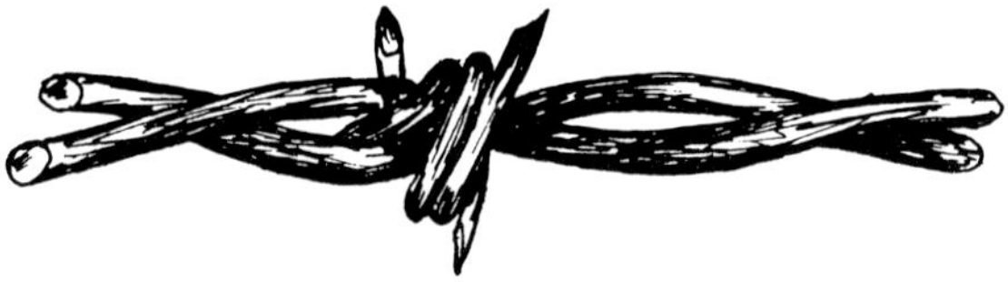

PLASTIC COATED MODERN WIRE
(COMES IN DIFFERENT COLORS)

360

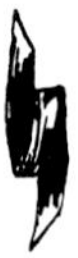

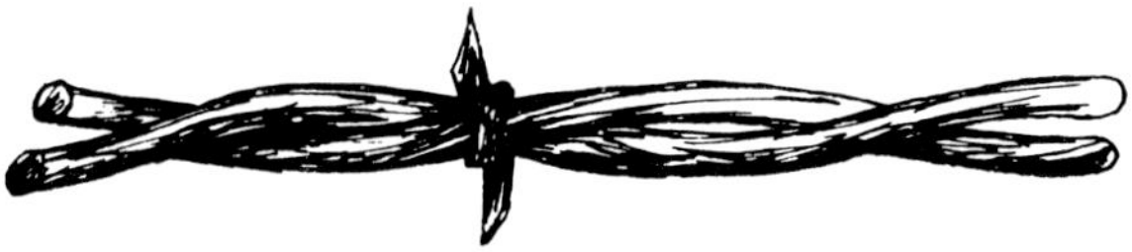

361

BAKER BARB AROUND BOTH LINES
(MODERN WIRE)

PAT. G. C. BAKER, FEB. 27, 1883

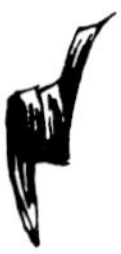

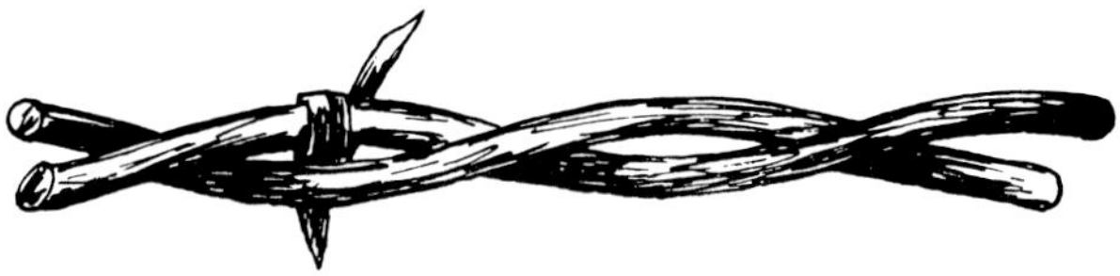

362

GLIDDEN COILED ON BARB
(VERY OLD, RUSTY, APPROXIMATELY 1880's)
ONE LINE NEARLY FLAT

PAT. NOV. 24, 1874

363

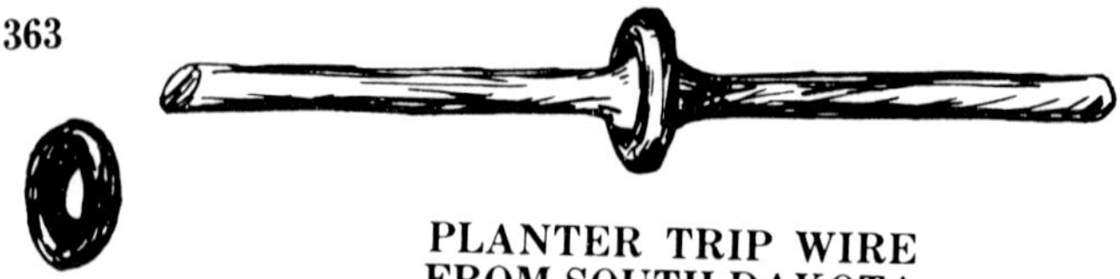

PLANTER TRIP WIRE
FROM SOUTH DAKOTA

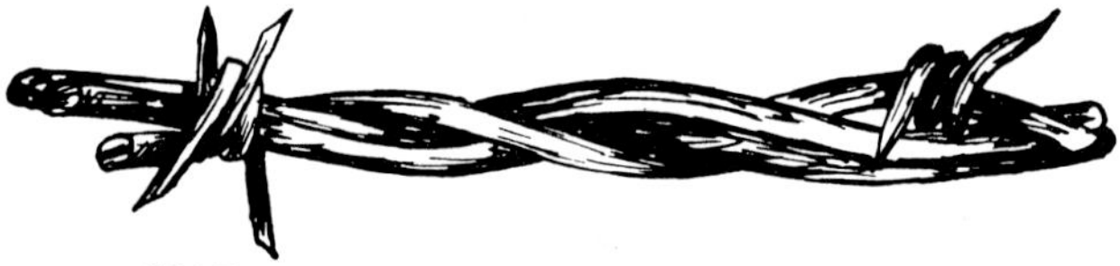

364

GLIDDEN AND ROSS COMBINATION
THREE LINE
(APPEARS TO BE FACTORY MANUFACTURE)
PAT. NOV. 24, 1874

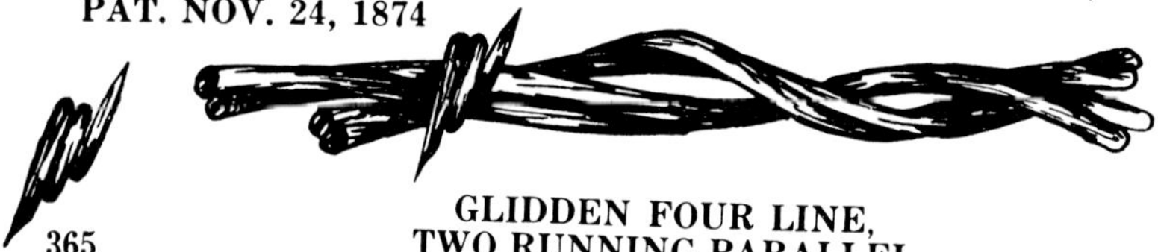

365

GLIDDEN FOUR LINE,
TWO RUNNING PARALLEL
WITH TWO WRAPPED AROUND

PAT. NOV. 24, 1874

366

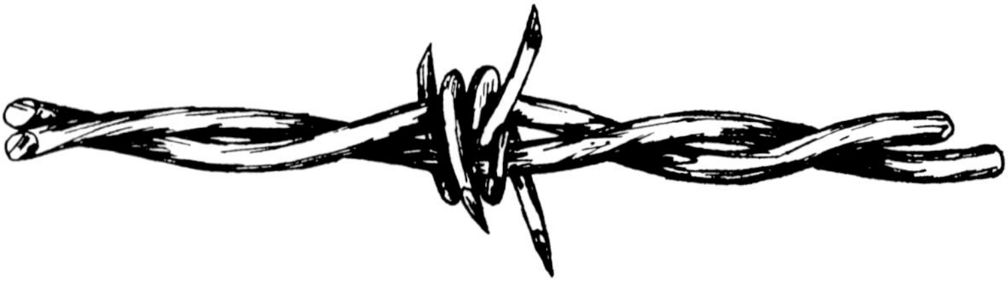

CURTIS FOUR POINT
PATENTED BY J. D. CURTIS, MARCH 28, 1893
HAS HALF ROUND BARB MATERIAL

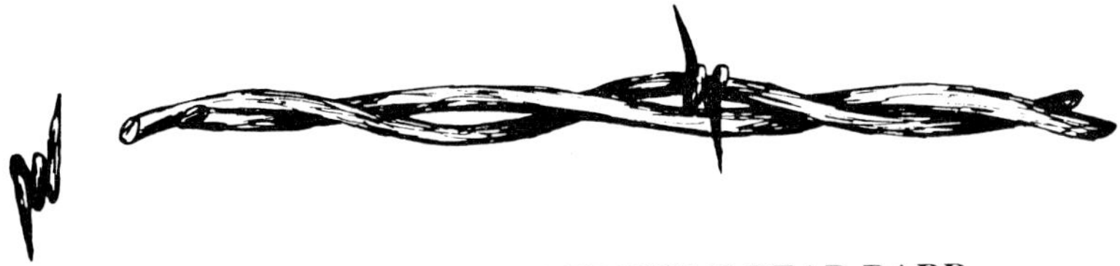

367 ROSE WIRE WITH PRICKLY-PEAR BARB
PATENTED H. M. ROSE, DECEMBER 25, 1877

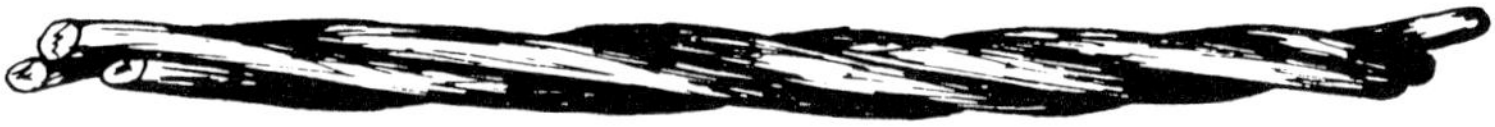

368 THREE LINE HORSE WIRE
(ONE LINE SOLID COPPER)

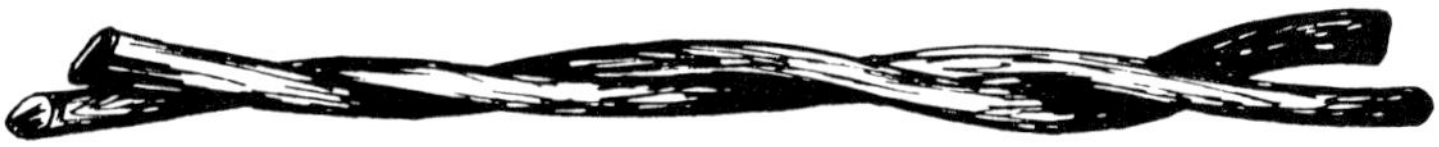

369 R. R. HORSE HOLDING PEN WIRE
FLAT AND OVAL LINE WIRES
(STAYS APPROXIMATELY ONE FOOT APART)

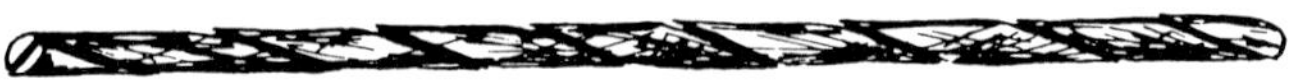

370 USED ROCK SAW WIRE (HIGH TINSEL)

371 NEW ROCK SAW WIRE (HIGH TINSEL)

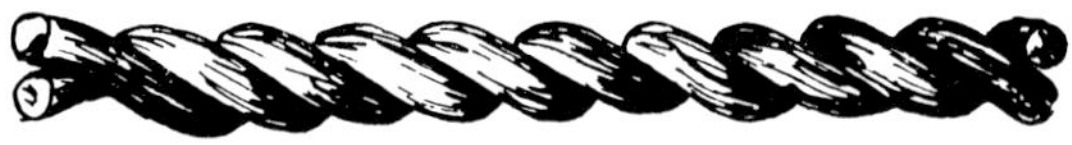

372 NEW TWO LINE ROCK SAW WIRE
(HIGH TINSEL)

373

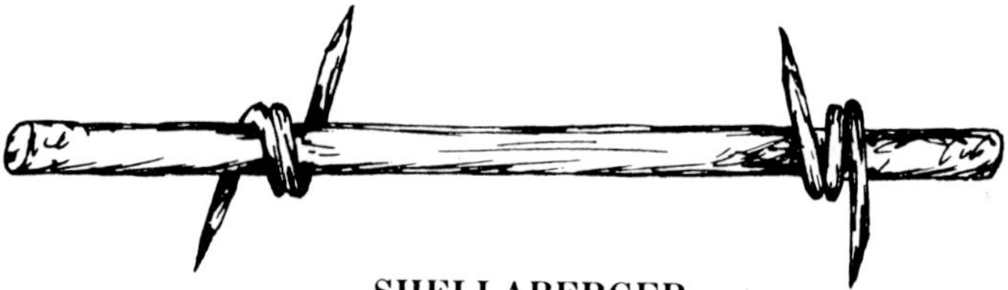

SHELLABERGER
PATENTED FEBRUARY 13, 1883

374

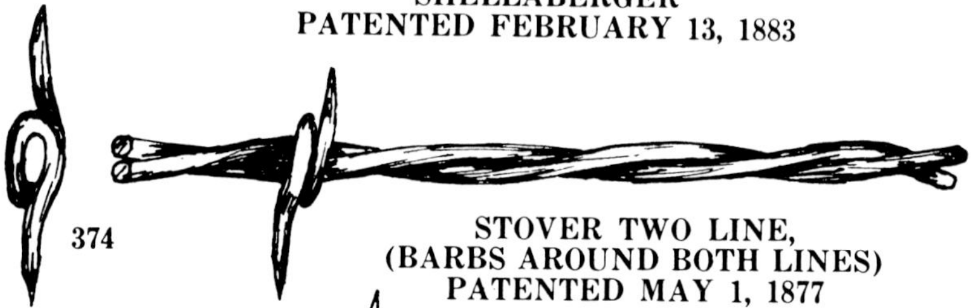

STOVER TWO LINE,
(BARBS AROUND BOTH LINES)
PATENTED MAY 1, 1877

375

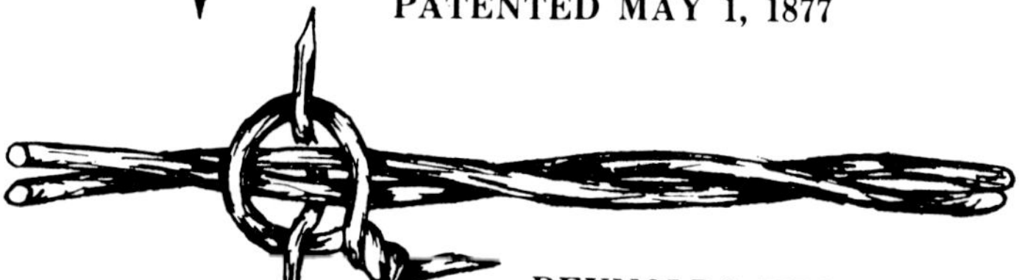

REYNOLDS NECKTIE
(BARB AROUND BOTH LINES)

PATENTED BY H. REYNOLDS, MAY 14, 1878

376

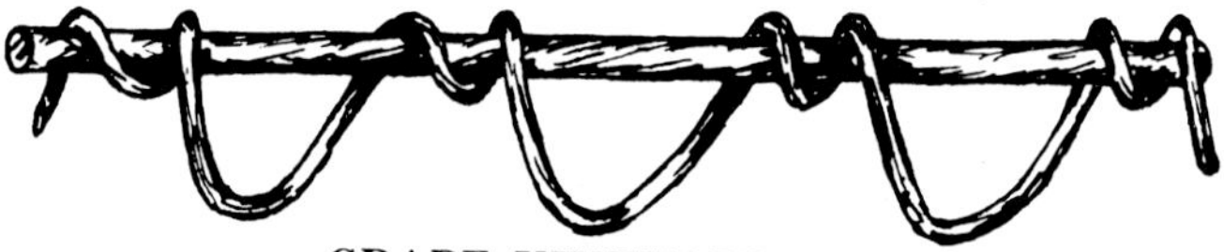

GRAPE VINEYARD WIRE
(LATER USED FOR FENCING)

377

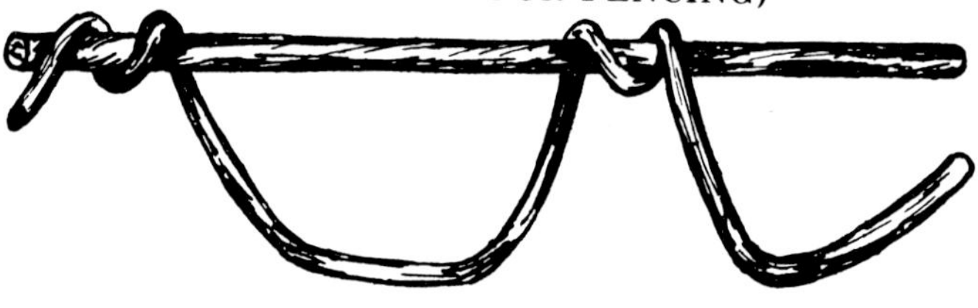

LARGE VINEYARD WIRE
(USED FOR FENCING)

PAT. DATUS C. SMITH

MAR. 2, 1897

378

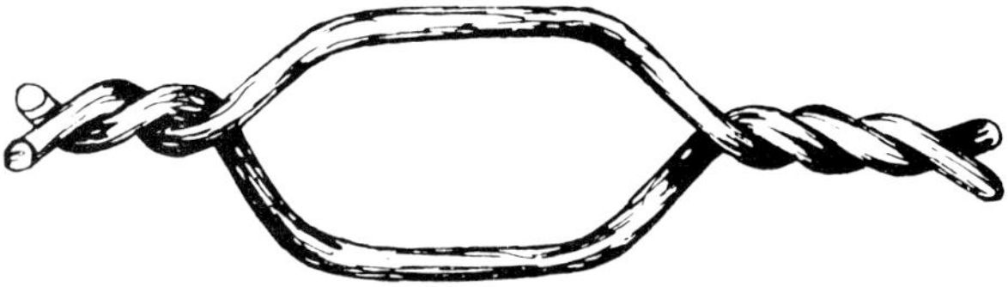

PICKET FENCE WIRE
PATENTED BY R. C. EATON, SEPTEMBER 2, 1884
(USED FOR FENCING AFTER PICKETS ROTTED OUT)

379

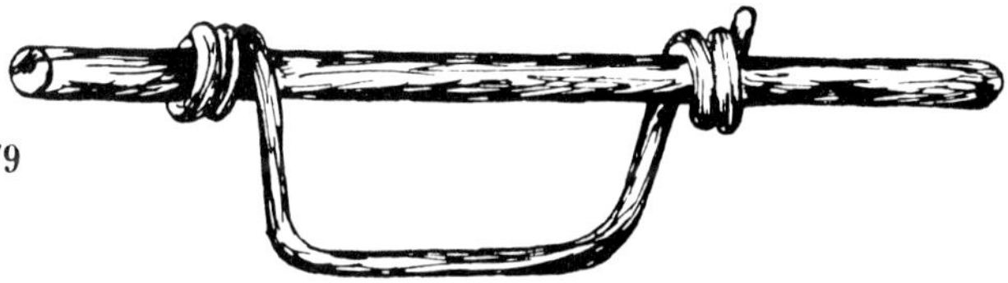

PAT. DATUS C. SMITH
MAR. 2, 1897

SNOW FENCE PICKET WIRE
(EARLY MANUFACTURE)
(USED FOR HORSE FENCE AFTER PICKETS ROTTED)

380

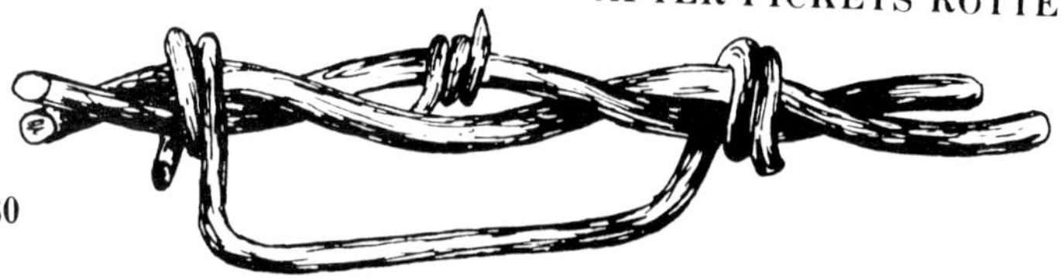

HOG GLIDDEN SLAT WIRE
(PROBABLY HOME-MADE)

381

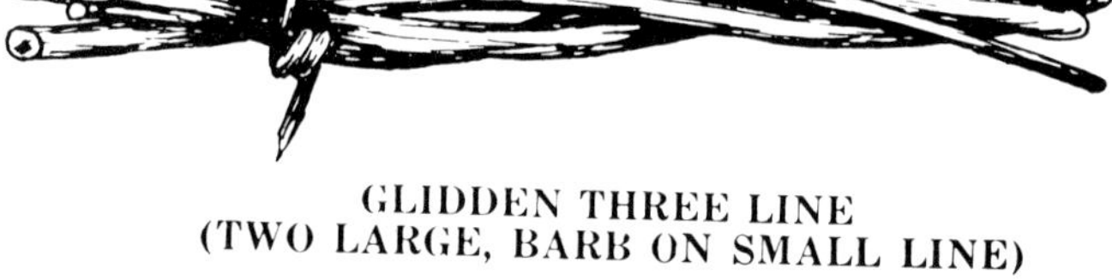

GLIDDEN THREE LINE
(TWO LARGE, BARB ON SMALL LINE)

PAT. NOV. 24, 1874

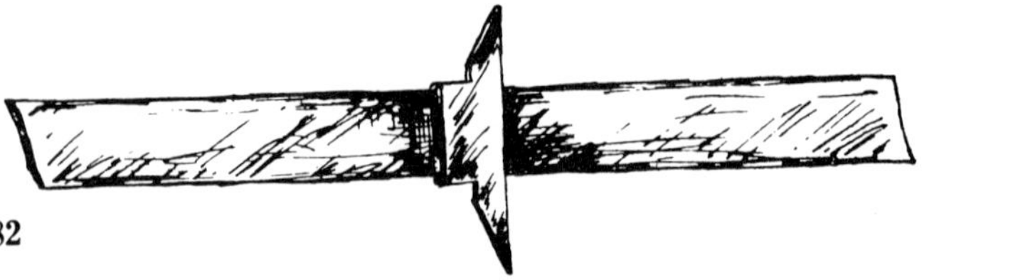

382

J. BRINKERHOFF
PATENTED MAY 17, 1881

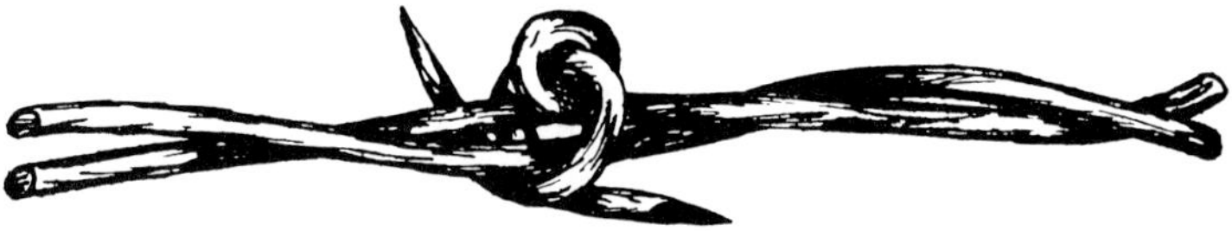

383

J. E. PAGE
PATENTED DECEMBER 7, 1875

384

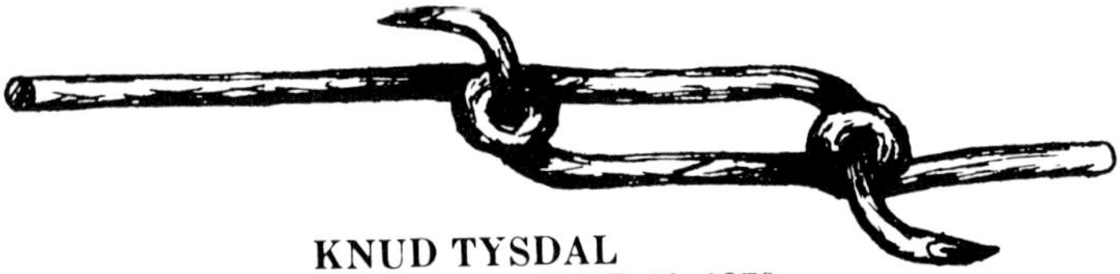

KNUD TYSDAL
PATENTED SEPTEMBER 10, 1878

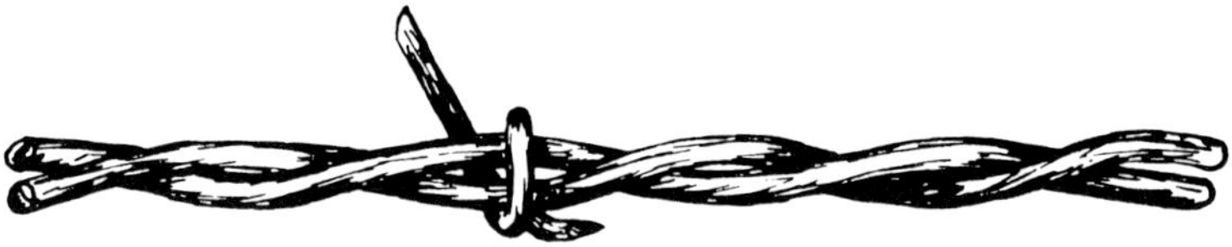

WILLIAM EDENBORN
PATENTED MARCH 17, 1885

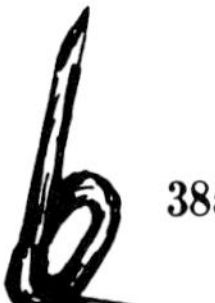

385

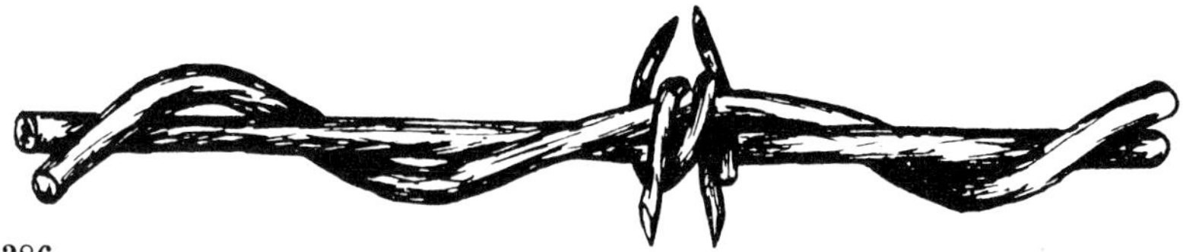

386

BURNELL ON STRAIGHT LINE WITH
ONE LINE TWISTED AROUND
PATENTED JUNE 19, 1877

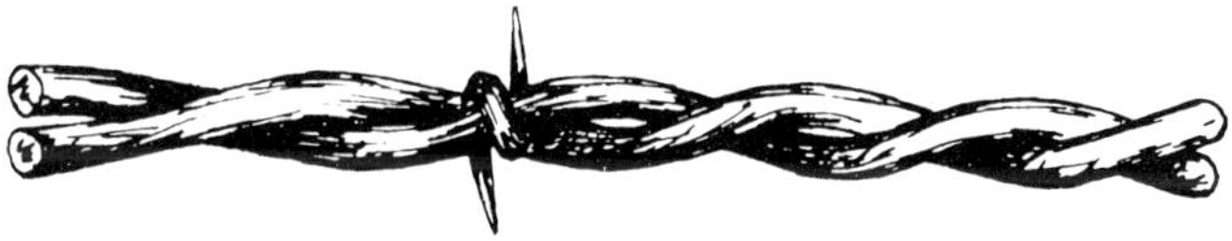

387

J. D. CURTIS
PATENTED MARCH 15, 1892

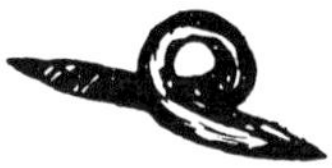

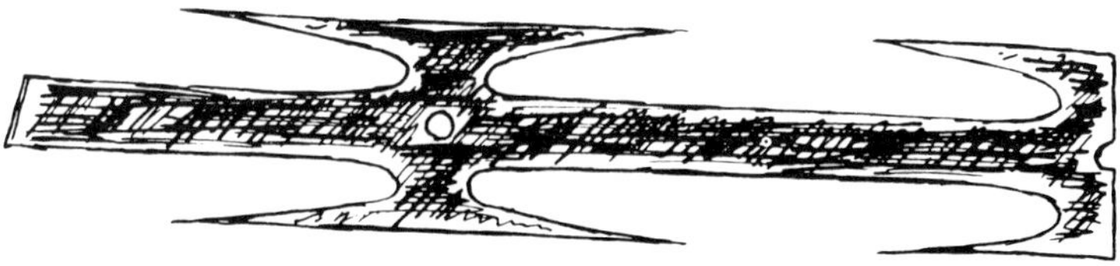

388 GENERAL PURPOSE BARBED TAPE OBSTACLE WIRE.
FOR COMPOUND ENCLOSURE, BATTLEFIELDS

389

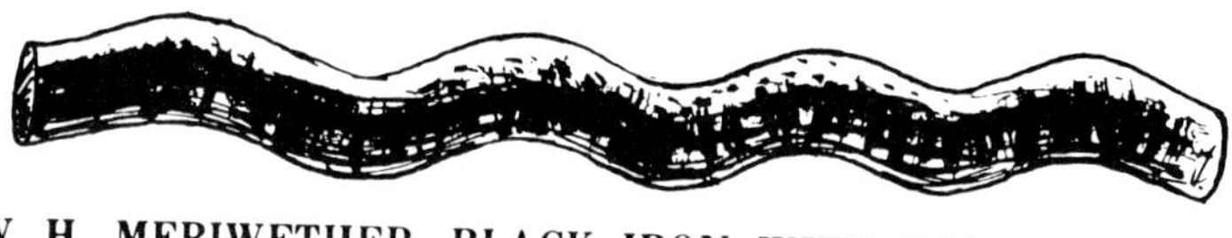

W. H. MERIWETHER, BLACK IRON WITH FASTER CURL.
PAT. NOV. 8, 1853 NEW BRAUNFELS, TEXAS

390

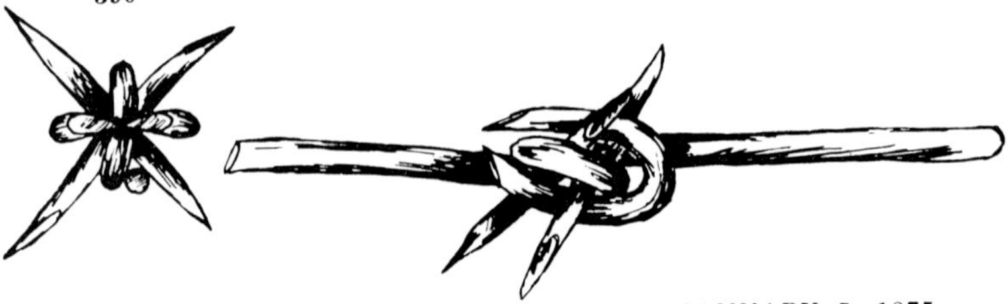

BARB & STAPLE, F. F. WILSON PAT. JANUARY 5, 1875

391

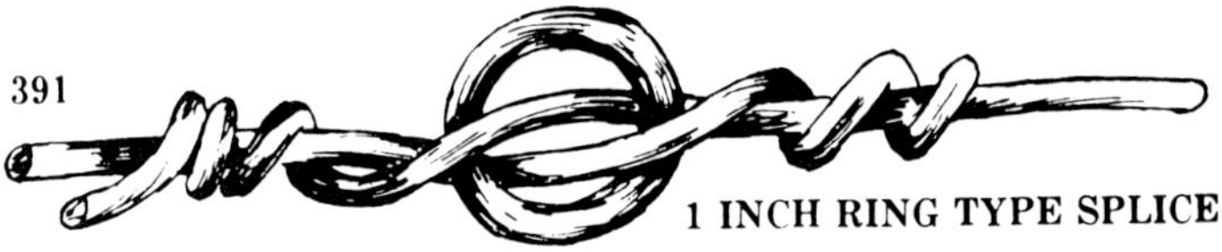

1 INCH RING TYPE SPLICE

392

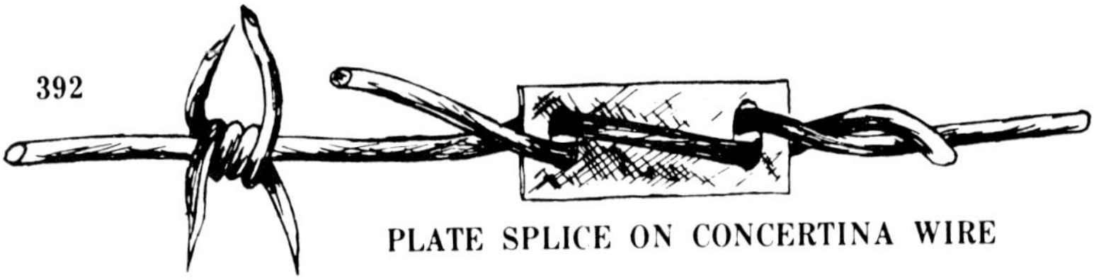

PLATE SPLICE ON CONCERTINA WIRE

393

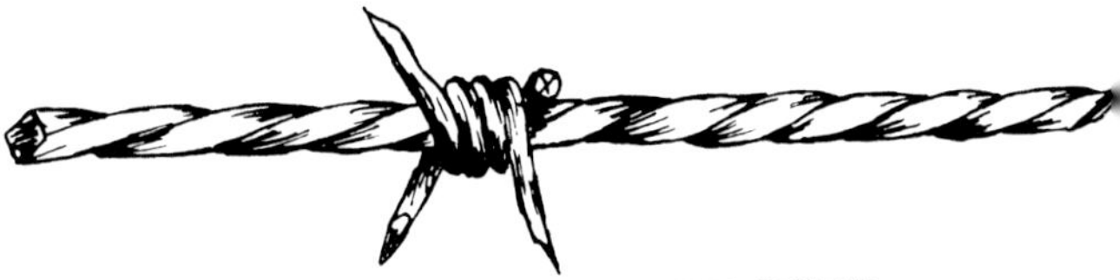

CURTIS SQUARE LINE TWIST
PAT. J. D. CURTIS MARCH 28, 1893

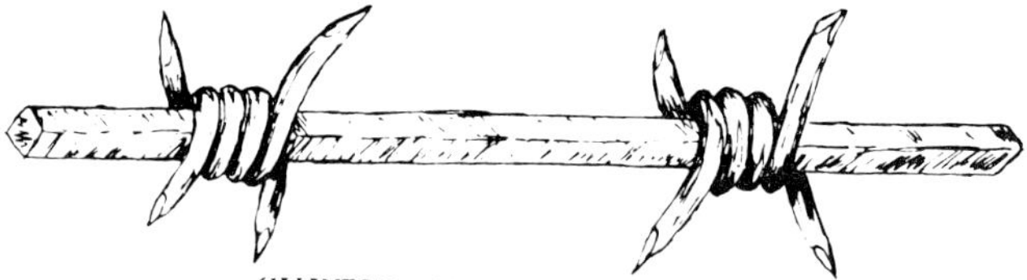

394

CURTIS SQUARE LINE
J. D. CURTIS PAT. MARCH 28, 1893

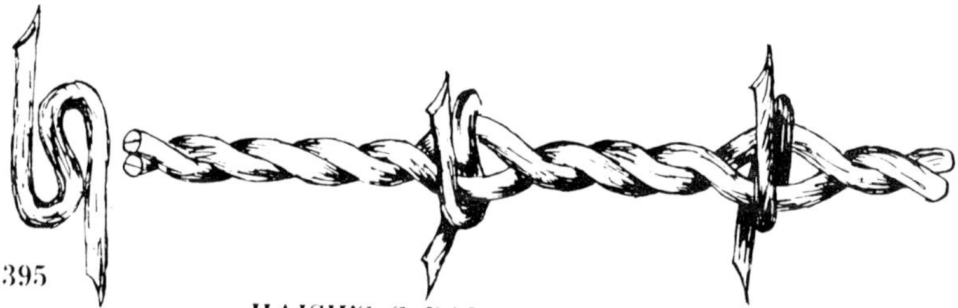

395

HAISH'S S BARB HOG WIRE.
BARBS 1 INCH APART. PAT. AUG. 31, 1875

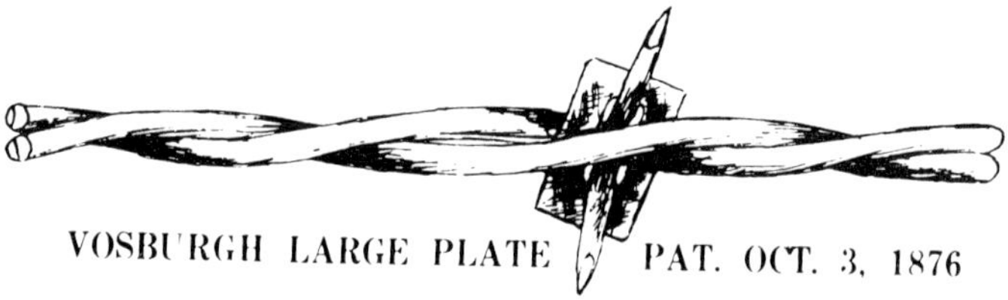

396

VOSBURGH LARGE PLATE PAT. OCT. 3, 1876

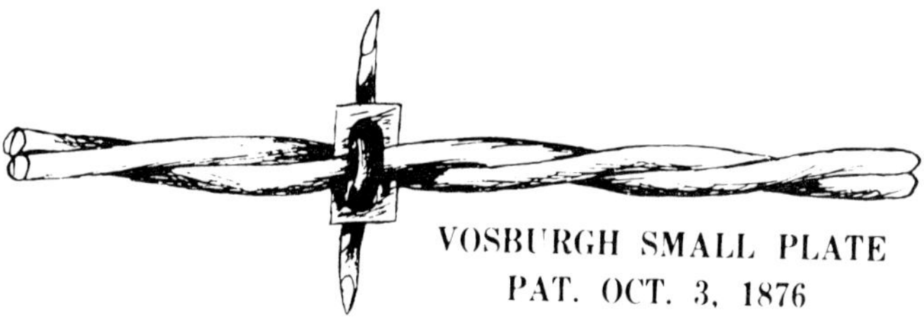

397

VOSBURGH SMALL PLATE
PAT. OCT. 3, 1876

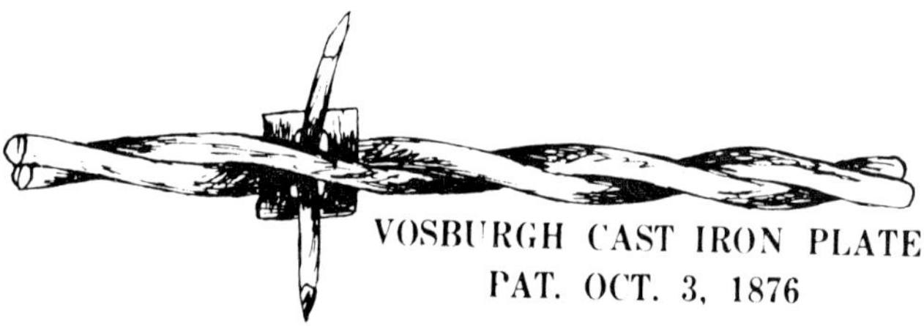

398

VOSBURGH CAST IRON PLATE
PAT. OCT. 3, 1876

399

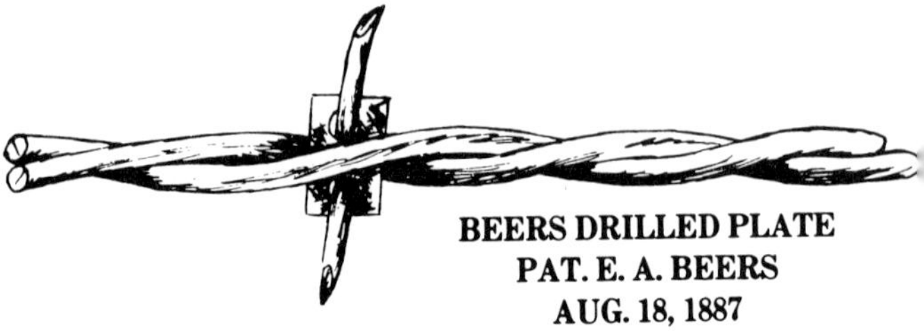

BEERS DRILLED PLATE
PAT. E. A. BEERS
AUG. 18, 1887

400

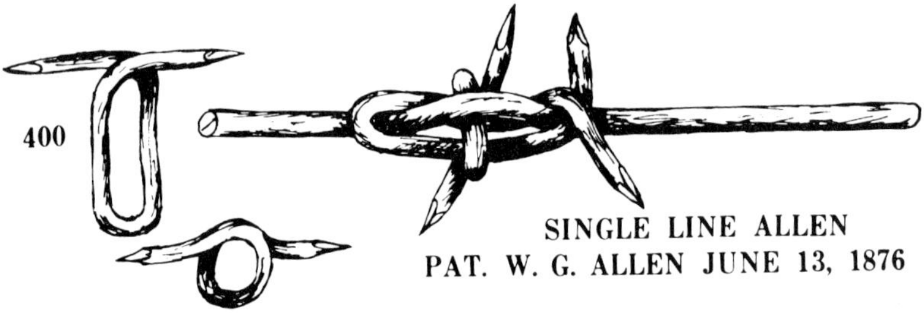

SINGLE LINE ALLEN
PAT. W. G. ALLEN JUNE 13, 1876

401

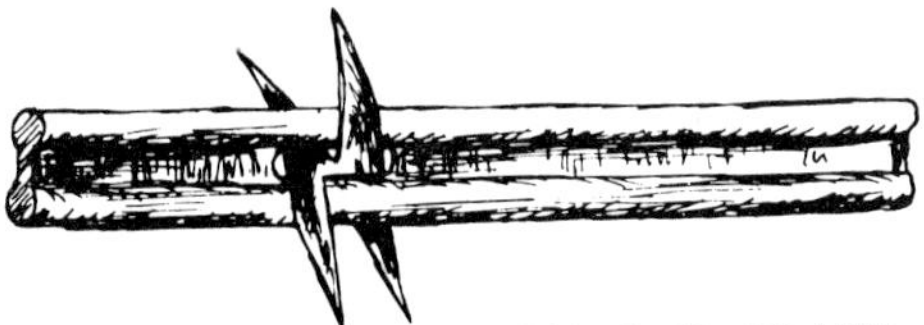

DOUBLE BAR RIBBON & H PLATE
JOHN WESTGARTH PAT. MARCH 22, 1881

402

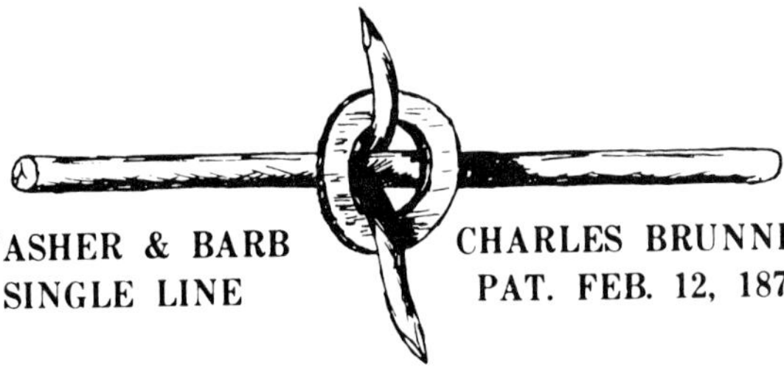

WASHER & BARB
SINGLE LINE

CHARLES BRUNNER
PAT. FEB. 12, 1878

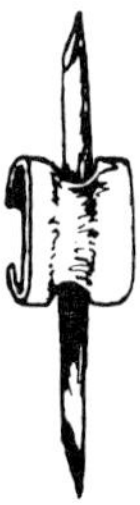

403

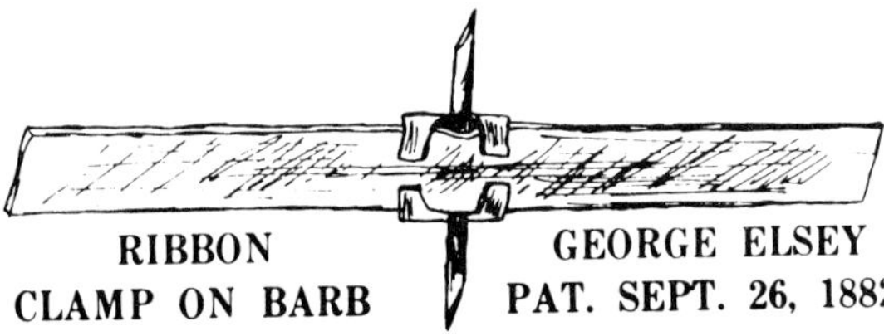

RIBBON
CLAMP ON BARB

GEORGE ELSEY
PAT. SEPT. 26, 1882

PUNCH PRESS SCRAP ORNAMENTAL

404 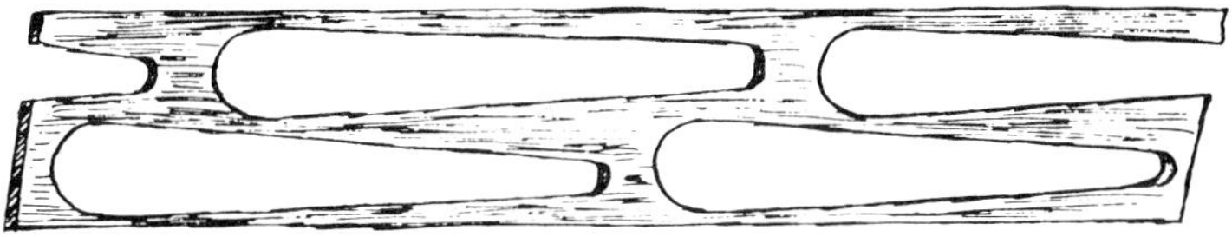

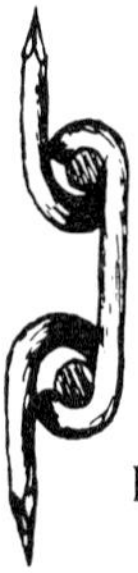

405 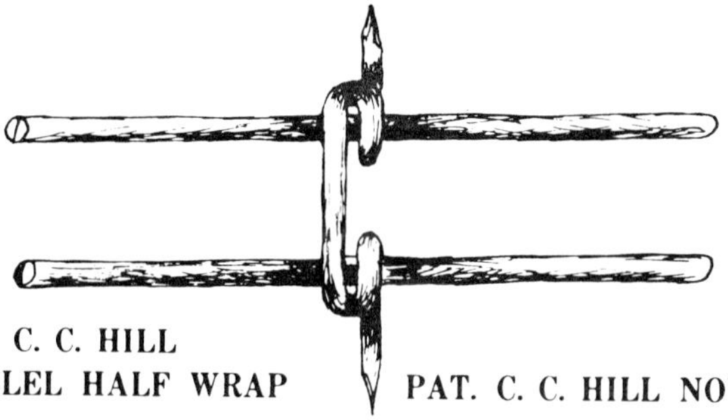

C. C. HILL
PARALLEL HALF WRAP

PAT. C. C. HILL NOV. 24, 1885

406 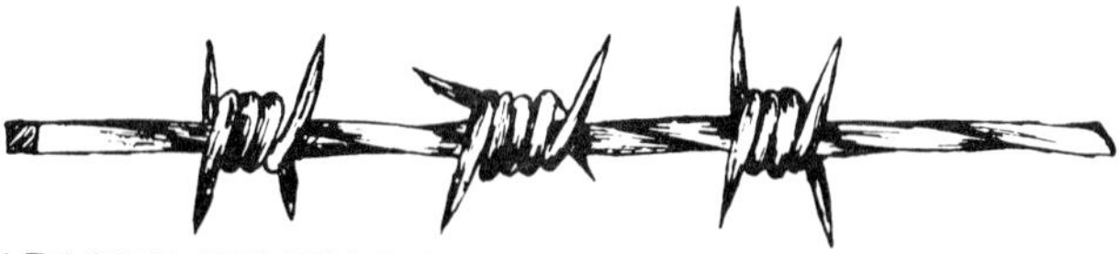

APACHE STOCKADE WIRE. ARMY CONSTRUCTED.
BARBS 1 INCH APART. V. MERRILL PAT. SEPT. 29, 1871

407 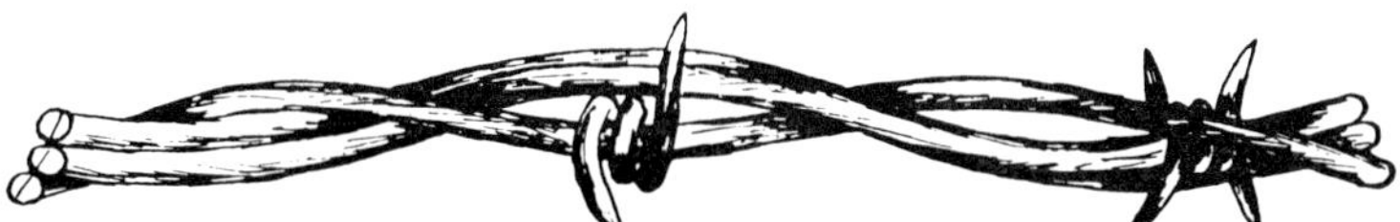

THREE LINE ROSE & ROSS
TWO LINES RUNNING PARALLEL

PAT. H. M. ROSE, DEC. 25, 1877

408 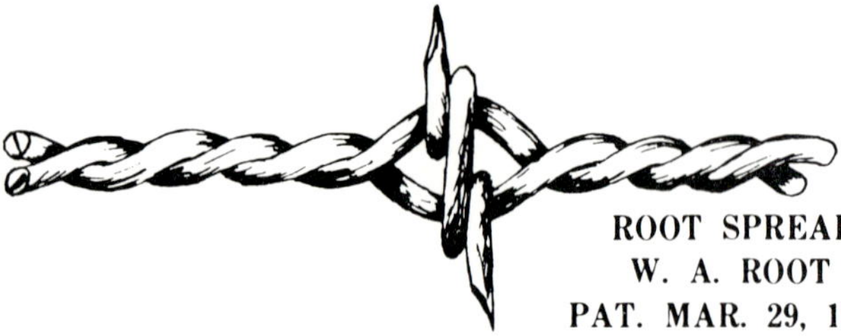

ROOT SPREAD.
W. A. ROOT
PAT. MAR. 29, 1881

409 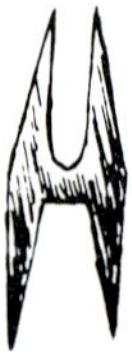

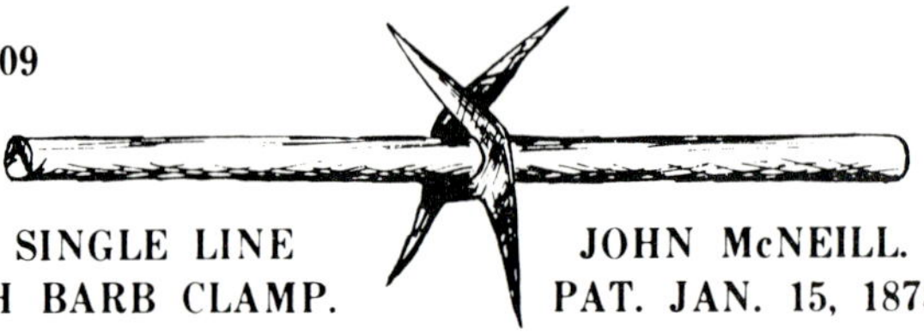

SINGLE LINE
H BARB CLAMP.

JOHN McNEILL.
PAT. JAN. 15, 1878

410 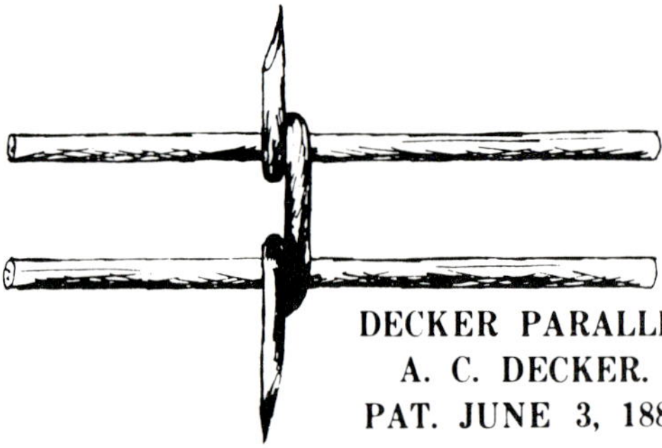

DECKER PARALLEL.
A. C. DECKER.
PAT. JUNE 3, 1884

411 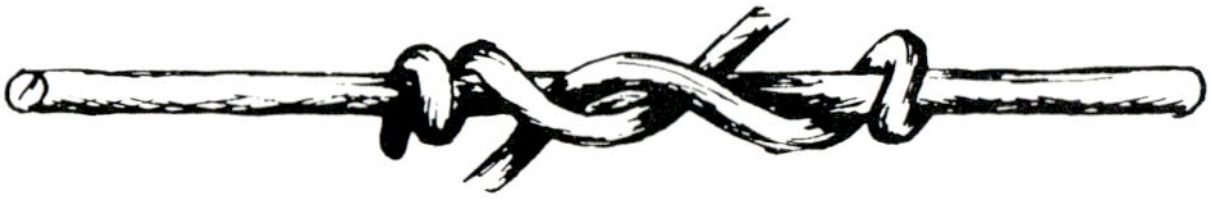

NET FENCE TIE

412 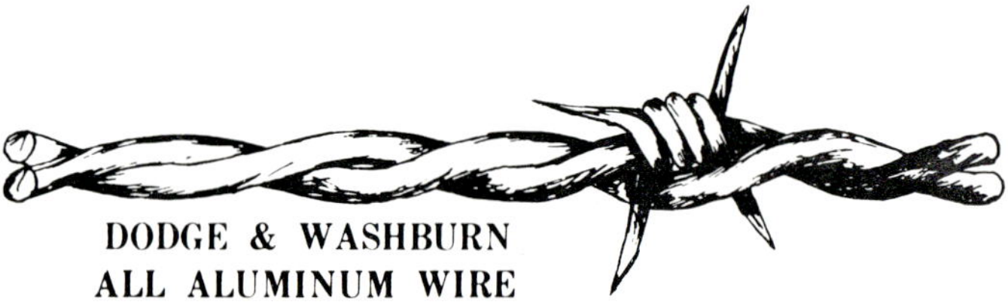

DODGE & WASHBURN
ALL ALUMINUM WIRE

413 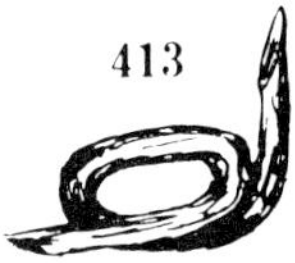 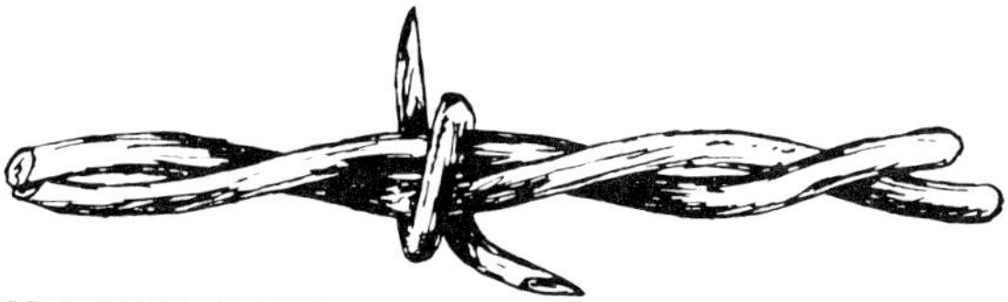

RUNNING BARB. SLIDES UP & DOWN LINE.
PAT. WILLIAM EDENBORN. MAR. 17, 1885

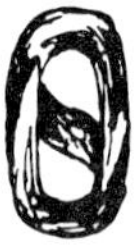 414 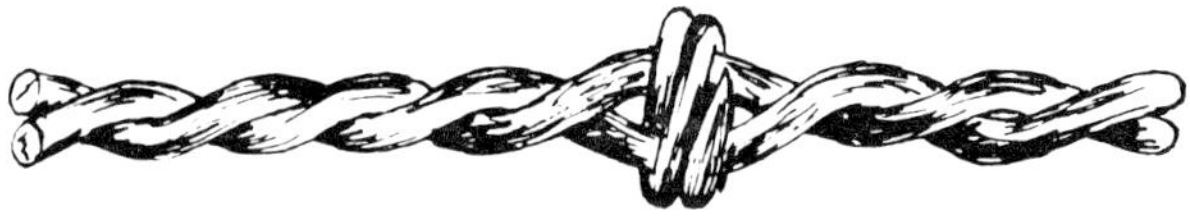

HAISH'S WALKING CANE WIRE. PAT. AUG. 31, 1875

415  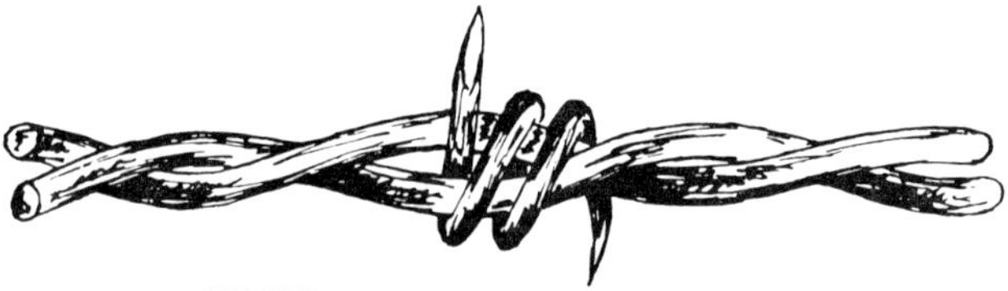

BROTHERTON OPEN WRAP.
J. BROTHERTON. PAT. SEPT. 3, 1887

416 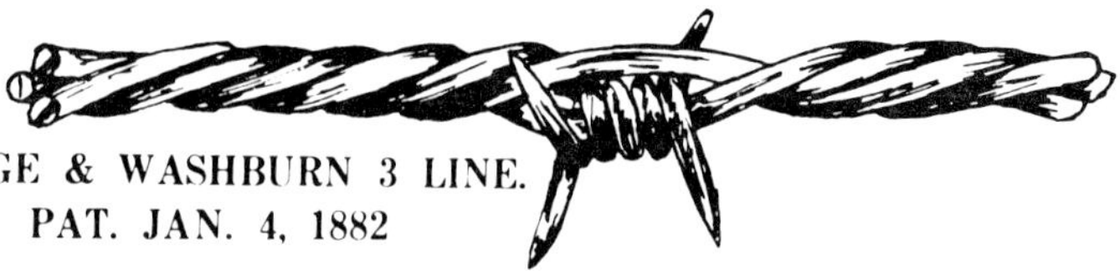

DODGE & WASHBURN 3 LINE.
PAT. JAN. 4, 1882

417

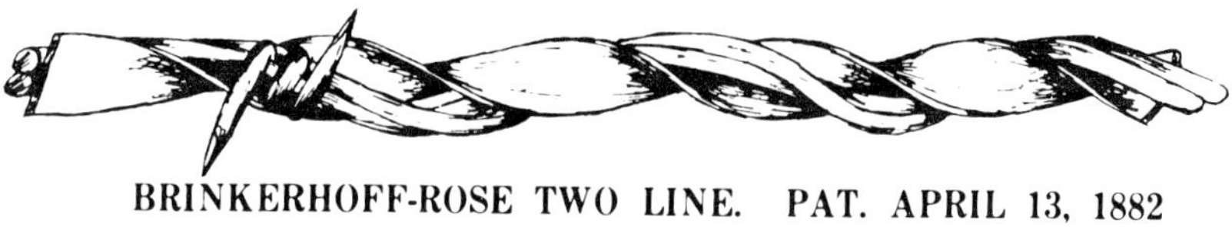

BRINKERHOFF-ROSE TWO LINE. PAT. APRIL 13, 1882

418 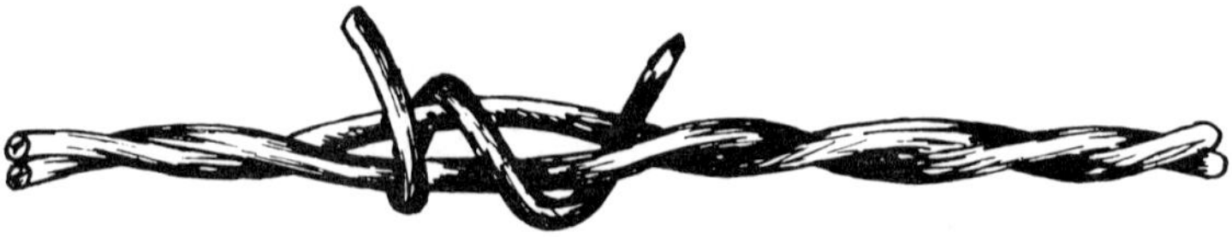

NET FENCE TIE

419 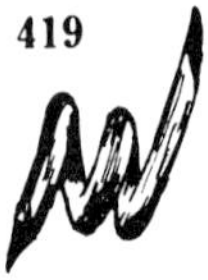 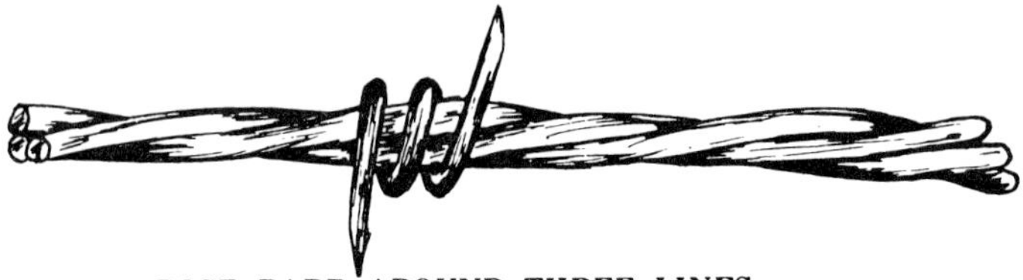

ROSE BARB AROUND THREE LINES.
H. M. ROSE. PAT. DEC. 25, 1877

420 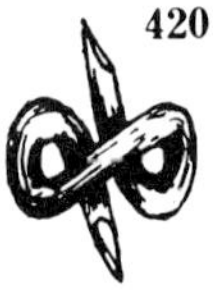 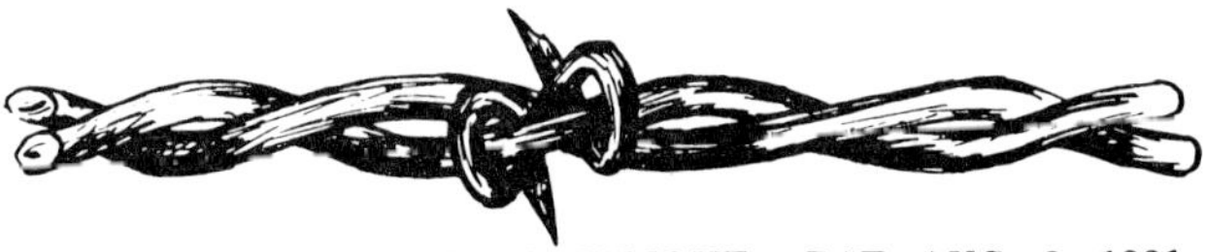

FIGURE 8 BARB. IVY E. WRIGHT. PAT. AUG. 2, 1881

421 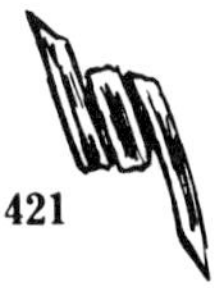 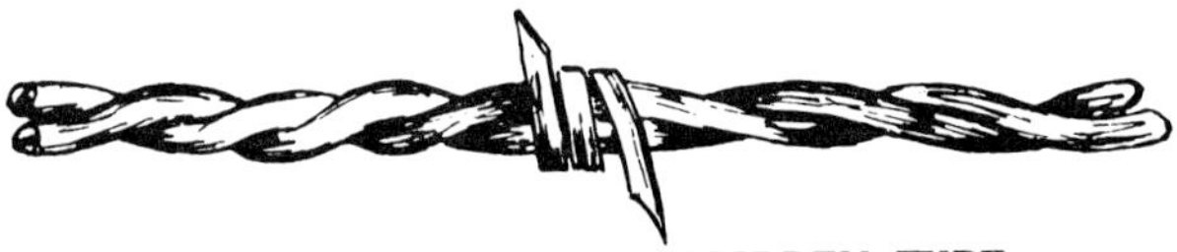

MODERN FLAT BARB GLIDDEN TYPE.
LINE TWIST BACK & FORTH

422 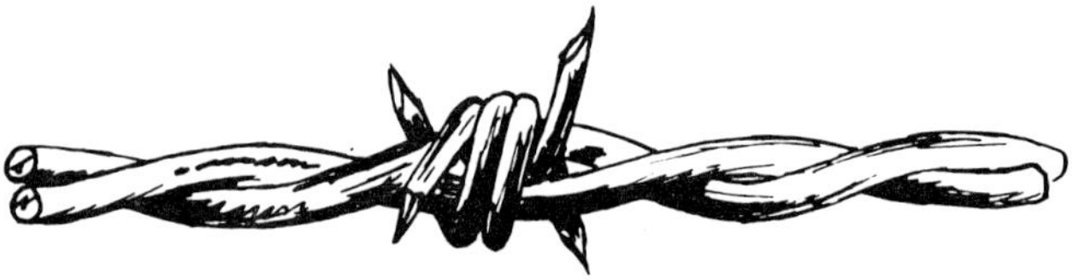

ROSS COPPER BARB & IRON LINE WIRES.
N. G. ROSS. PAT. JUNE 10, 1879

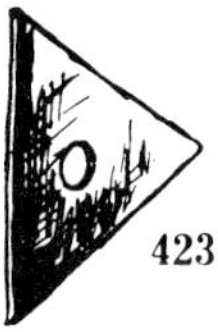

423

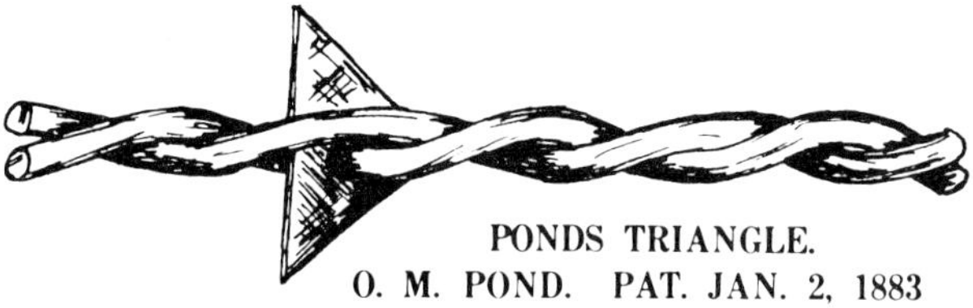

PONDS TRIANGLE.
O. M. POND. PAT. JAN. 2, 1883

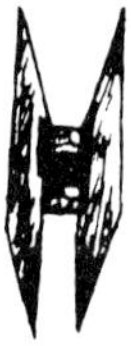

424

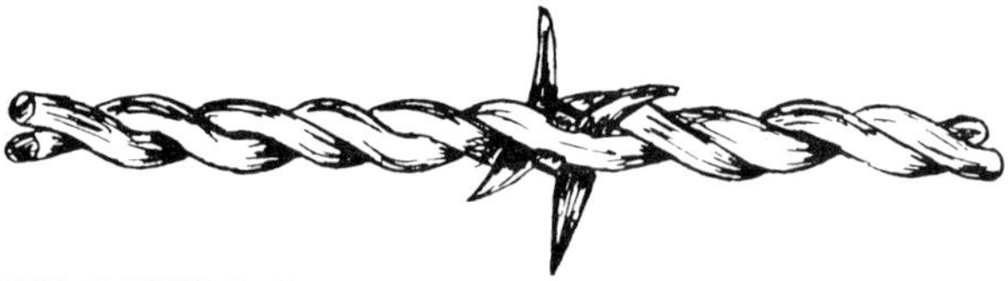

SCUTT'S DOUBLE CLIP. H. B. SCUTT. PAT. JUNE 18, 1878

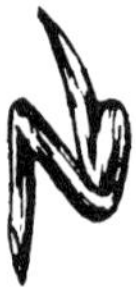

425

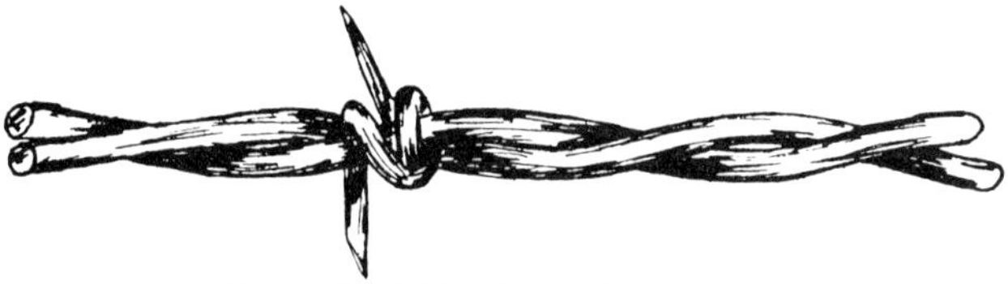

E. A. BEERS PAT. MAY 25, 1880
"FLOP OVER BARB"

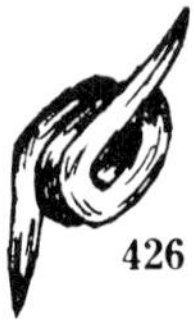

426

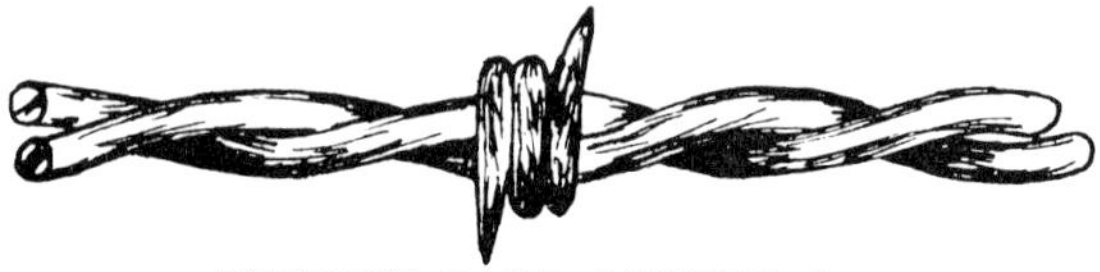

GLIDDEN BARB AROUND 2
J. F. GLIDDEN PAT. NOV. 24, 1874

427

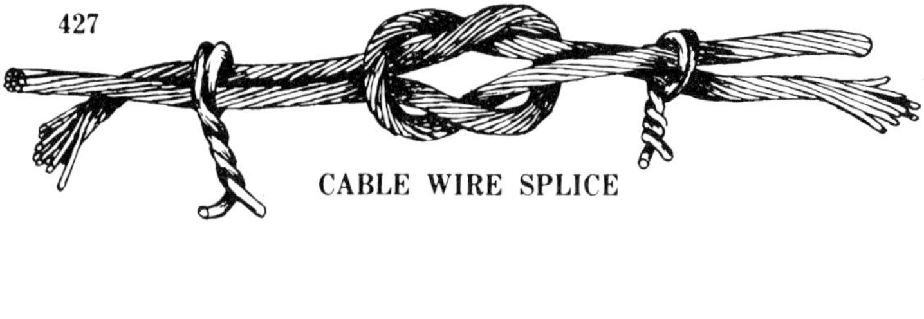

CABLE WIRE SPLICE

428

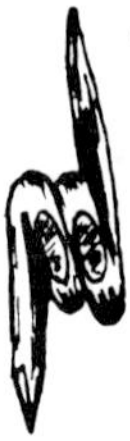

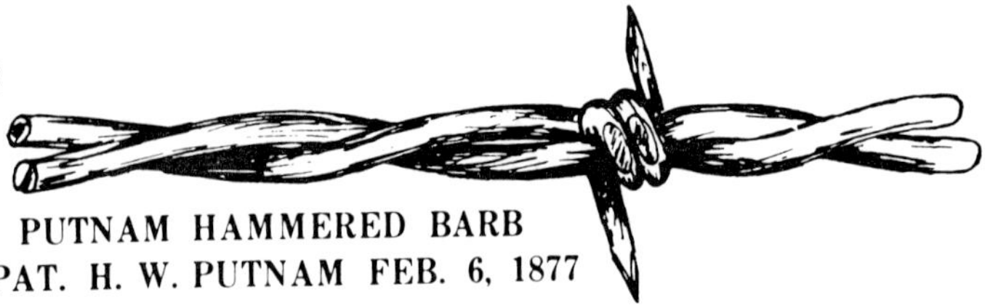

PUTNAM HAMMERED BARB
PAT. H. W. PUTNAM FEB. 6, 1877

429

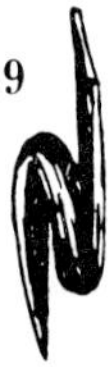

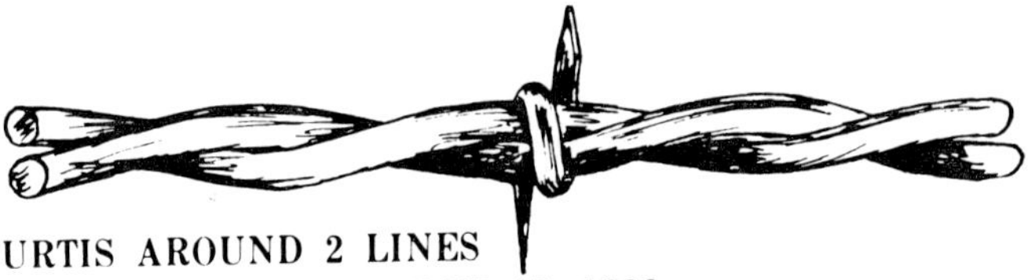

CURTIS AROUND 2 LINES
J. D. CURTIS PAT. MARCH 15, 1892

430

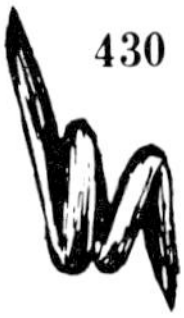

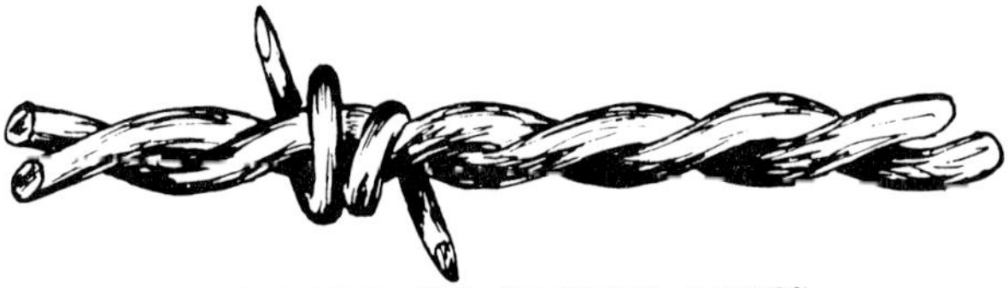

ROSE BARB ON COPPER LINES
PAT. DEC. 25, 1877

431

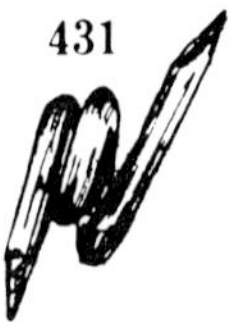

ROSE ON PARALLEL, 2 TWISTED LINES AROUND
H. M. ROSE PAT. DEC. 25, 1877

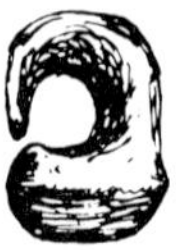

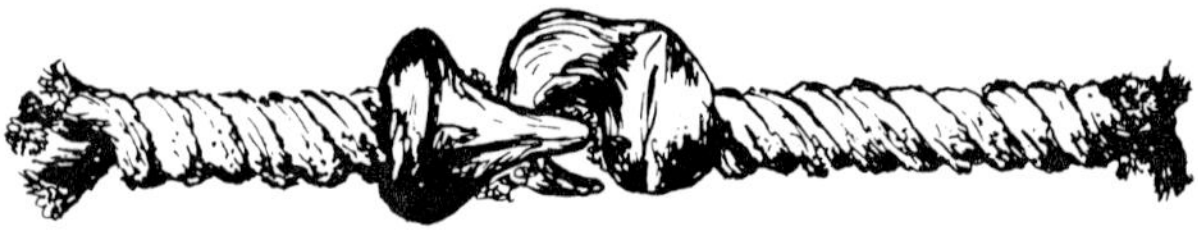

432

ROPE PLANTER WIRE TRIP

433

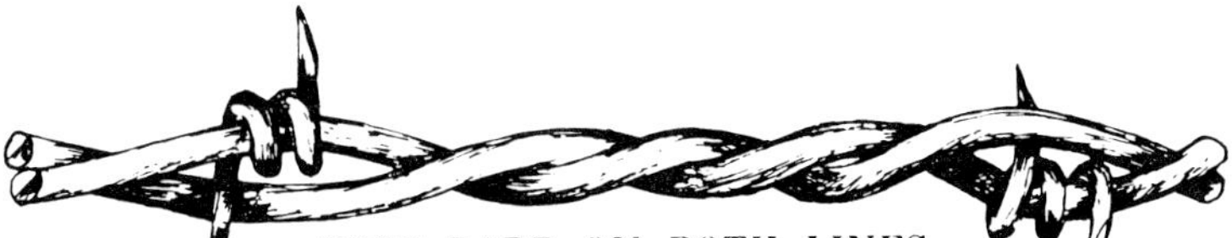

ROSE BARB ON BOTH LINES
H. M. ROSE PAT. DEC. 25, 1877

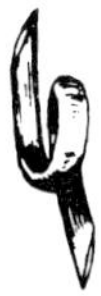

434

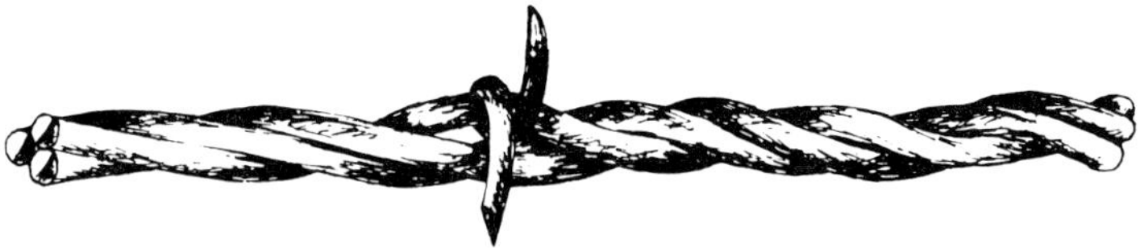

CURTIS THREE LINE
J. D. CURTIS PAT. MARCH 15, 1892

435

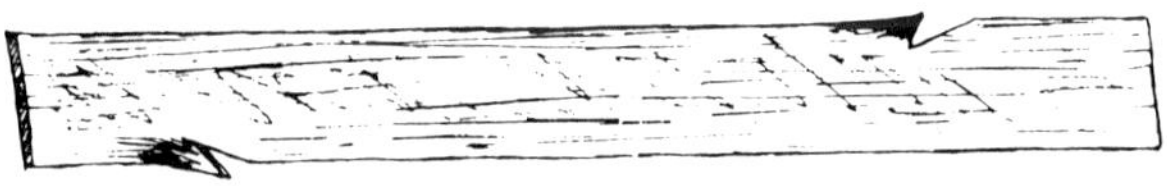

CANADIAN JUDSON PAT. ALBERT JACKSON, JULY 31, 1911

436

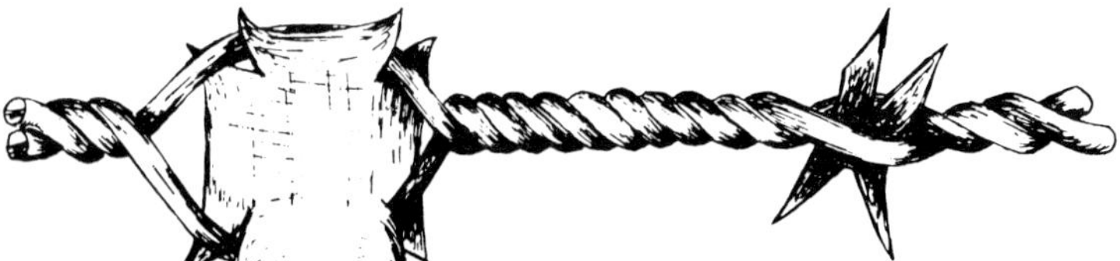

SCUTT'S ARROW PLATE WITH STUBBE PLATE
PAT. OCT. 23, 1883

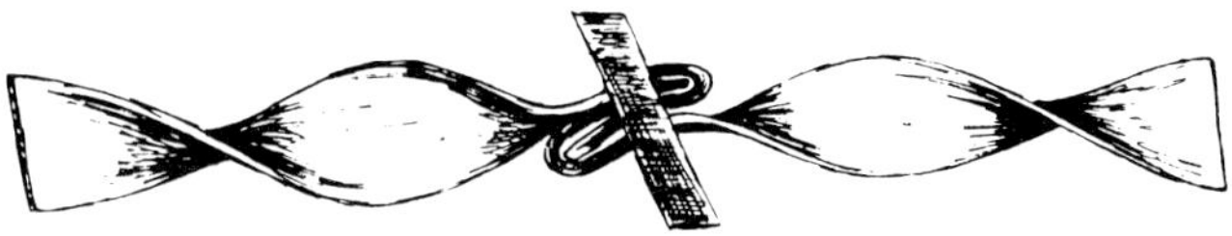

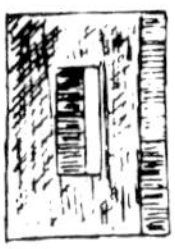

437

RIBBON WIRE CONNECTOR

PAT. J. BRINKERHOFF, MAY 17, 1881

438

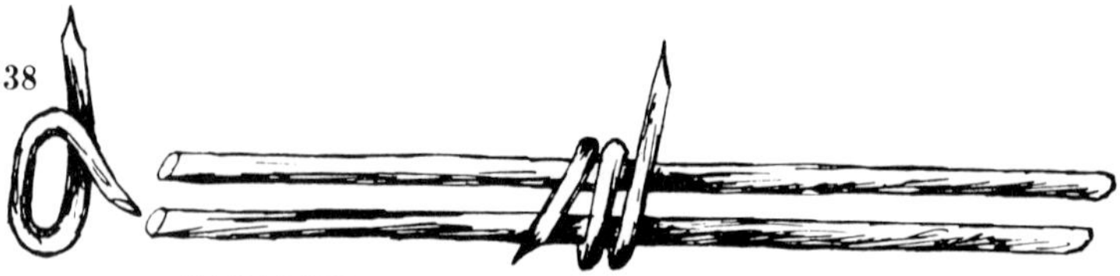

PARALLEL ROSE H. M. ROSE PAT. MARCH 5, 1874

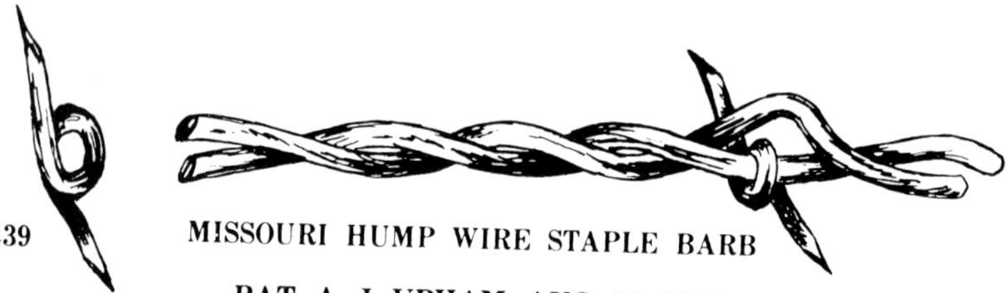

439 MISSOURI HUMP WIRE STAPLE BARB

PAT. A. J. UPHAM, AUG. 29, 1876

440

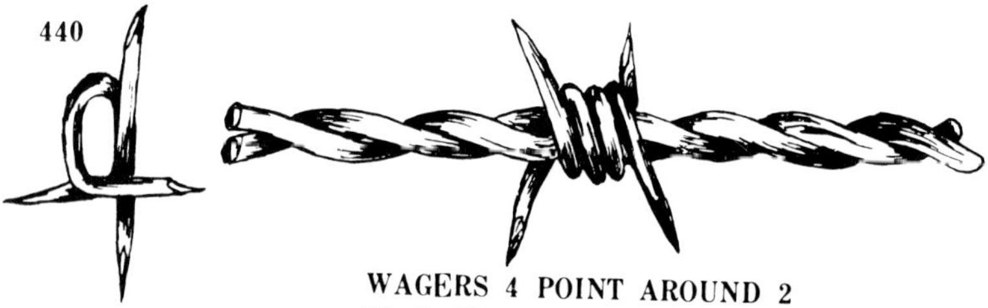

WAGERS 4 POINT AROUND 2
W. H. WAGOR APRIL 8, 1879

441

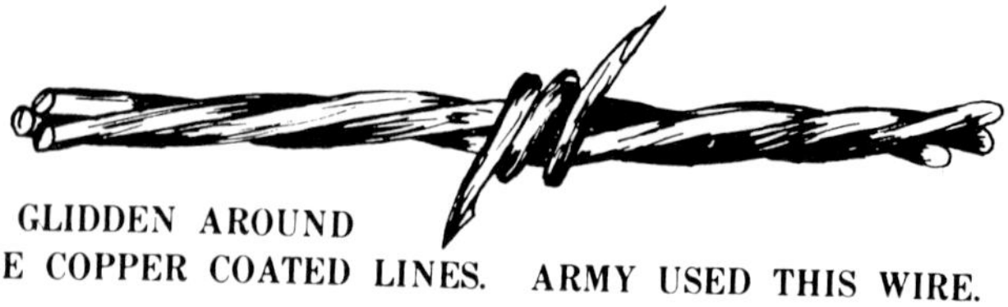

GLIDDEN AROUND
THREE COPPER COATED LINES. ARMY USED THIS WIRE.

442

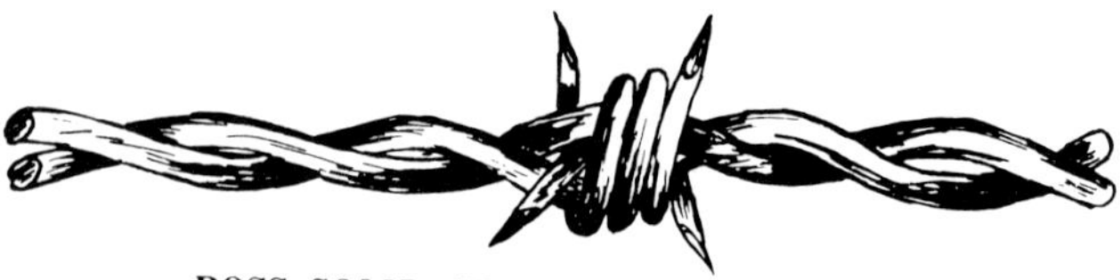

ROSS SOLID STAINLESS STEEL

443

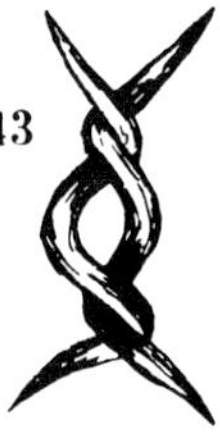

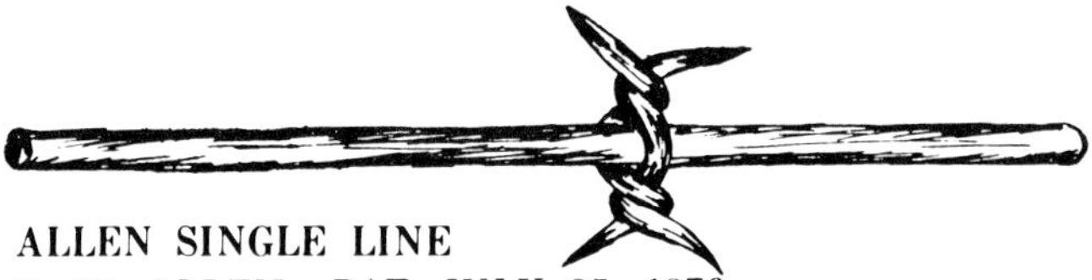

ALLEN SINGLE LINE
G. W. ALLEN PAT. JULY 25, 1876

444

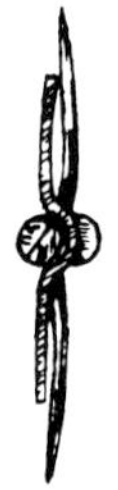

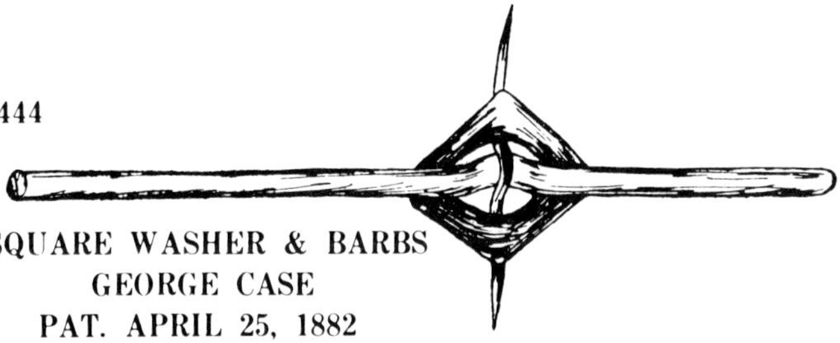

SQUARE WASHER & BARBS
GEORGE CASE
PAT. APRIL 25, 1882

445

T BAR HORSE PEN WIRE

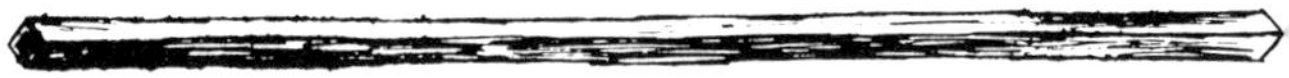

446

SQUARE COPPER LINE
USED FOR FENCES IN THE CALIF. DESERT

447

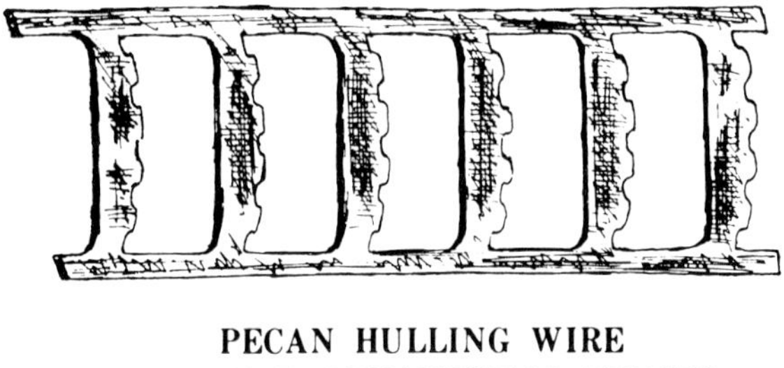

PECAN HULLING WIRE
USED FOR ORNAMENTAL FENCES

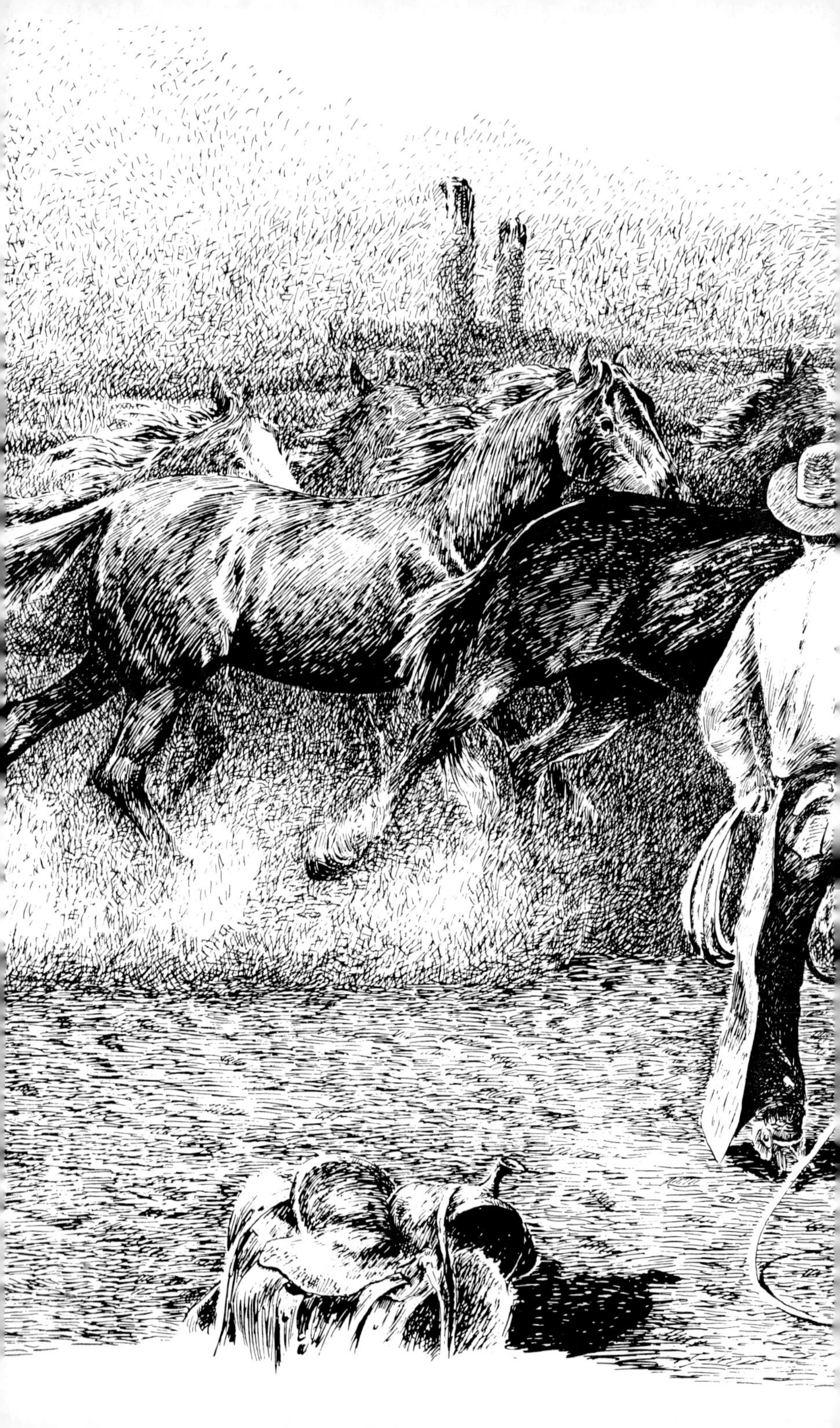

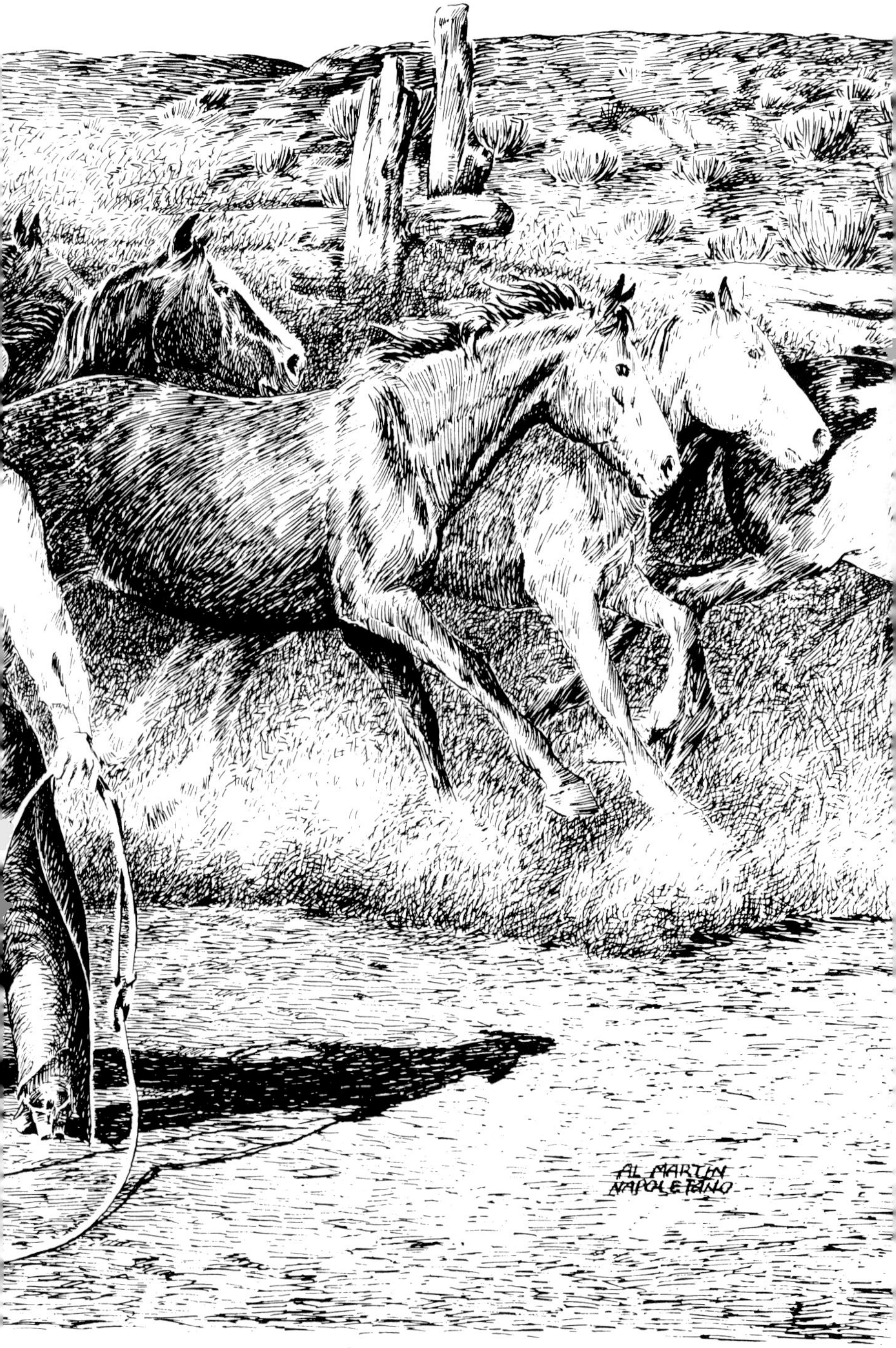
AL MARTIN
NAPOLETANO

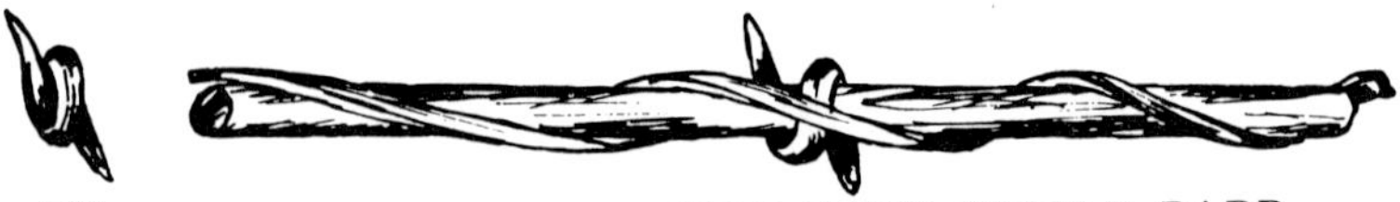

448

LARGE ROD WIRE WITH STAPLE BARB
WITH A SMALL LINE BARB HOLDER
PAT. J. D. CURTIS, MAR. 15, 1892

449

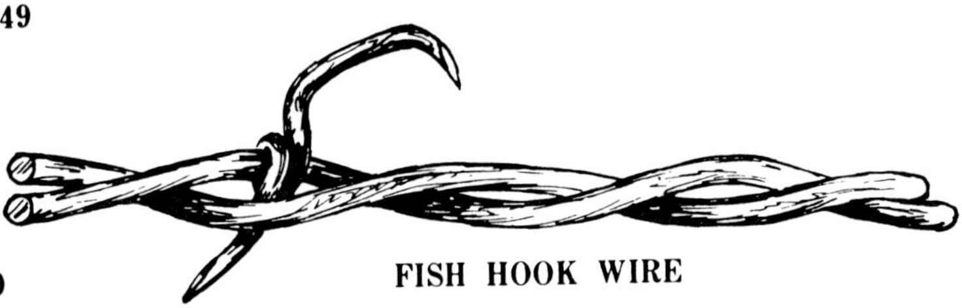

FISH HOOK WIRE

PAT. A. J. UPHAMS, AUG. 1876

450

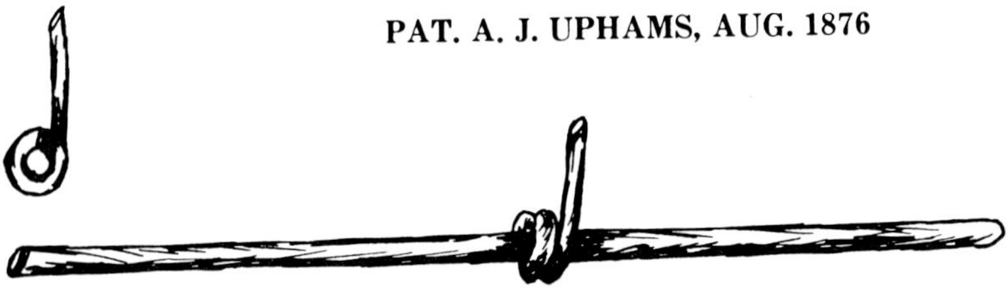

SINGLE LINE, HIGH TENSILE WIRE.
USED IN THE 1870'S

451

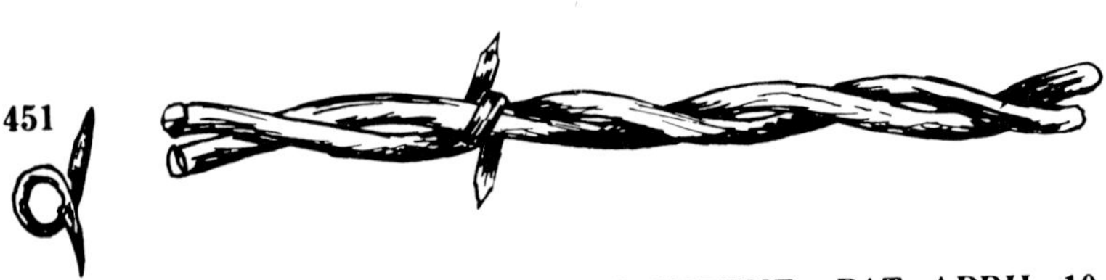

NOTCH CENTER BARB MERRITT GREENE PAT. APRIL 10, 1888

452 HALF ROUND DOUBLE BAR.
USED BY RAILROADS FOR EARLY FENCING

PAT. JACOB HAISH, DEC. 15, 1885

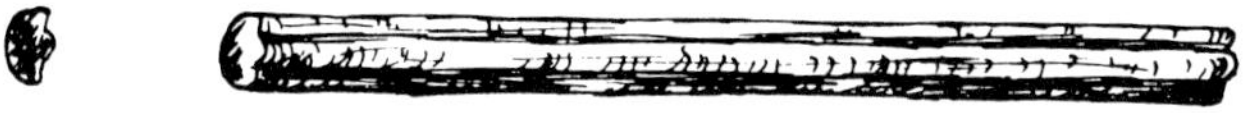

453 HALF ROUND RAISED CENTER
RAILROAD WIRE

PAT. JACOB HAISH, DEC. 15, 1885

454 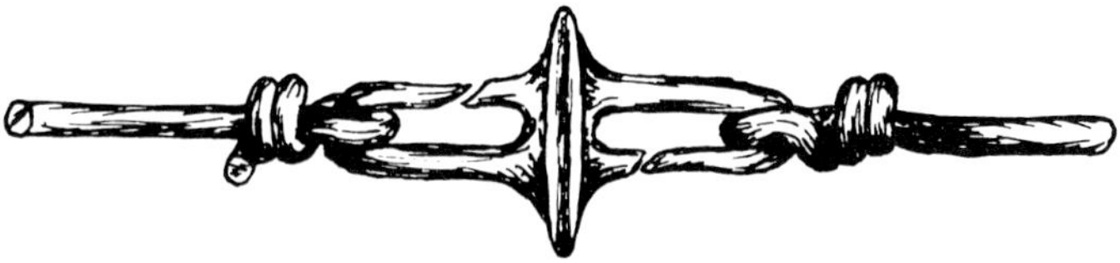

PLANTER WIRE TRIP

455 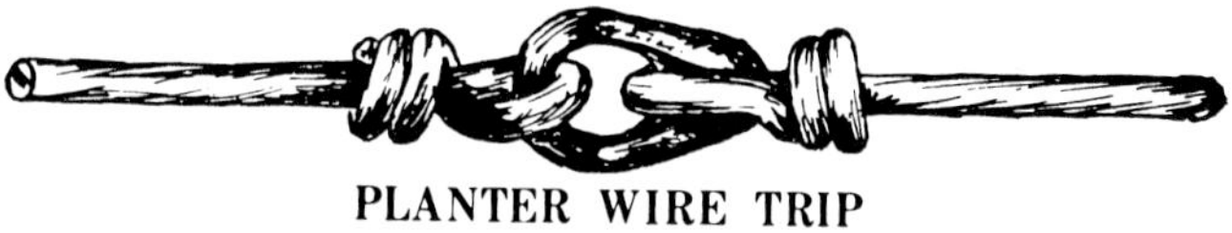

PLANTER WIRE TRIP

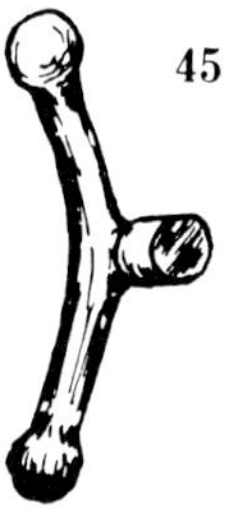

456

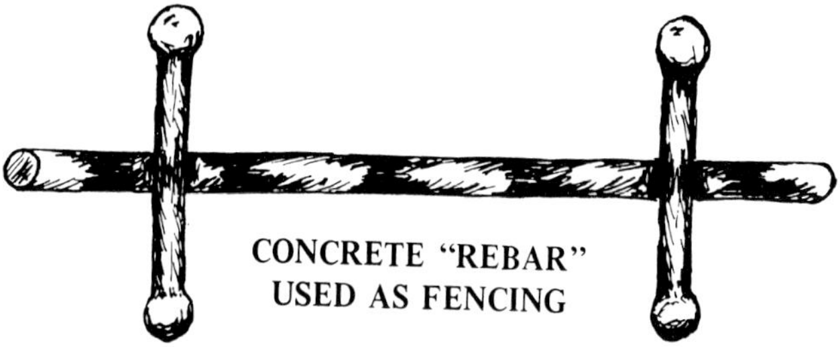

CONCRETE "REBAR"
USED AS FENCING

457 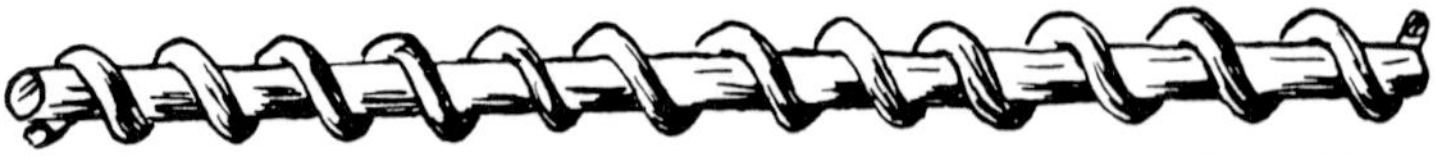

GRAPE VINEYARD WIRE. USED FOR LIVESTOCK ALSO.

PAT. M. M. SHELLABERGER, DEC. 15, 1891

458 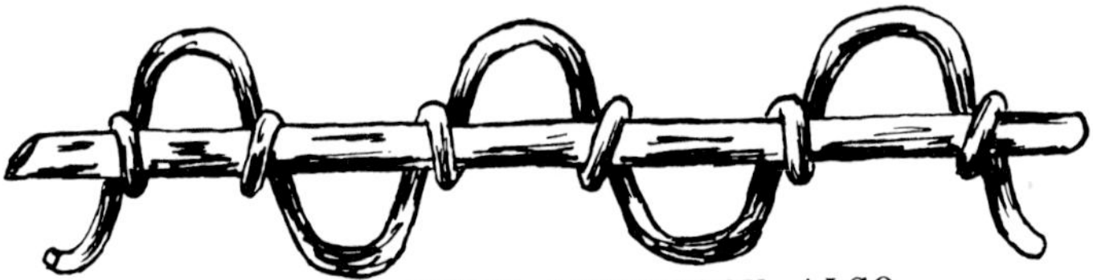

VINEYARD WIRE. LIVESTOCK ALSO.
J. B. CLEVELAND PAT. MAY 24, 1892

459 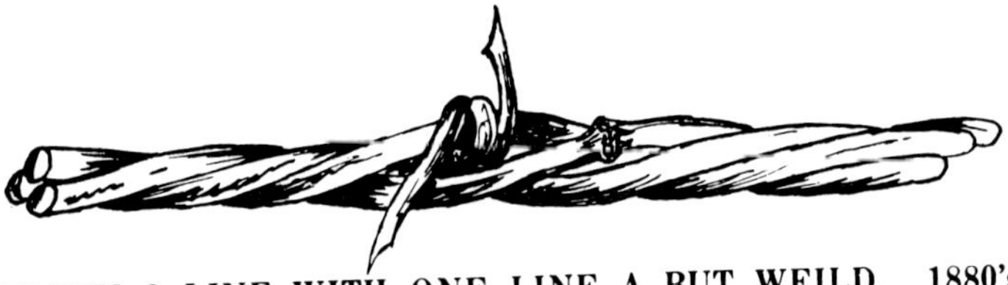

GLIDDEN 3 LINE WITH ONE LINE A BUT WEILD. 1880's

460 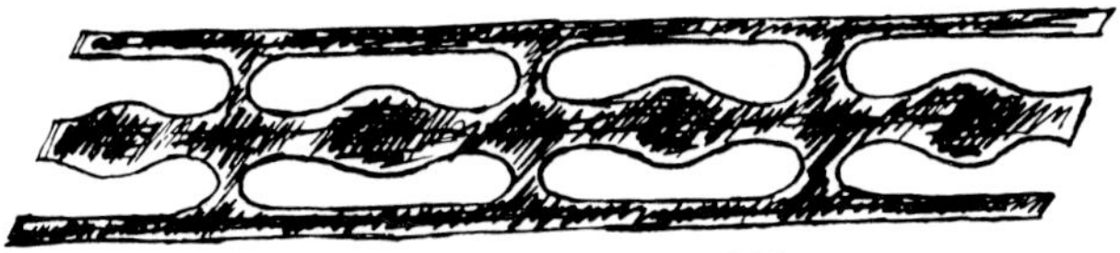

ORNAMENTAL WIRE
PUNCH PRESS SCRAP

461 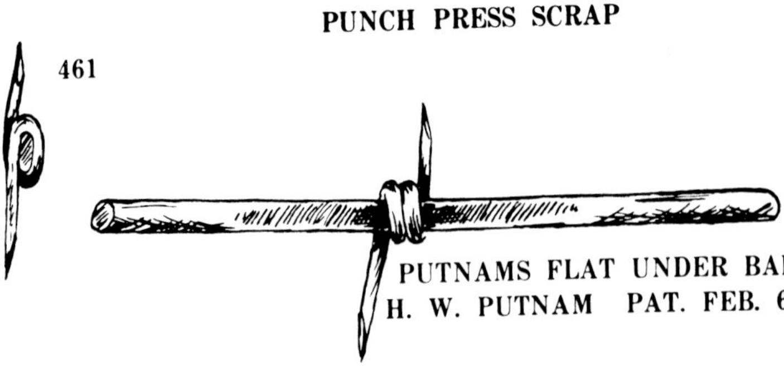

PUTNAMS FLAT UNDER BARB.
H. W. PUTNAM PAT. FEB. 6, 1877

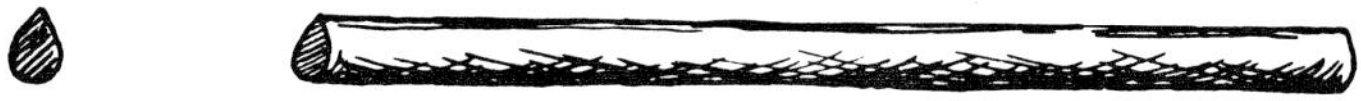

462

TEAR DROP DRAWN WIRE

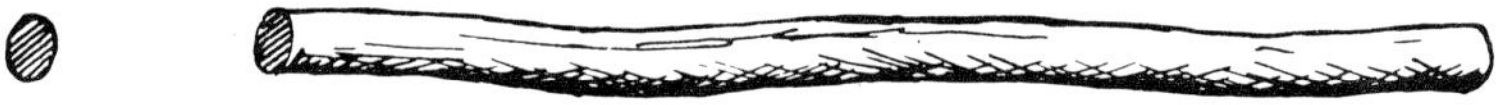

463

OVAL DRAWN

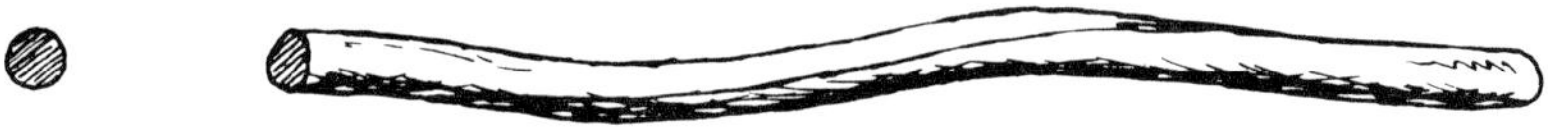

464

ROUND DRAWN

465

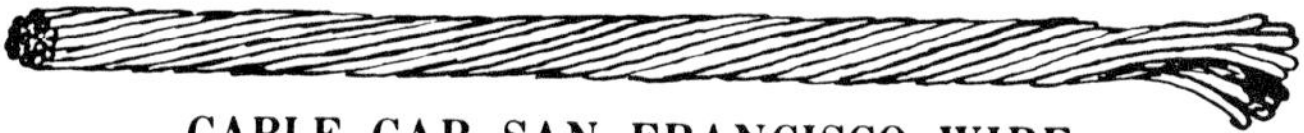

CABLE CAR SAN FRANCISCO WIRE

466

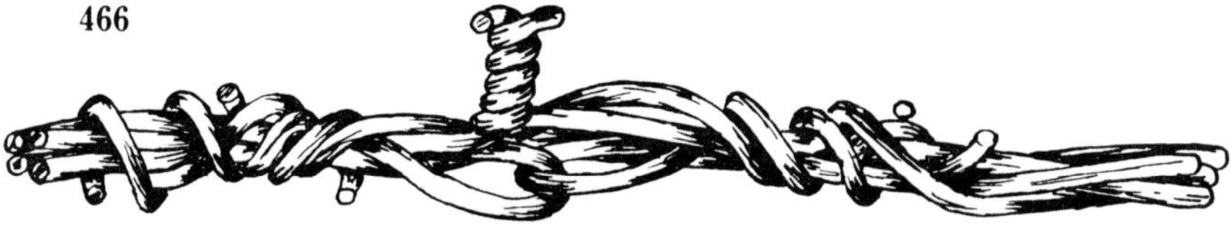

BAILING WIRE PLATTED WIRE

467 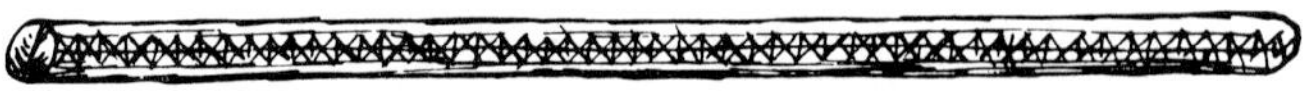

TERRYS RANA RANCH HORSE PEN WIRE.
PUT UP IN THE 1870's.

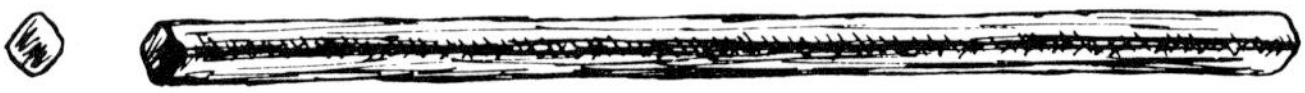

468 DIAMOND SHAPED COPPER HAMMER MILL WIRE
USED FOR LIVESTOCK IN CALIF.

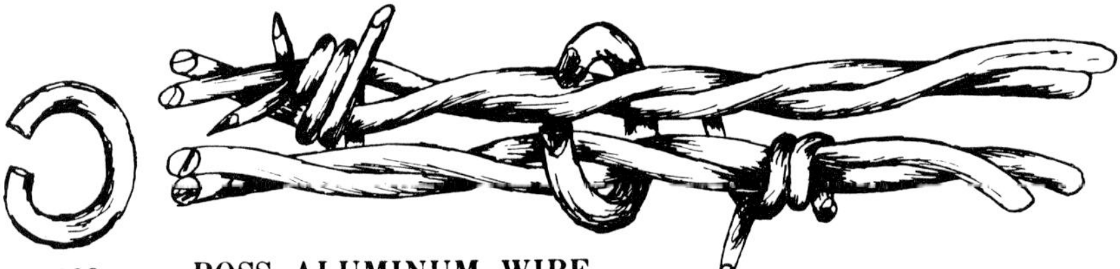

469 ROSS ALUMINUM WIRE
USED FOR RIOT & COMPOUND CONTROL. MODERN

470 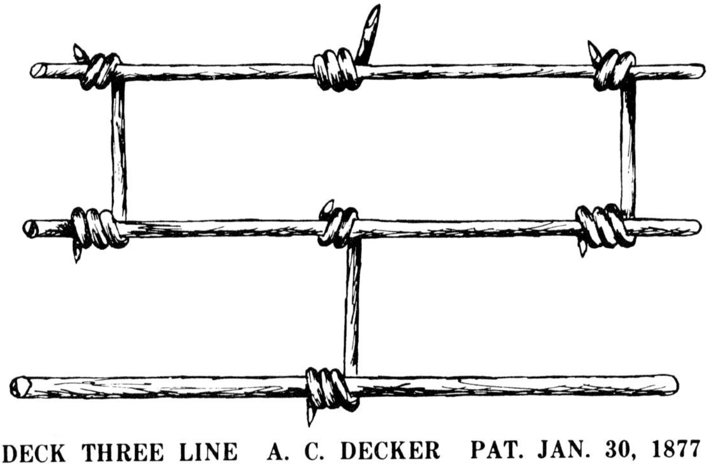

DECK THREE LINE A. C. DECKER PAT. JAN. 30, 1877

471

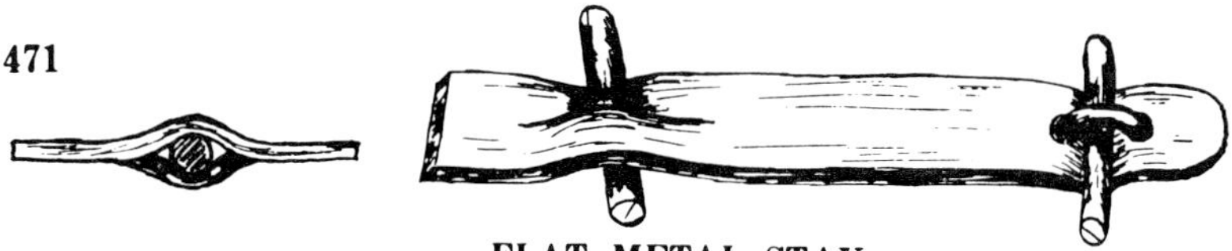

FLAT METAL STAY

472

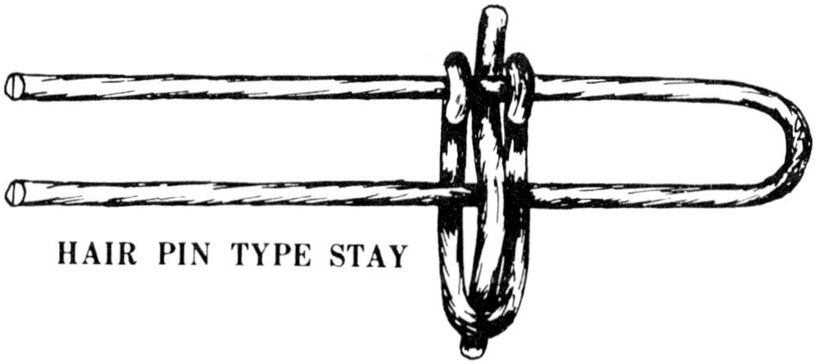

HAIR PIN TYPE STAY

473

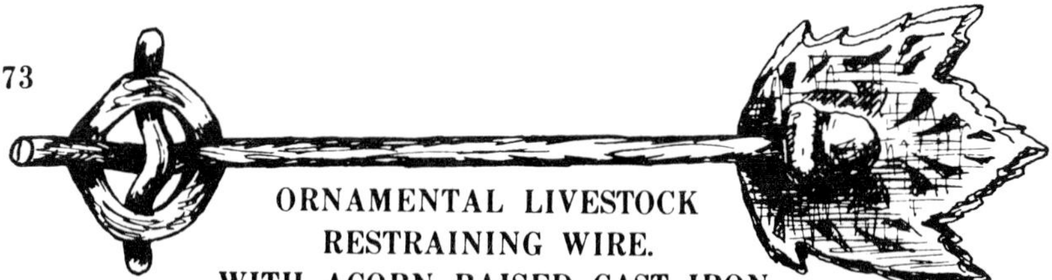

ORNAMENTAL LIVESTOCK
RESTRAINING WIRE.
WITH ACORN RAISED CAST IRON

474

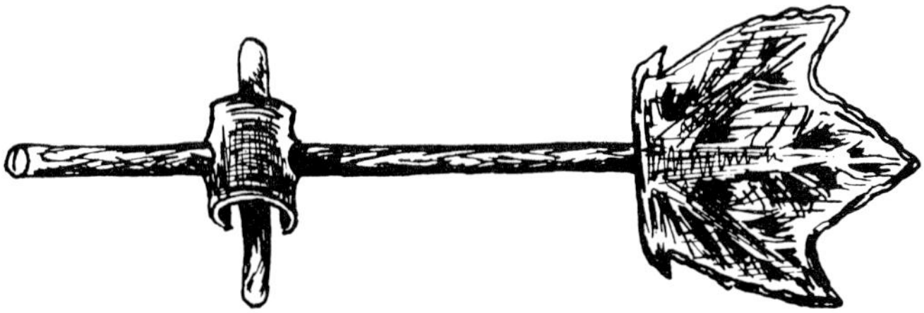

CAST IRON WITH NO ACORN

475

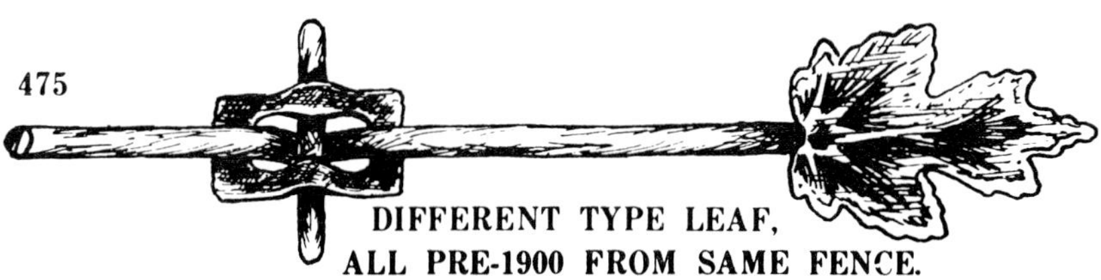

DIFFERENT TYPE LEAF,
ALL PRE-1900 FROM SAME FENCE.

476

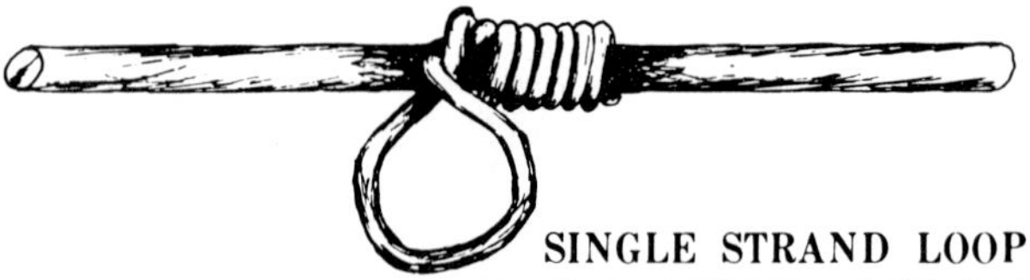

SINGLE STRAND LOOP WIRE.
M. F. McNELLY PAT. MAY 14, 1895

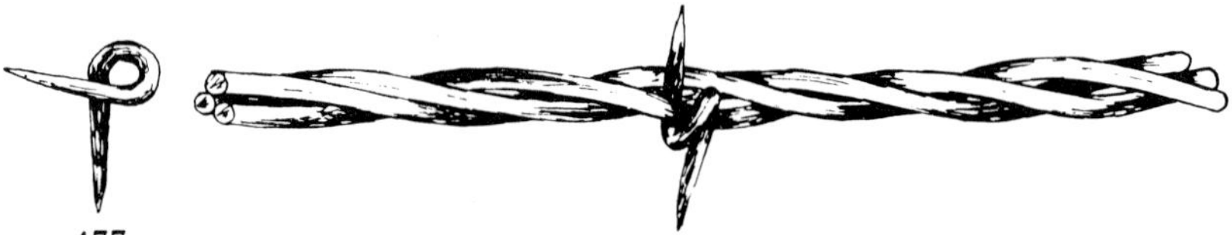

477

THREE LINE STAPLE BARB.
PURCHES MILES PAT. OCT. 8, 1878

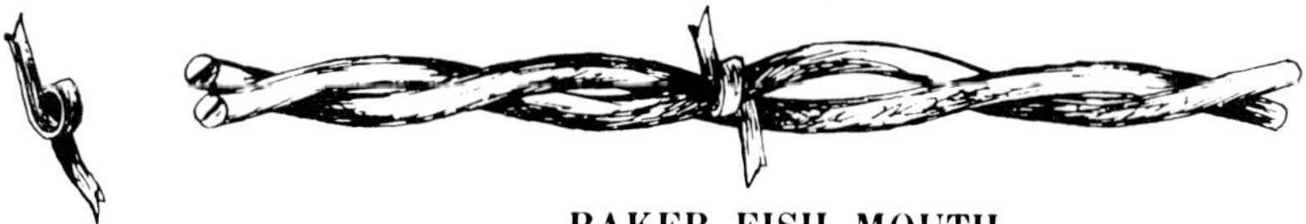

478

BAKER FISH MOUTH
PAT. G. C. BAKER, FEB. 27, 1883

479

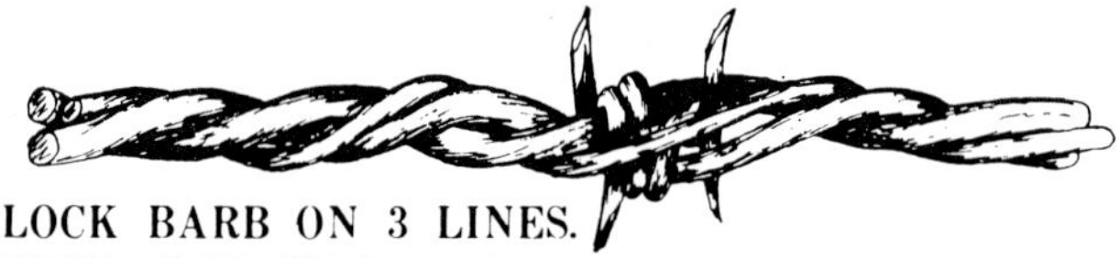

SHINN LOCK BARB ON 3 LINES.
M. C. SHINN PAT. MAR. 1, 1881

480

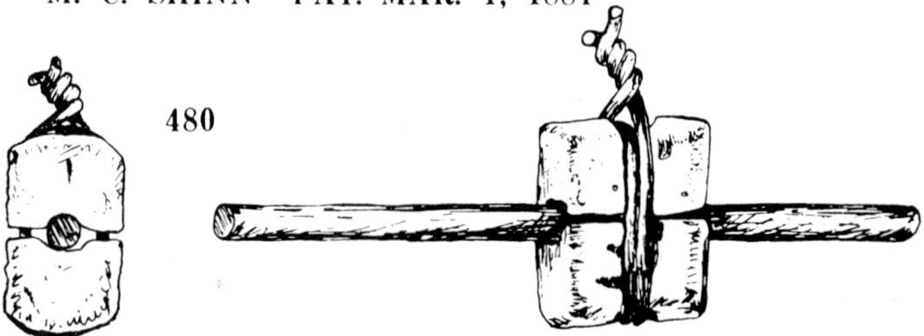

EARLY DAY FENCE PHONE INSULATOR

481 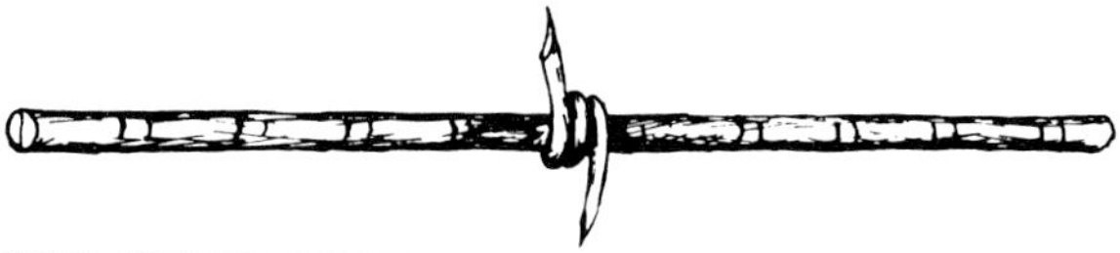

HOXIE KNOT LINE. V. HOXIE PAT. DEC. 14, 1909

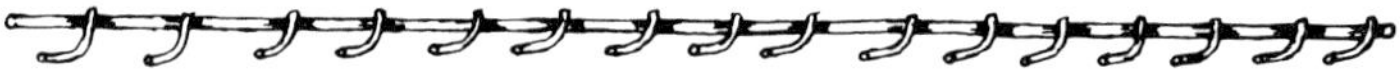

WELDED PIN STAY

482

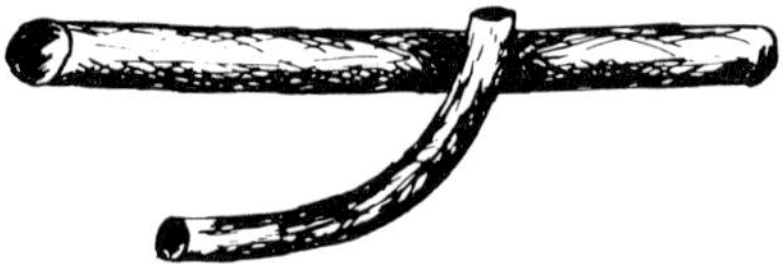

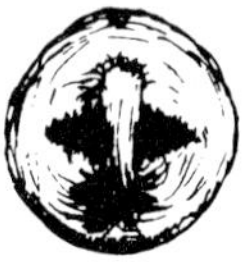

483 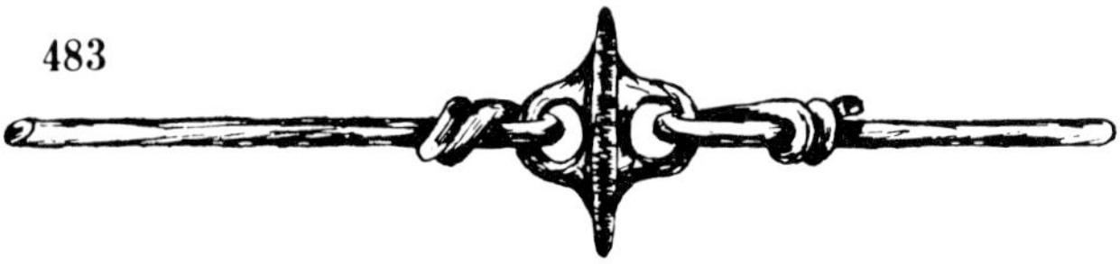

PLANTER TRIP WIRE

484 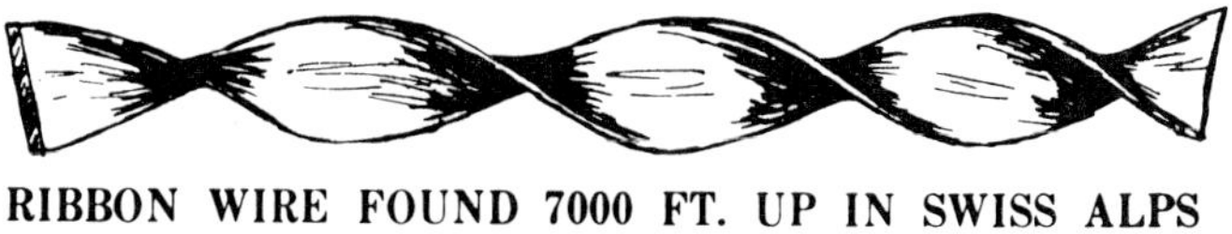

RIBBON WIRE FOUND 7000 FT. UP IN SWISS ALPS
BY JOE LEWIS

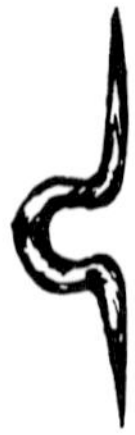

485

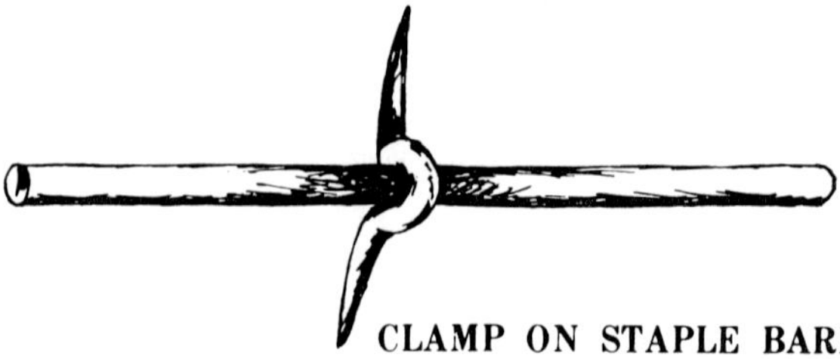

CLAMP ON STAPLE BARB.
JACOB HAISH PAT. JUNE 15, 1875

486

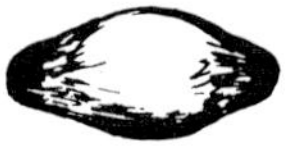

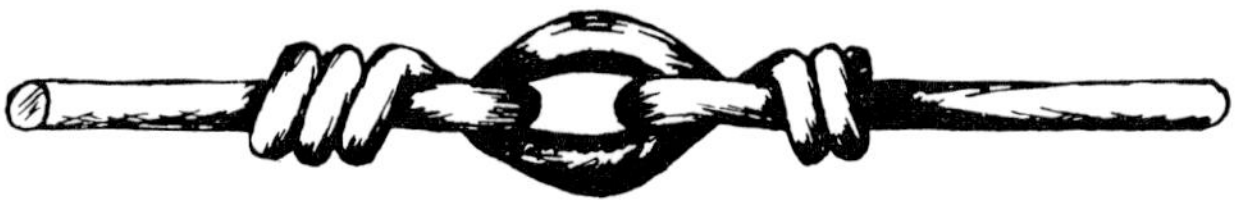

PLANTER TRIP WIRE

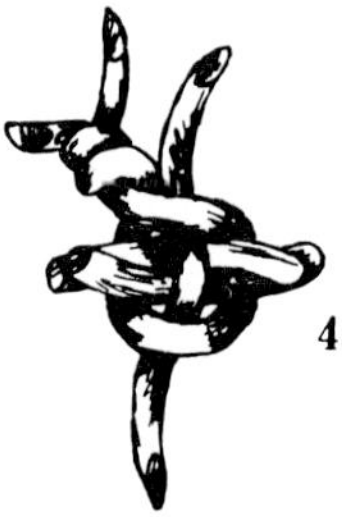

487

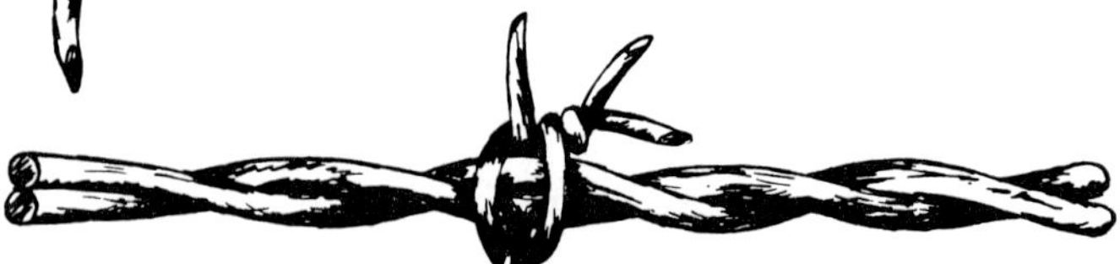

HOME MADE WIRE BARB
ON 2 LINE HORSE WIRE

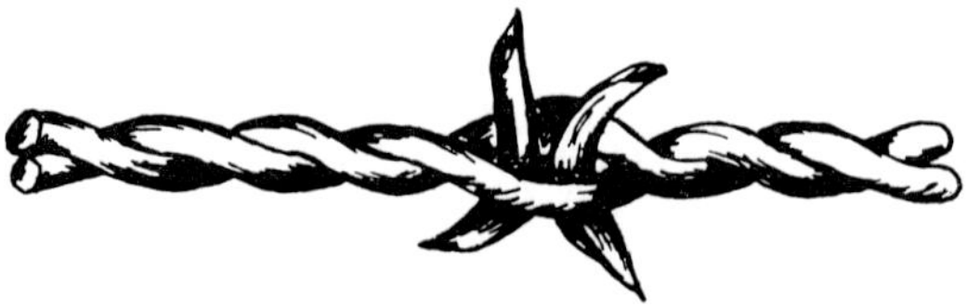

SCUTTS WEBBED RIBBON BARB.
VARIATION OF JOHN SCUTT.
JOHN SCUTT PAT. SEPT. 12, 1882

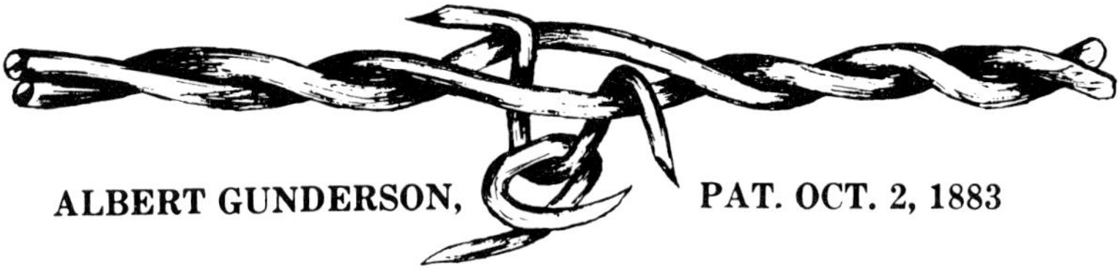

ALBERT GUNDERSON, PAT. OCT. 2, 1883

489

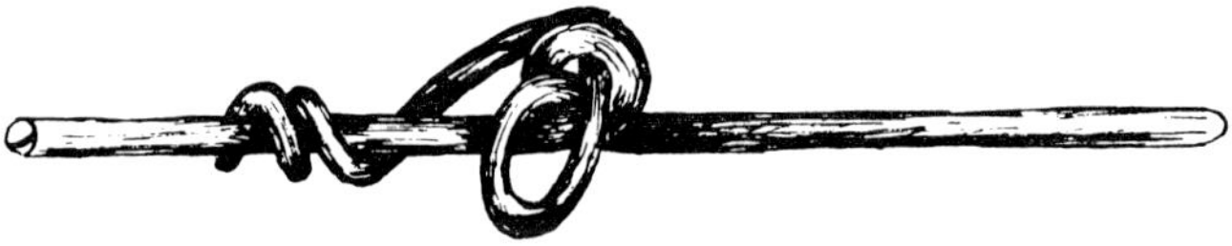

490

M. KELLEY LOOP NET WIRE

PAT. JULY 14, 1885

491

PAT. BY O. O. KITTLESON, MAY 7, 1878

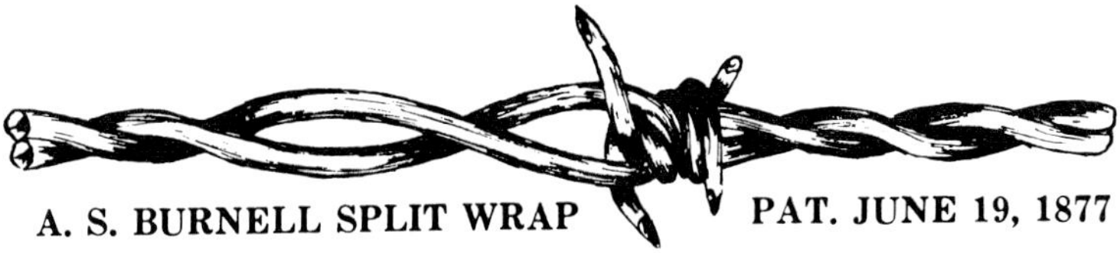

492

A. S. BURNELL SPLIT WRAP PAT. JUNE 19, 1877

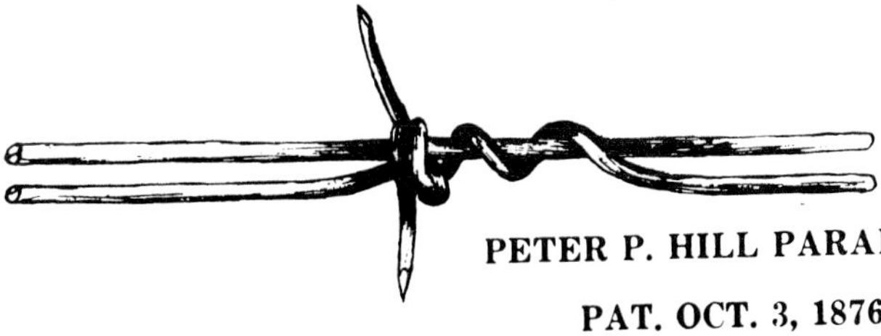

493

PETER P. HILL PARALLEL

PAT. OCT. 3, 1876

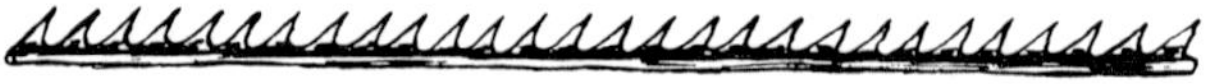

494 TEXTILE CARDING WIRE USED AS FENCING

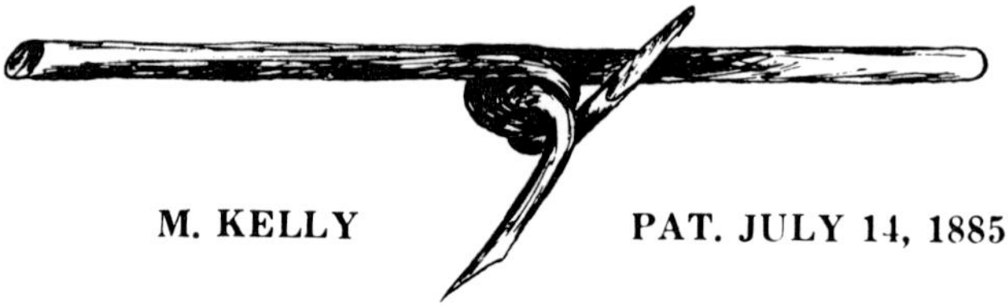

495

M. KELLY PAT. JULY 14, 1885

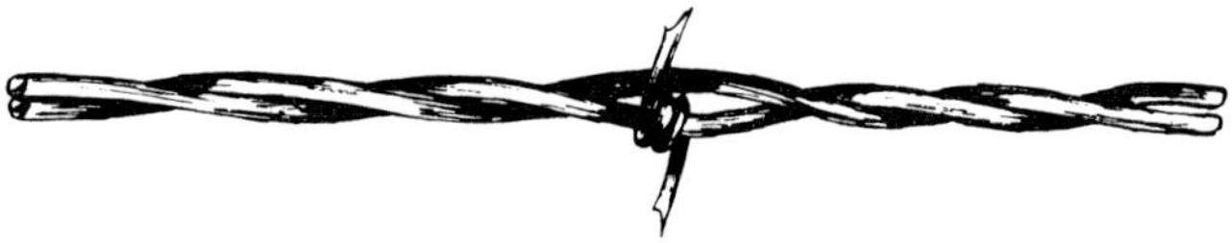

496

GLIDDEN COPPER LINE & BARB FISH MOUTH

J. GLIDDEN PAT. NOV. 24, 1874

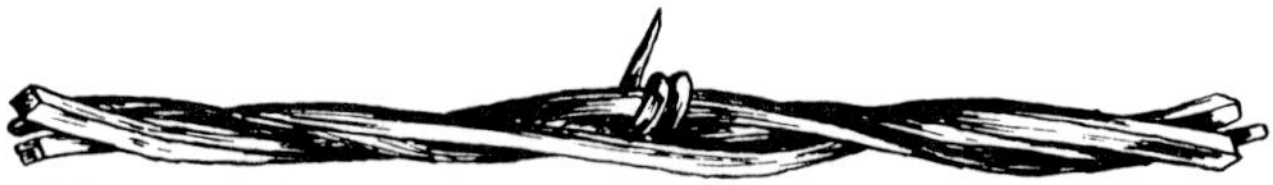

497 GLIDDEN R.R. THREE STRAND, TWO SQUARE

J. GLIDDEN PAT. NOV. 24, 1874

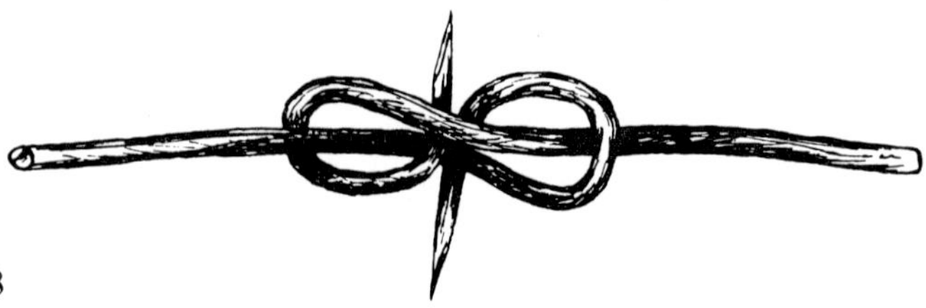

498

SINGLE LINE BRONSON

PAT. A. E. BRONSON, APRIL 24, 1877

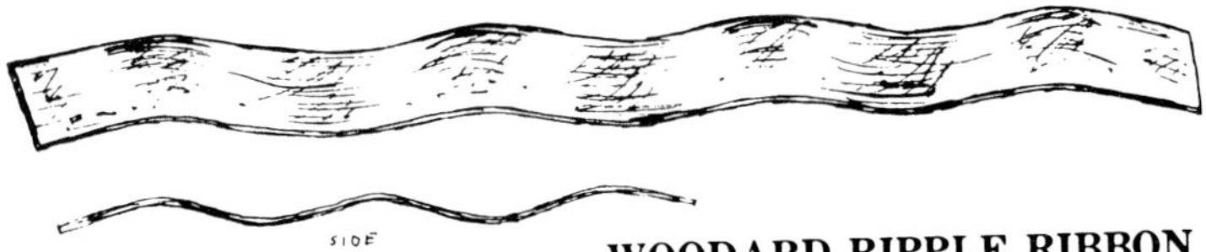

499 WOODARD RIPPLE RIBBON

PAT. A. B. WOODARD, NOV. 10, 1891

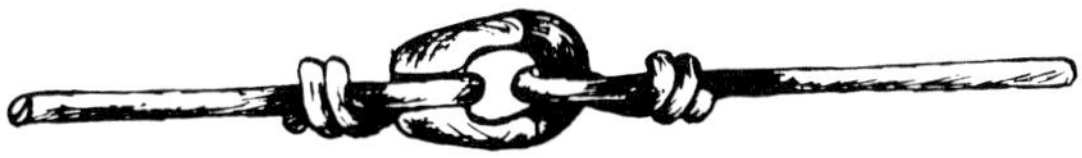

500 PLANTER CHECK WIRE

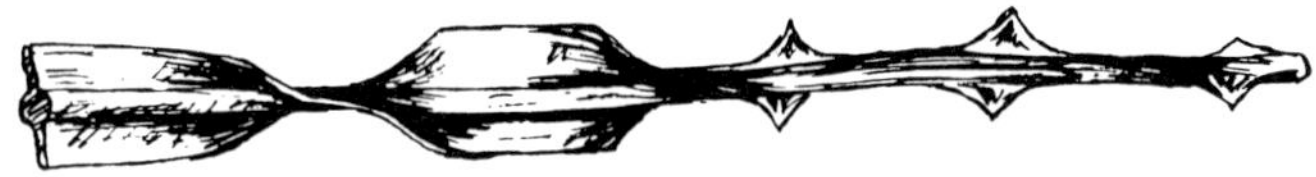

501 FACTORY FREAK

T. V. ALLIS PAT. NOV. 30, 1880

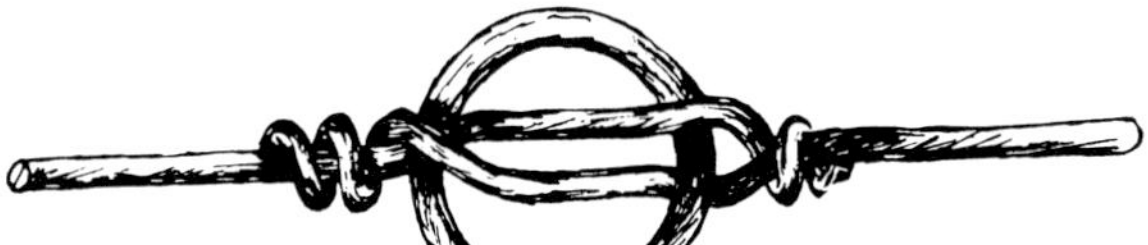

502 SINGLE LINE EXPANTION RING 2 INCH

PAT. T. NEELY, DEC. 20, 1881

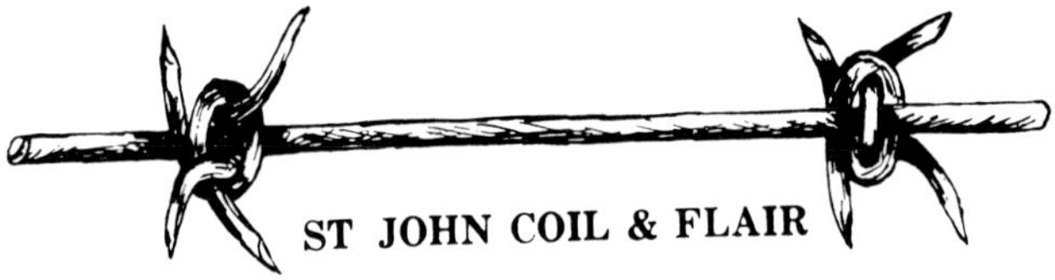

503 ST JOHN COIL & FLAIR

PAT. S. H. ST. JOHN, JAN. 15, 1878

504
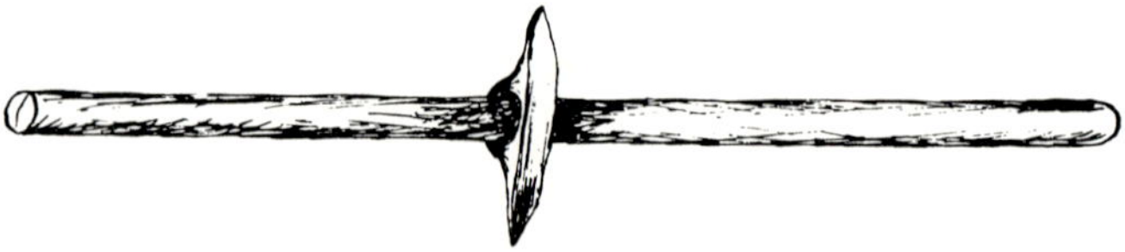

STOVER SINGLE LINE

PAT. D. C. STOVER, JUNE 29, 1875

505
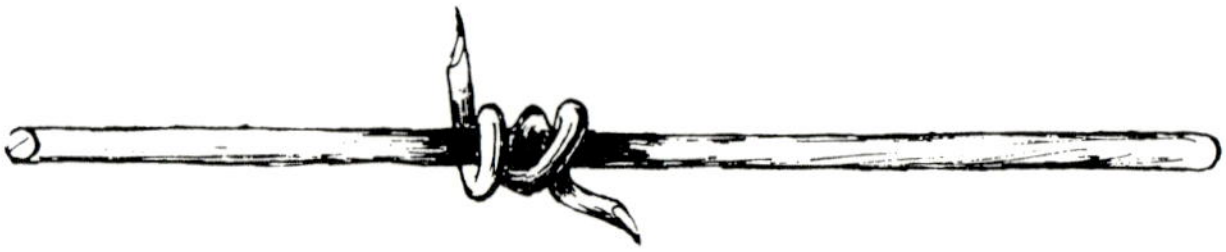

LENOX KNOT

PAT. E. S. LENOX, JUNE 24, 1884

506
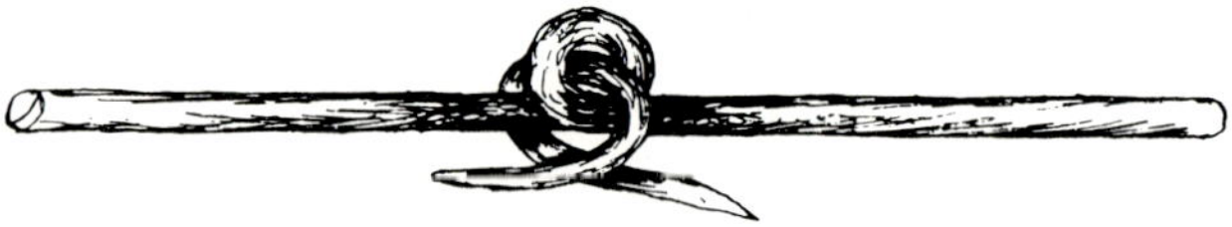

PAGE SINGLE LINE

PAT. J. E. PAGE, DEC. 7, 1875

507
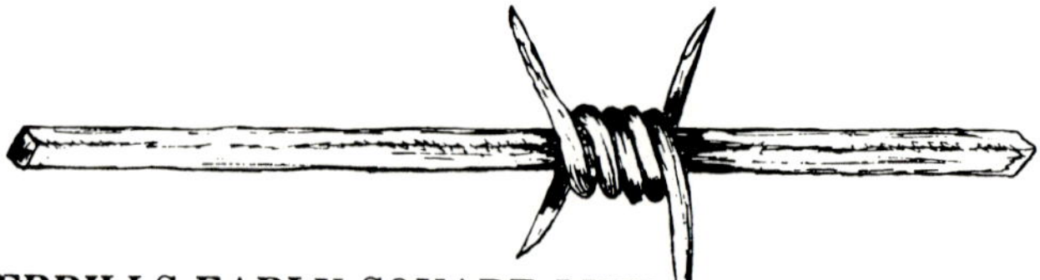

MERRILLS EARLY SQUARE LINE

PAT. J. & L. MERRILL, SEPT. 29, 1871

508
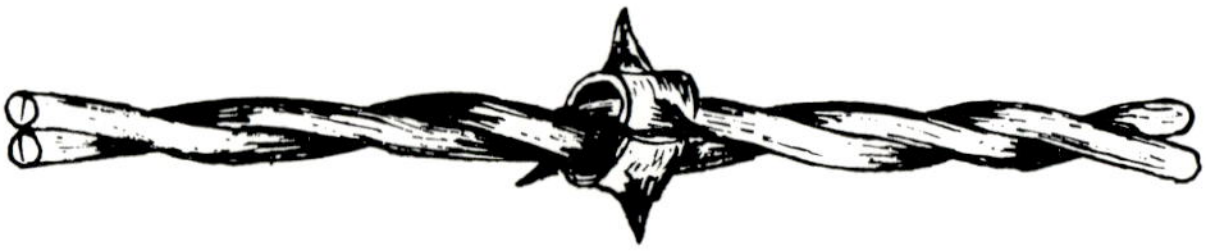

E. S. WHEELER SPIKE COLLAR

PAT. E. S. WHEELER, JUNE 30, 1885

509

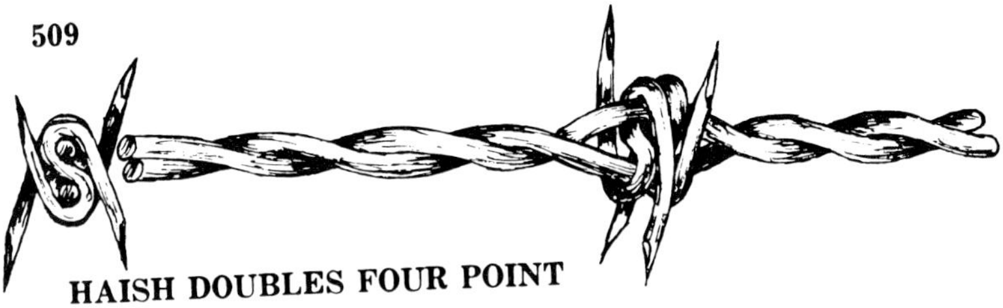

HAISH DOUBLES FOUR POINT

PAT. JACOB HAISH, AUG. 31, 1875

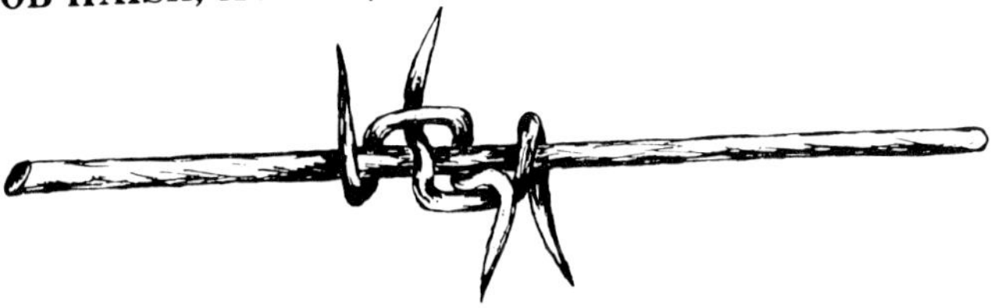

510

PETER P. HILL

PAT. PETER HILL, MAY 5, 1878

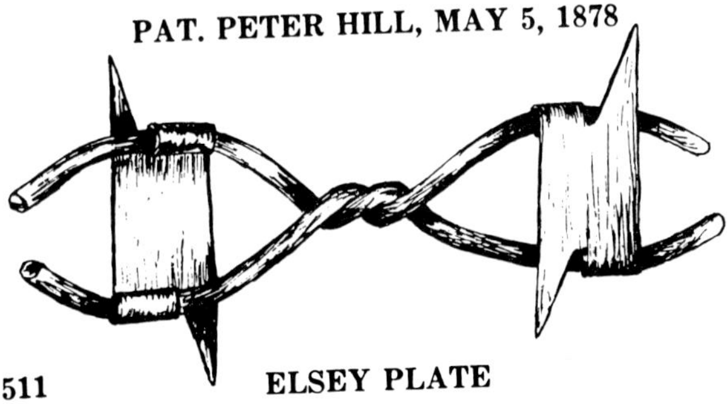

511 ELSEY PLATE

PAT. G. ELSEY, JULY 18, 1882

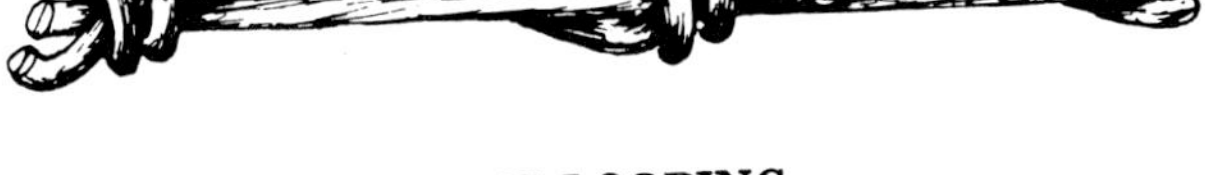

512 RITER'S LOOPING

PAT. J. L. RITER, OCT. 10, 1893

513

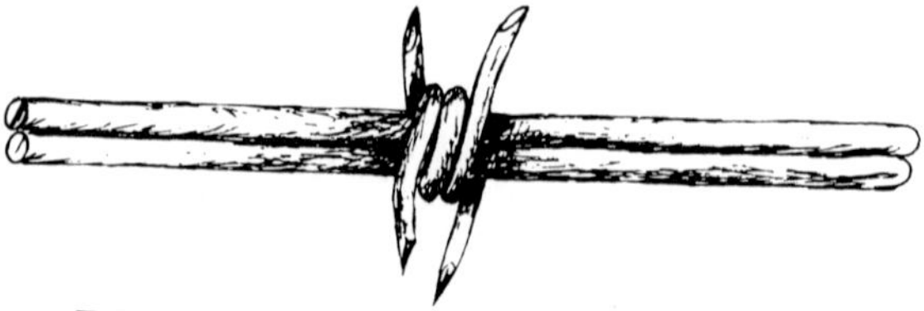

DODGE-WASHBURN PARALLEL

PAT. T. H. DODGE-C. G. WASHBURN
JAN. 24, 1882

514

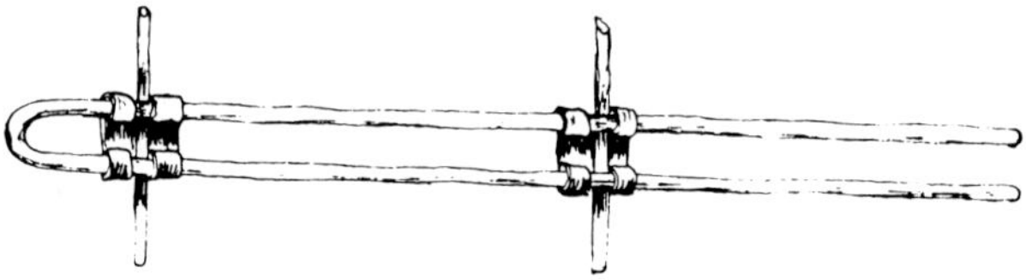

FENCING CLIPS FOR BUILDING THEIR OWN STYLE.

515

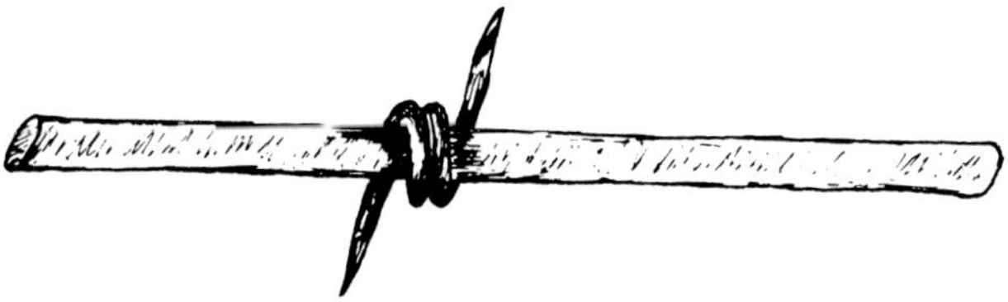

DECKER'S OVAL LINE

PAT. A. C. DECKER, JUNE 13, 1876

516

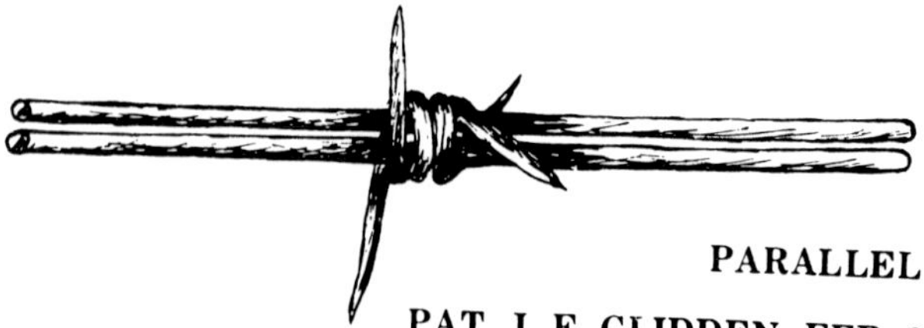

PARALLEL

PAT. J. F. GLIDDEN, FEB. 8, 1876
GLIDDEN'S REISSUE No. 6914

517

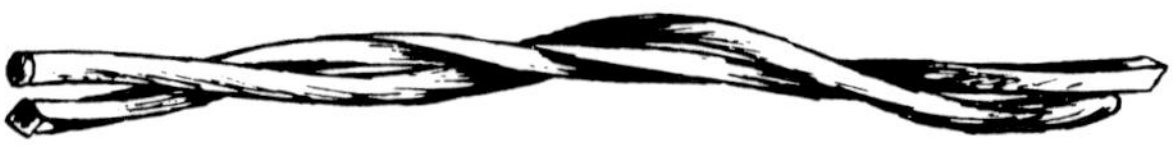

GLIDDEN R.R. NO BARB. PAT. J. F. GLIDDEN, NOV. 24, 1874

518

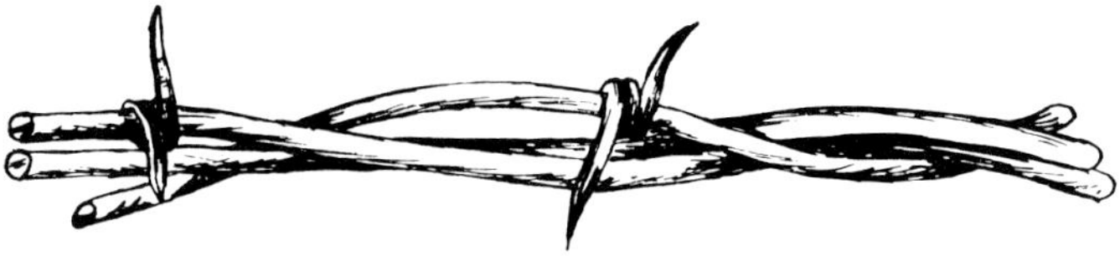

GLIDDEN & BAKER

GLIDDEN PAT. NOV. 24, 1874

519

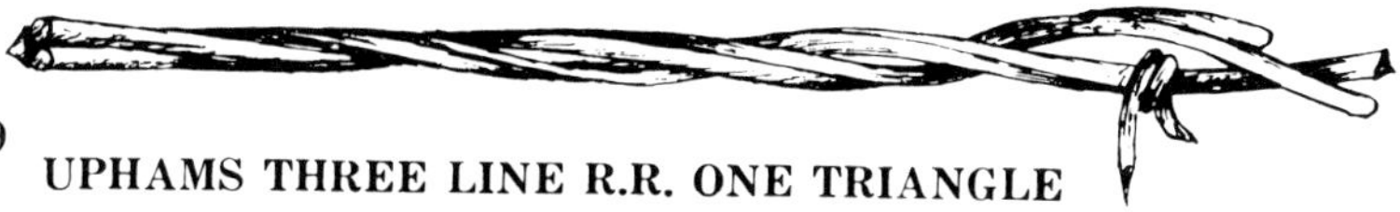

UPHAMS THREE LINE R.R. ONE TRIANGLE

PAT. A. J. UPHAM, AUG. 29, 1876

520

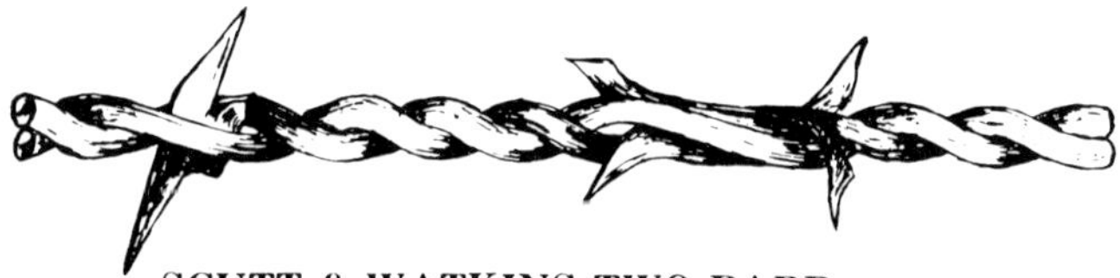

SCUTT & WATKINS TWO BARB

PAT. JUNE 18, 1878 H. B. SCUTT

521

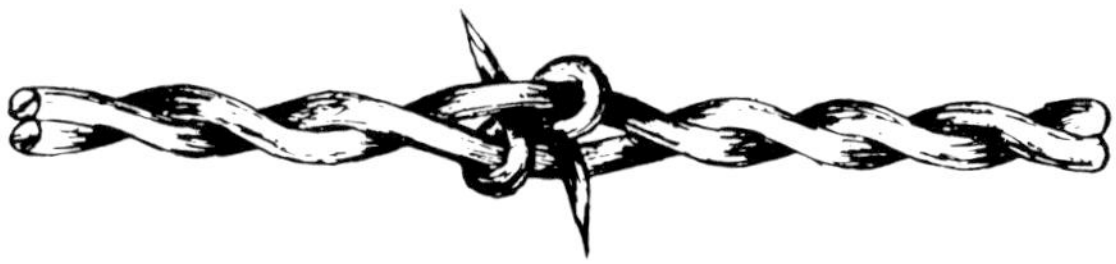

FIGURE EIGHT BARB

PAT. A. J. UPHAM, AUG. 29, 1876

522

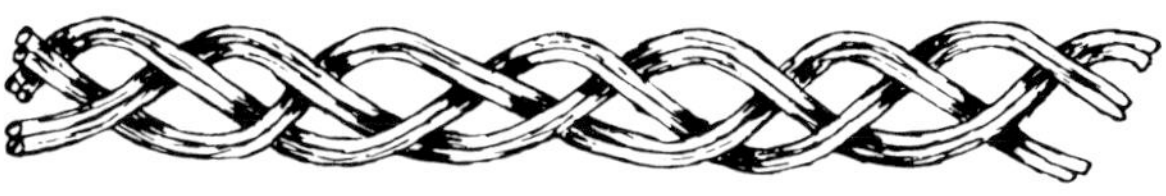

CLEAVELAND'S THEE LINE

PAT. J. B. CLEAVELAND, JULY 10, 1894

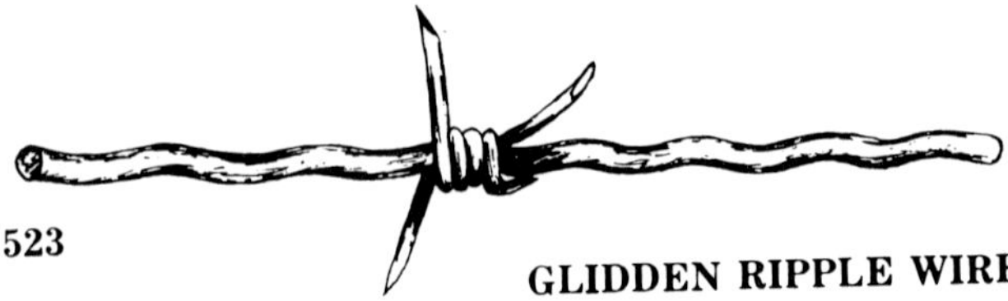

523

GLIDDEN RIPPLE WIRE

PAT. J. F. GLIDDEN, FEB. 8, 1876

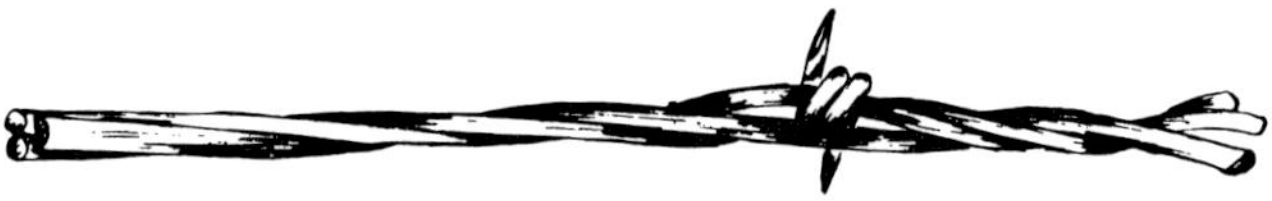

524

GLIDDEN 3 LINE 1 HALF ROUND

PAT. J. F. GLIDDEN, NOV. 24, 1874

525

HARBAUGH'S SLEEVE CONNECTORS

PAT. J. W. HARBAUGH, JUNE 7, 1881

526

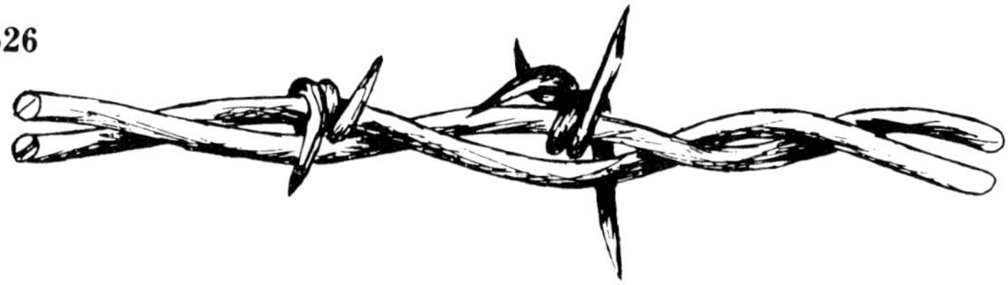

527

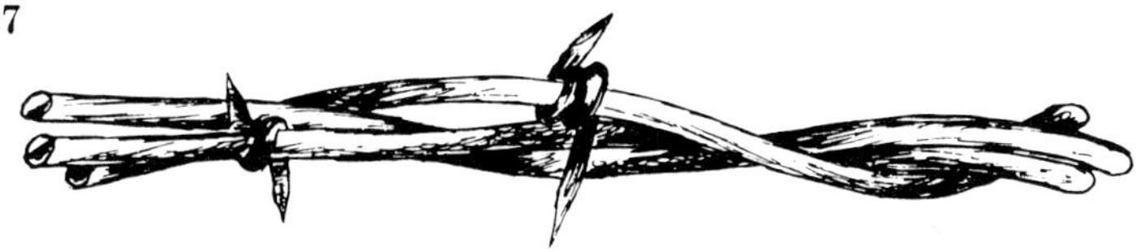

528

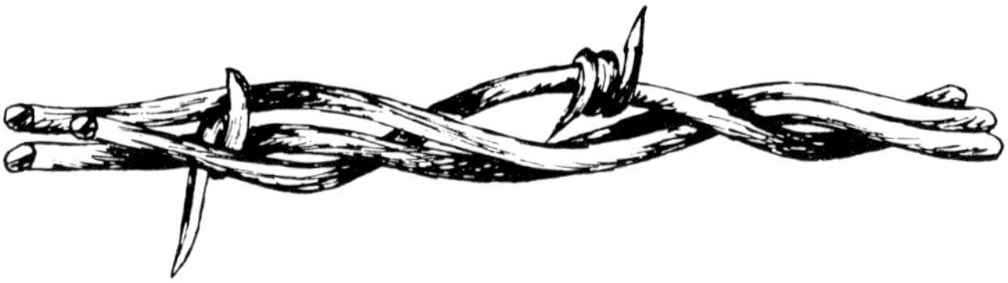

529

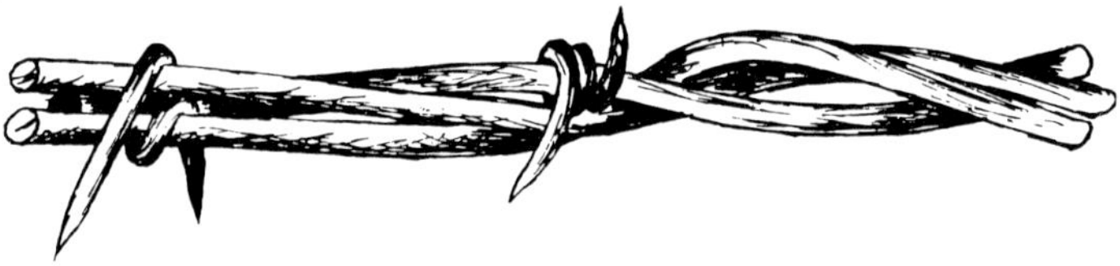

530

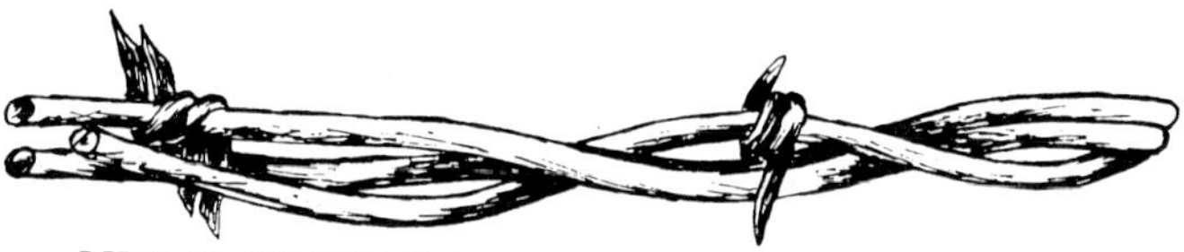

MULTI-COMPANY INDIAN ALLOTMENT FENCING WIRE. ARBUCKLE MT. OKLA. AREA.

531

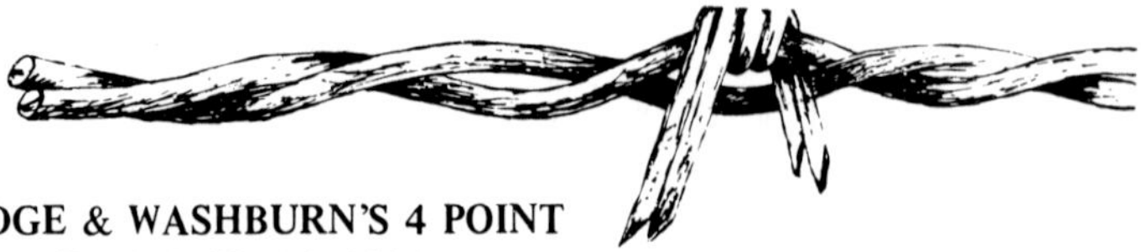

DODGE & WASHBURN'S 4 POINT
PAT. JAN. 24, 1882

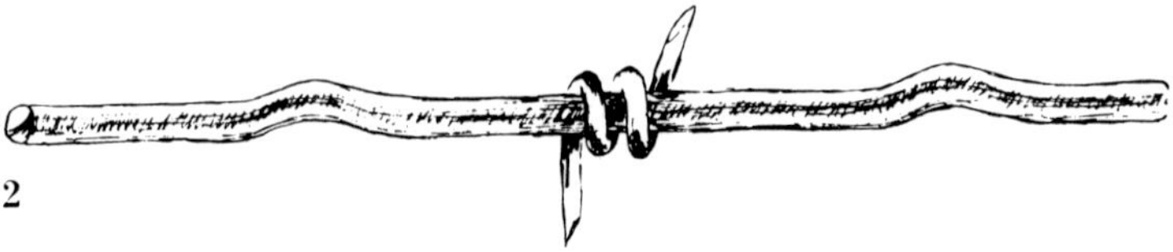

532

ROSE KINK LINE

PAT. H. M. ROSE, DEC. 25, 1877

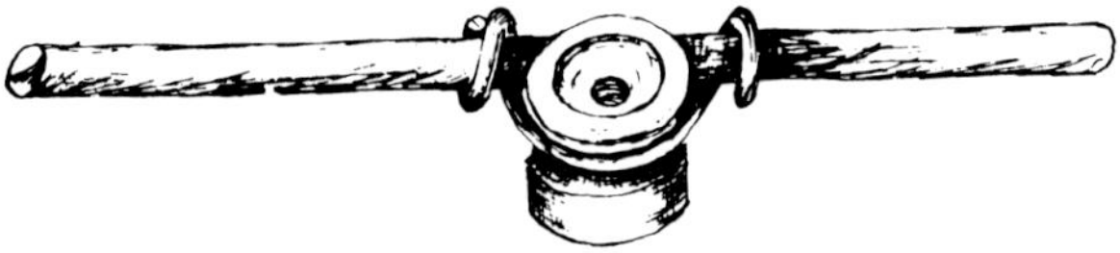

533

TELEPHONE FENCE LINE INSULATOR

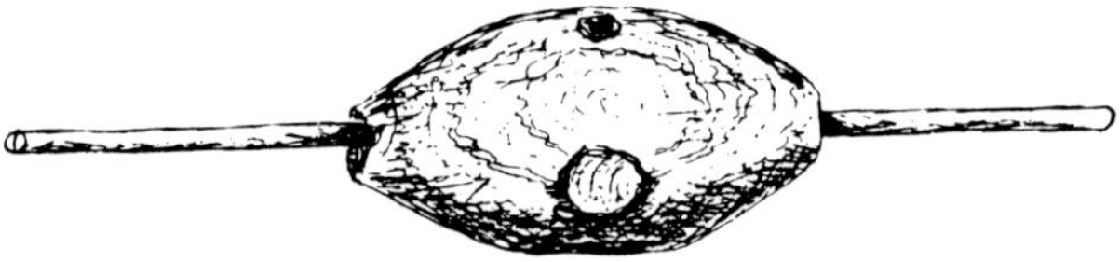

534

WOODEN OVAL PLANTER WIRE

535

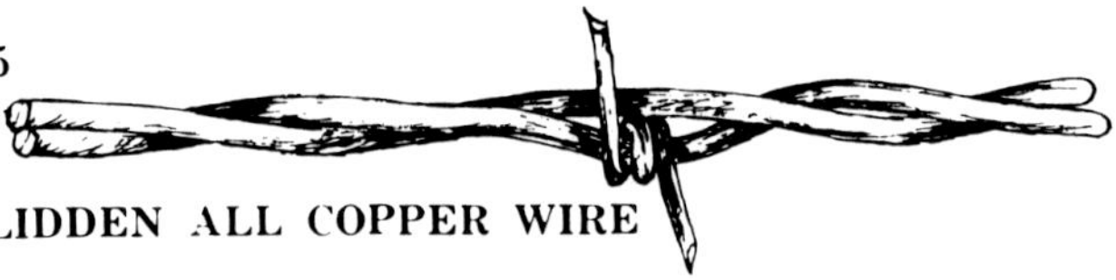

GLIDDEN ALL COPPER WIRE

PAT. J. F. GLIDDEN, NOV. 24, 1874

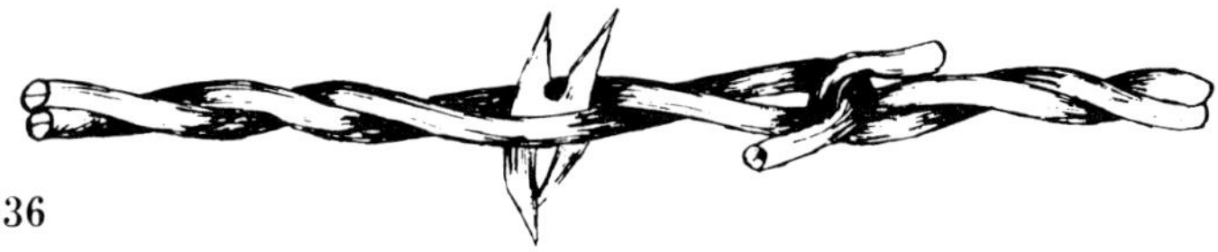

536

SCUTTS TURN BACK SPLICE

PAT. H. B. SCUTT, JUNE 18, 1878

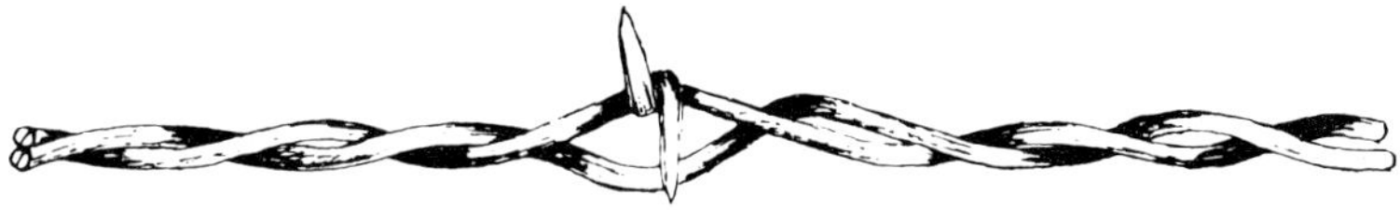

537

UPHAMS DOUBLE HUMP WIRE.

PAT. A. J. UPHAM, AUG. 29, 1876

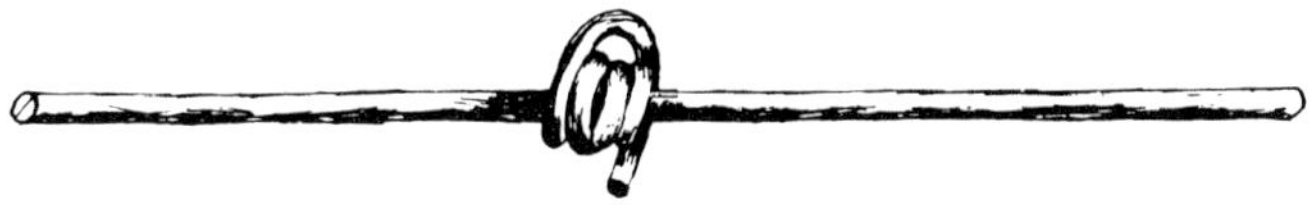

538

PLANTER KNOT WIRE

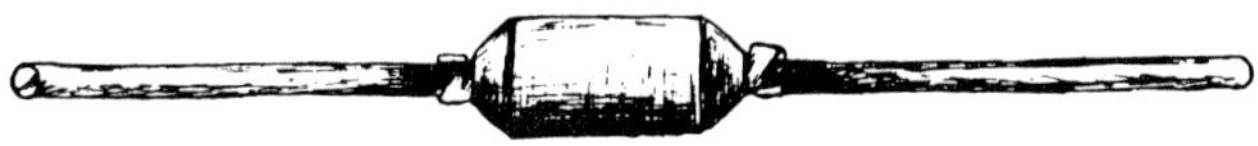

539

PLANTER LONG SLUG WIRE

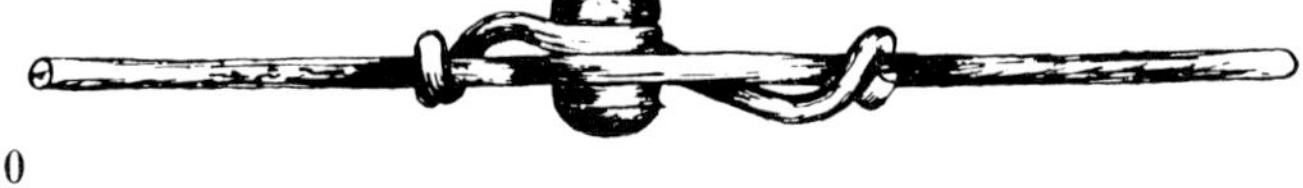

540

PLANTER SHORT SLUG WIRE

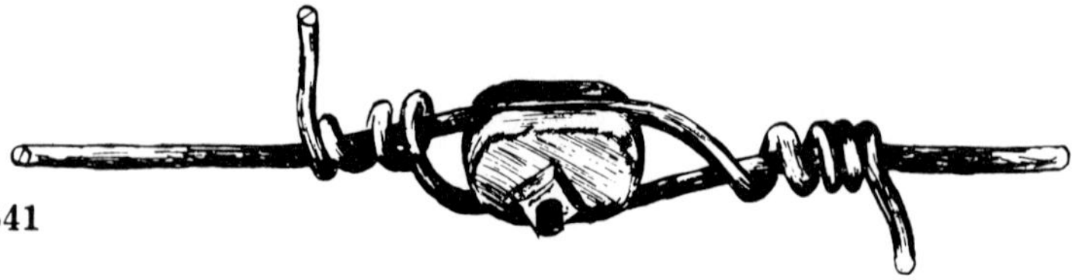

541

SUNDERLAND'S IRON BOLTED SPLICER

PAT. L. E. SUNDERLAND, AUG. 12, 1884

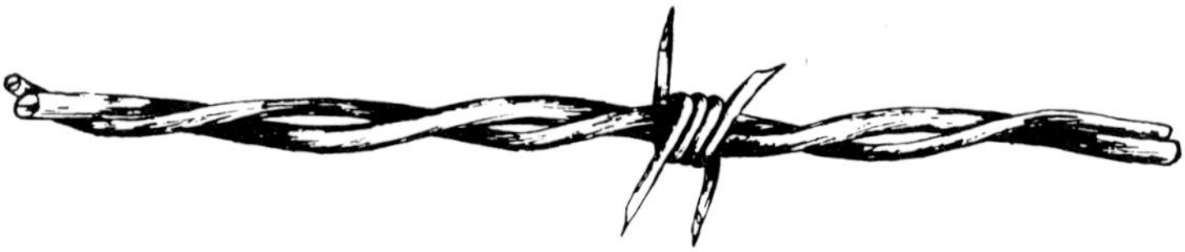

542

CURTIS, 2 AROUND 2, HALF ROUND BARB

PAT. J. D. CURTIS, MARCH 28, 1893

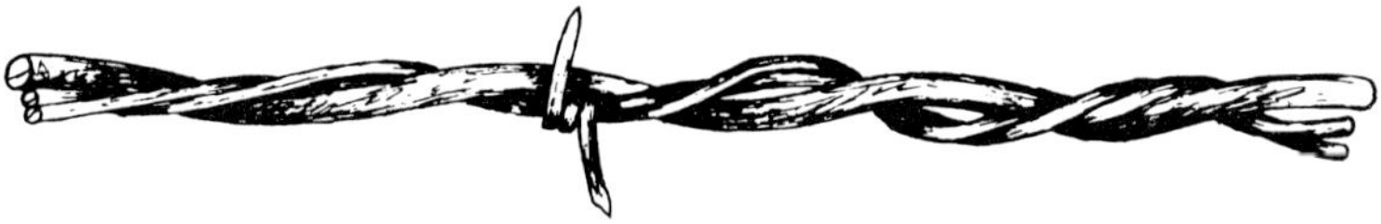

543

GLIDDEN ON 2 SMALL TWISTED WITH HEAVY LINE

PAT. J. F. GLIDDEN, NOV. 24, 1874

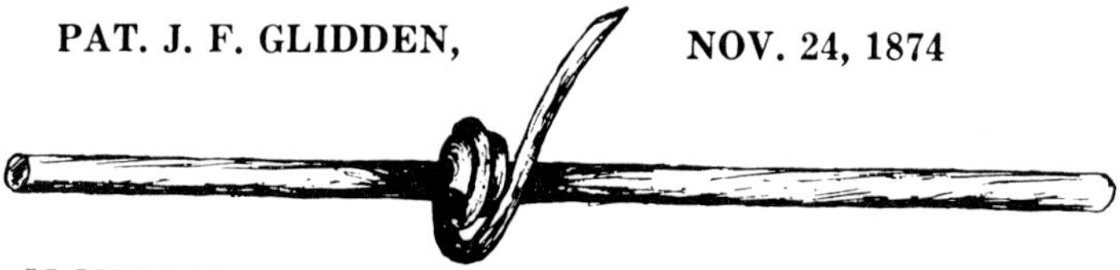

544

McNELLY'S NET FENCE WIRE KNOTS 12 IN. APART.

M. F. McNELLY, MAY 14, 1895

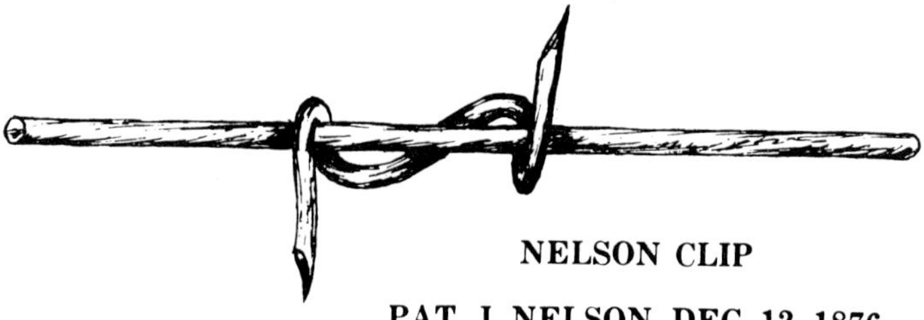

545

NELSON CLIP

PAT. J. NELSON, DEC. 12, 1876

546

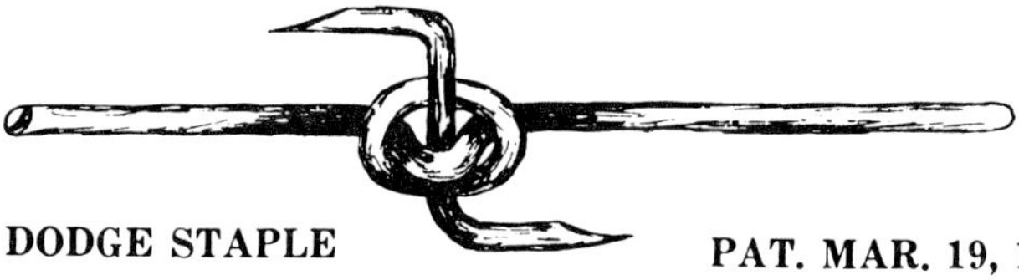

DODGE STAPLE PAT. MAR. 19, 1878

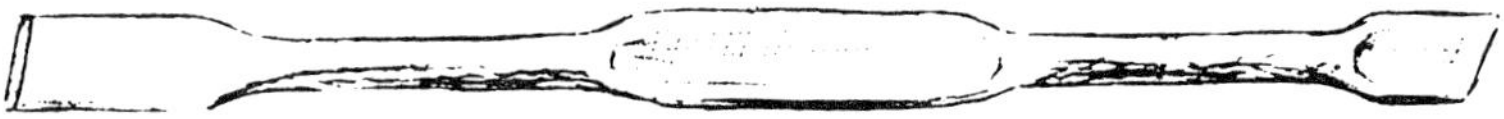

547 L. BAGGER

PAT. OCT. 31, 1876

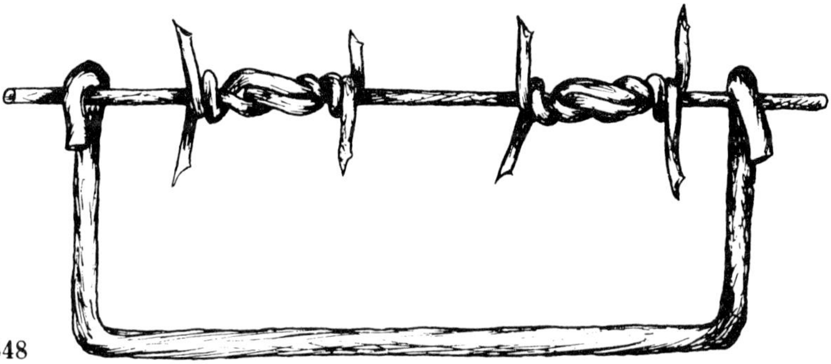

548

ABBOTT'S WIRE CARRYING HANDLE

PAT. H. A. ABBOTT, FEB. 13, 1939

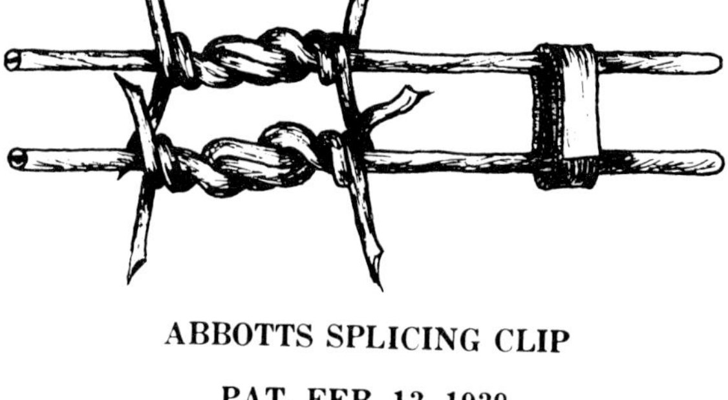

549

ABBOTTS SPLICING CLIP

PAT. FEB. 13, 1939

550 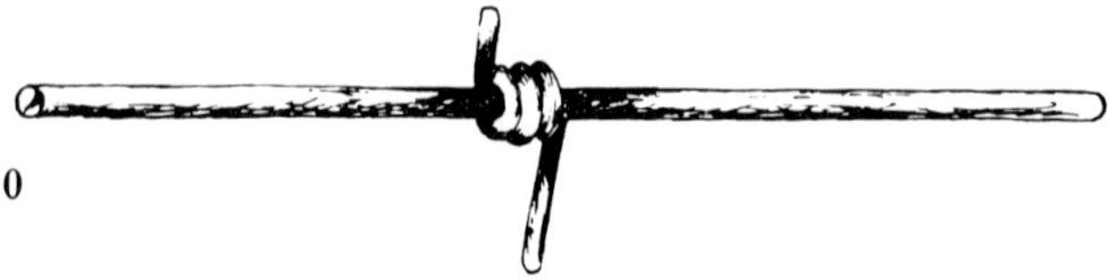

FENCE TIE

551 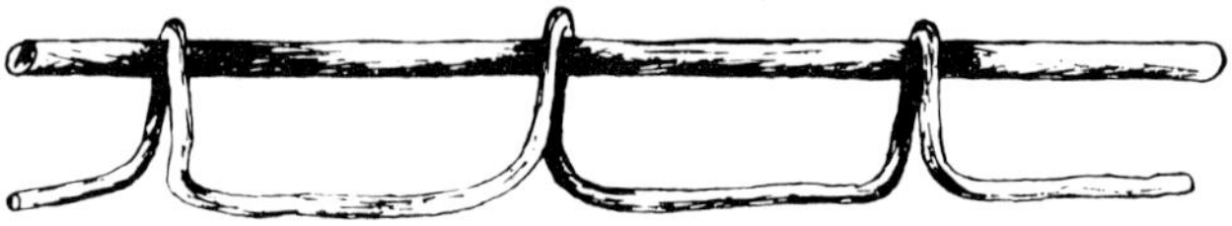

SMITH'S WEAVE PAT. D. C. SMITH, MAR. 2, 1897

552 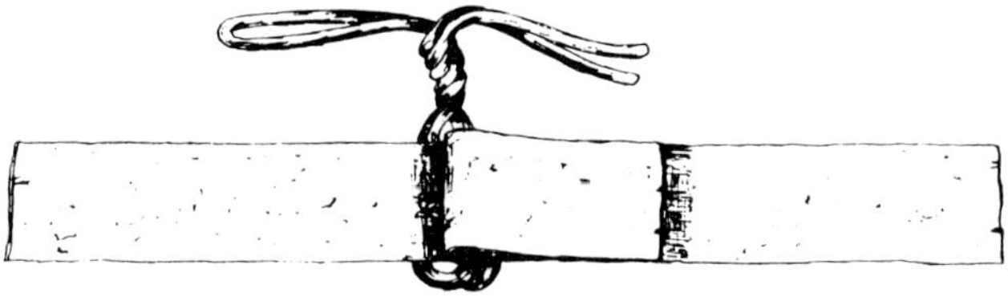

SPANISH LAND GRANT RIBBON
1880's PERIOD

553 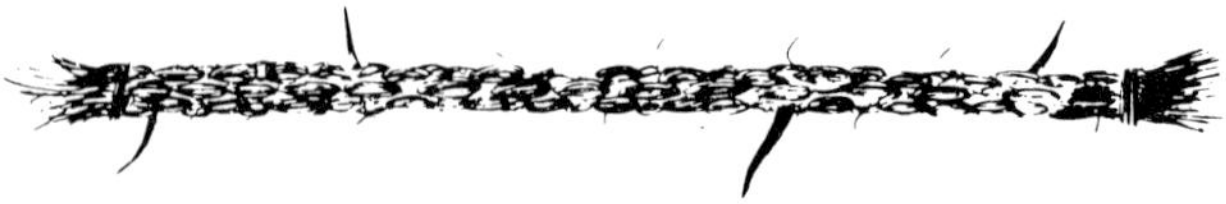

ROPE THORN WIRE
MADE IN PERSIAN DESERT FOR CENTURIES
BY NORMADIC TRIBES. 20 PCS. KNOWN.
ROPE MADE OF PLAITED HAIR WITH DESERT THORN.

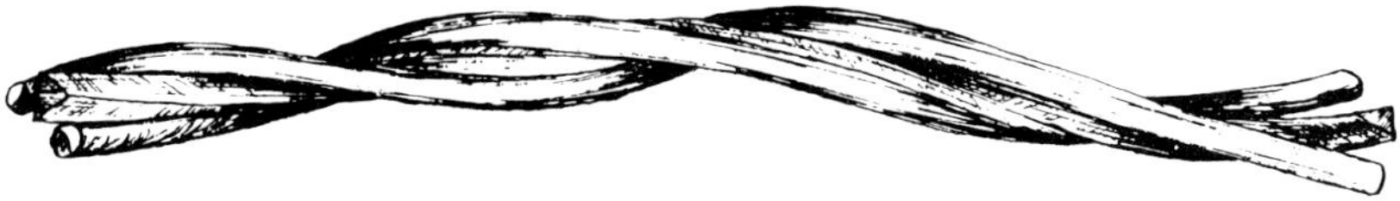

554

GLIDDEN 2 ROUND & 1 TRIANGLE RR WIRE.

PAT. J. F. GLIDDEN, NOV. 24, 1874

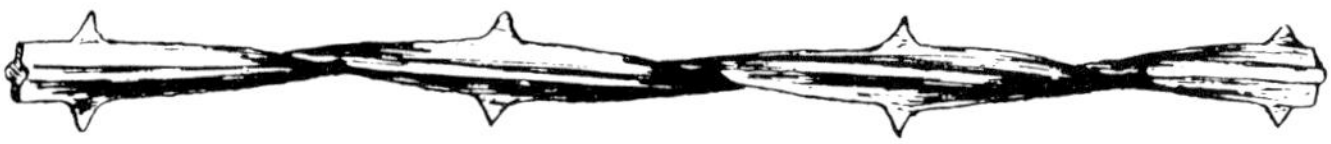

555

F. V. ALLIS PARALLEL SMALL LINE.

PAT. NOV. 30, 1880

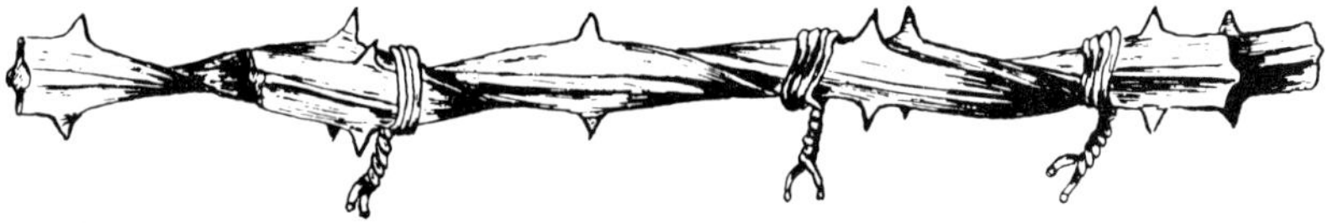

556

T. V. ALLIS FACTORY SPLICE

PAT. NOV. 30, 1880

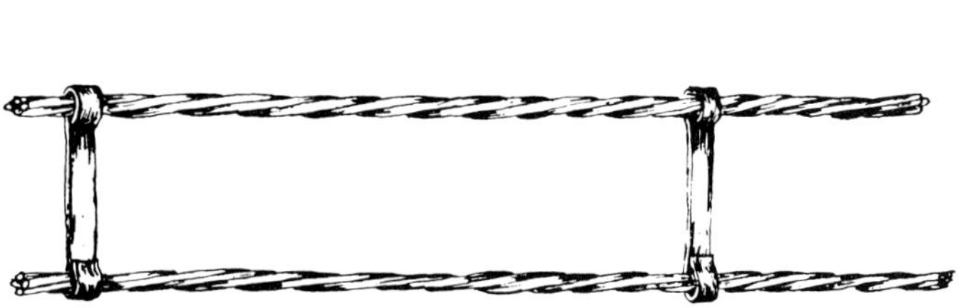

557

USED FOR FENCING

558 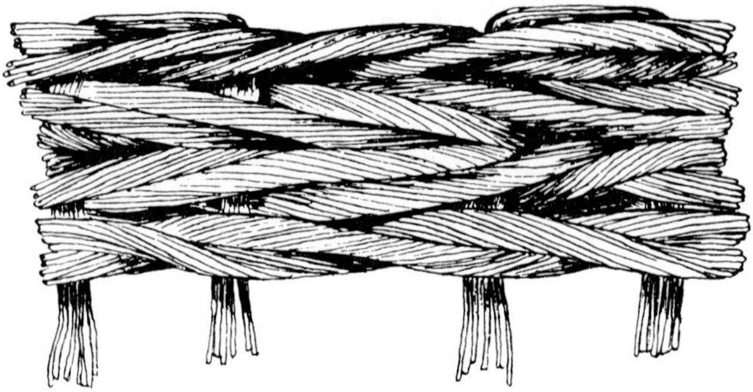

LARGE MINE CABLE WIRE

559 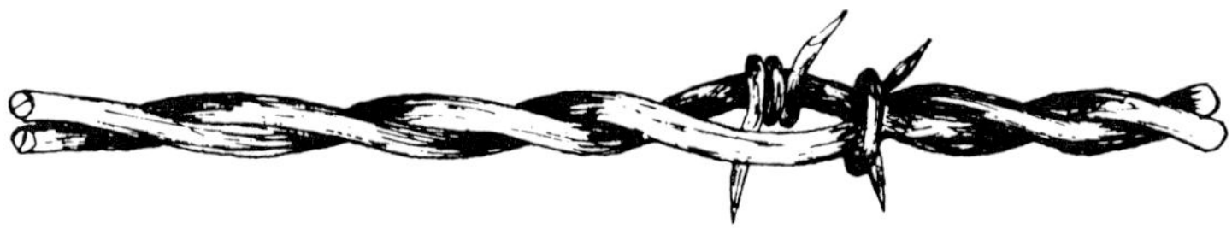

GLIDDEN WITH STAPLE BARB AROUND BOTH LINES.

PAT. J. F. GLIDDEN, NOV. 24, 1874

560 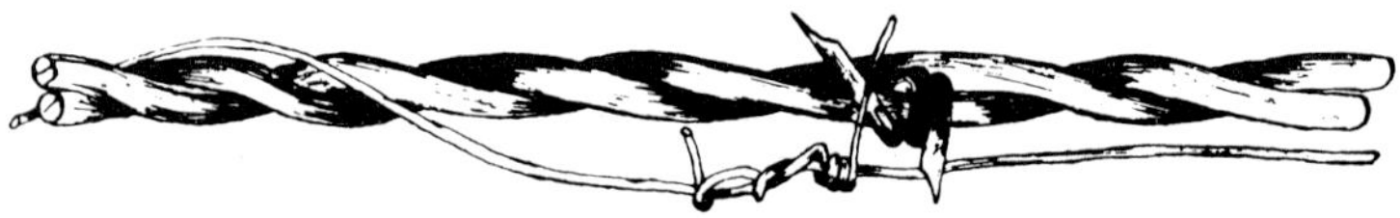

GLIDDEN WITH SMALL LINE WRAPPED AROUND

PAT. J. F. GLIDDEN, NOV. 24, 1874

561 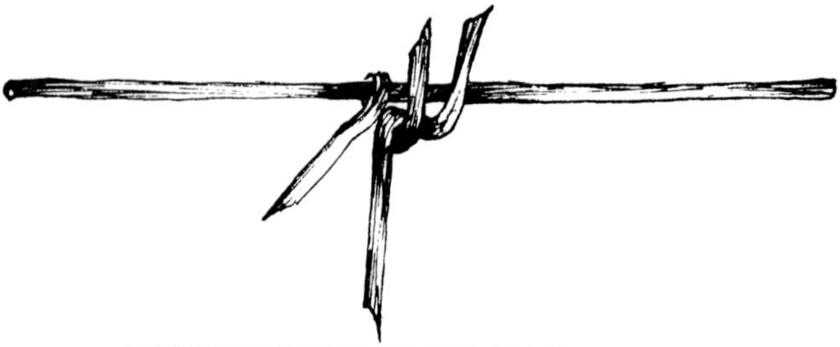

CALGARY SWALLOW TAIL

FROM CALGARY, ALBERTA, CANADA

562

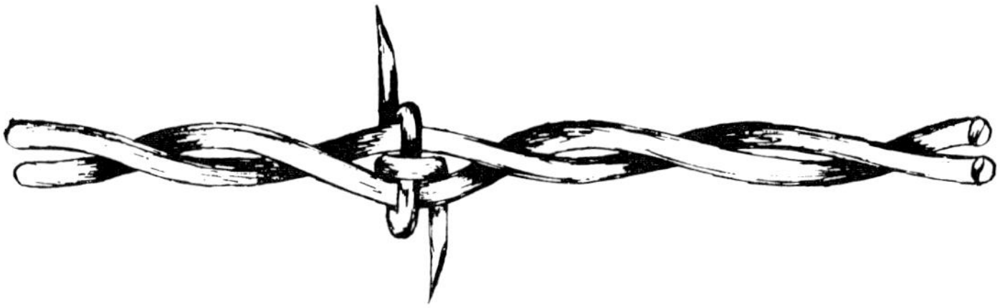

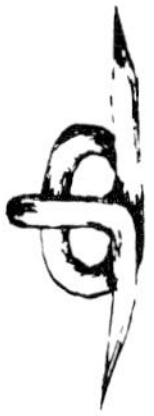

BRAINERDS BARB

PAT. F. W. BRAINERED, SEPT. 19, 1882

563

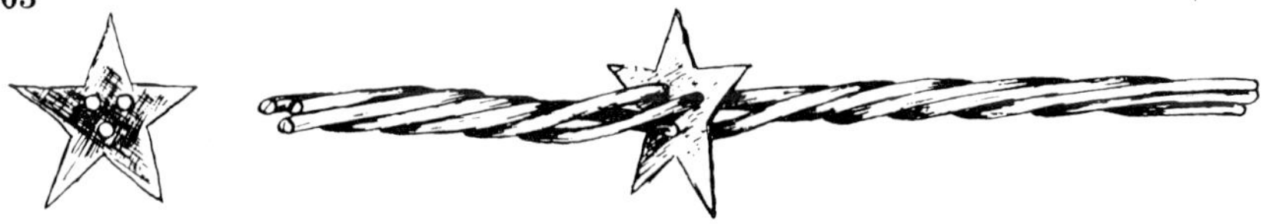

HART-McGLIN STAR

PAT. J. C. McGLIN & F. E. HART, SEPT. 12, 1876

564

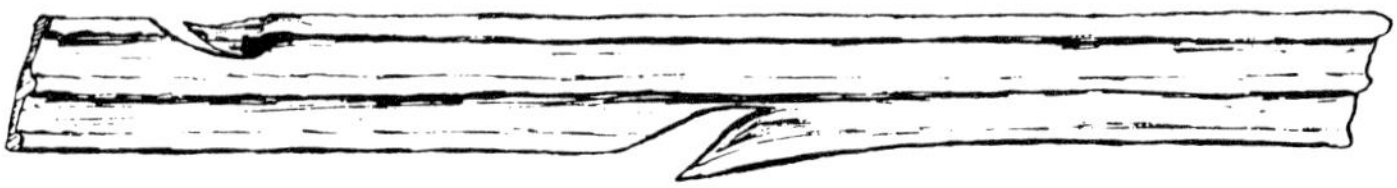

HARBAUGH'S CUT RIBBON

PAT. J. W. HARBAUGHS, JUNE 7, 1881

565 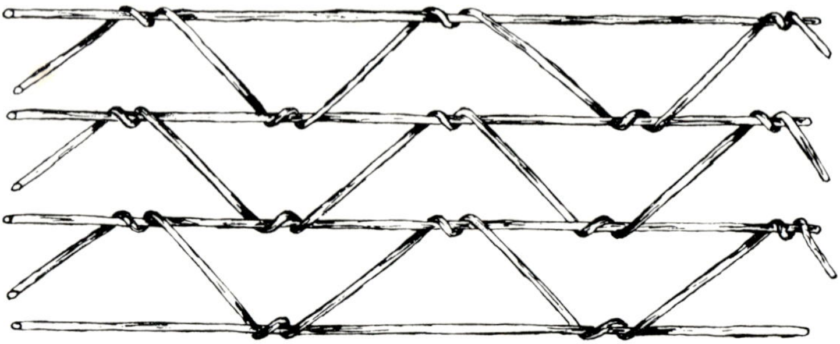

NEW ZEALAND HEDGE MENDING WIRE
ALUMINUM BARB ALSO

566 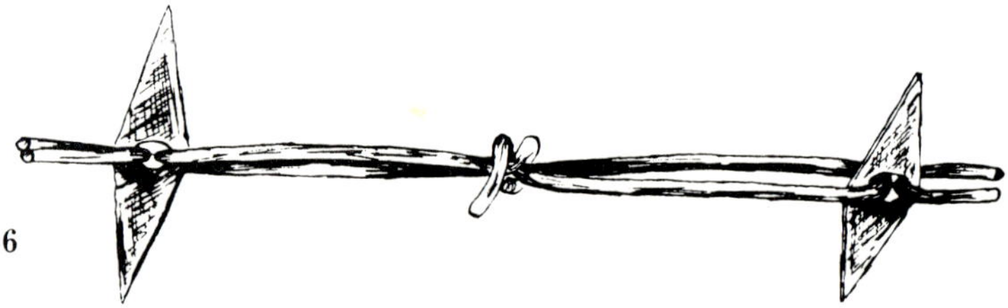

FREESE'S LINK BAR DIAMOND

PAT. P. C. FREESE MAY 29, 1888

567

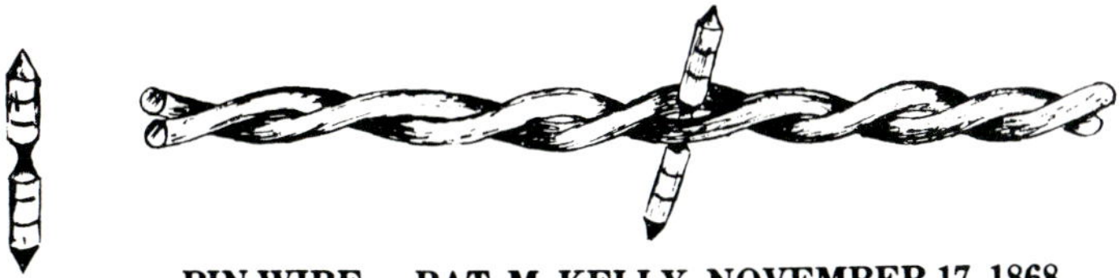

PIN WIRE — PAT. M. KELLY, NOVEMBER 17, 1868

568

HANGING STAR — PAT. G. E. BARKER, DECEMBER 27, 1881

569

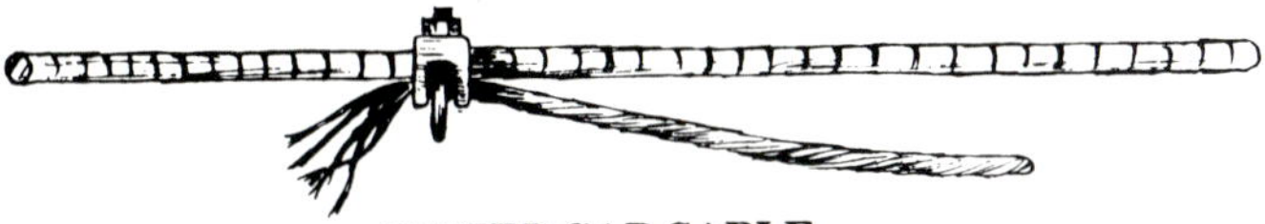

WATER GAP CABLE

570

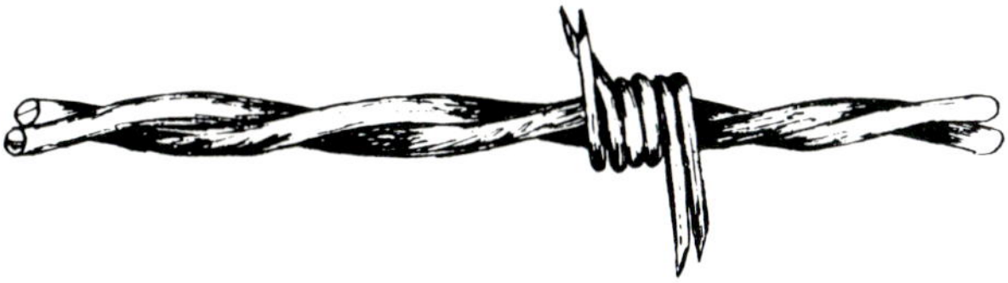

MERRILLS FOUR POINT AROUND BOTH LINES — PAT. L. & J. MERRILL, SEPTEMBER 29, 1871

571

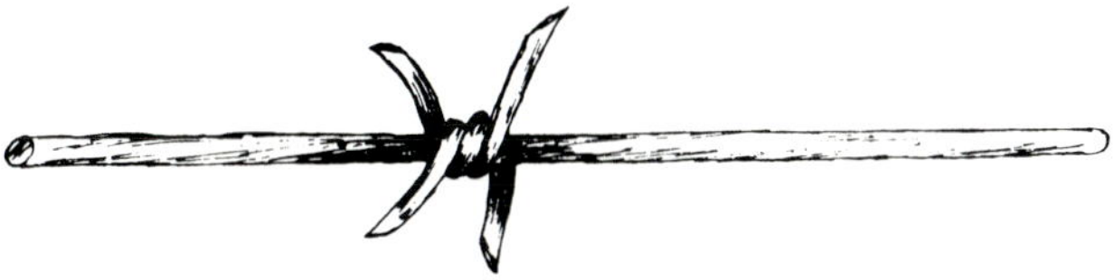

JAYNE & HILL SINGLE LINE — PAT. APRIL 11, 1876

572

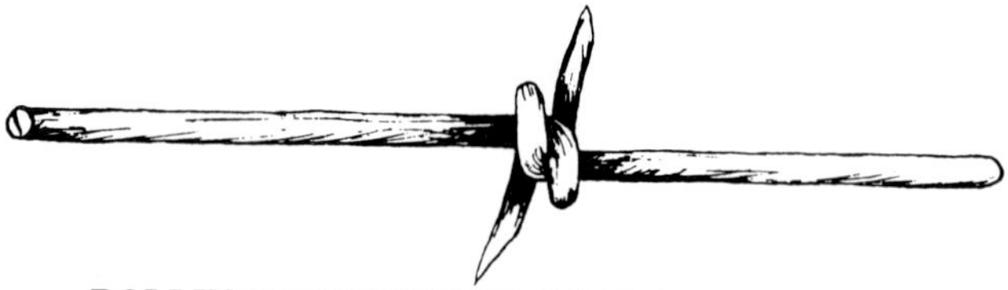

ROLLIN G. BROWN PAT. SEPTEMBER 19, 1876

573

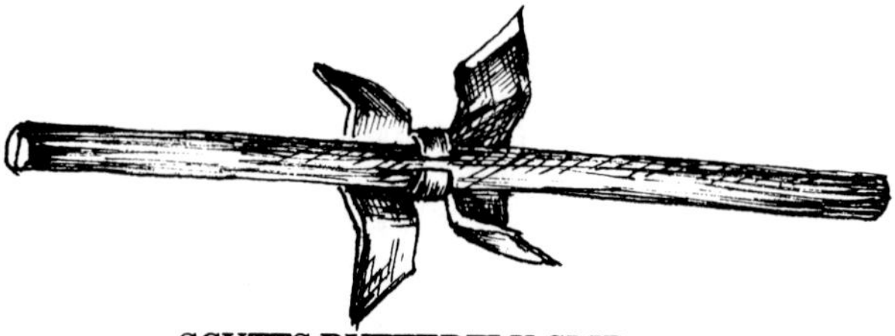

SCUTTS BUTTERFLY CLIP
H. B. SCUTT, FEBRUARY 10, 1878

574

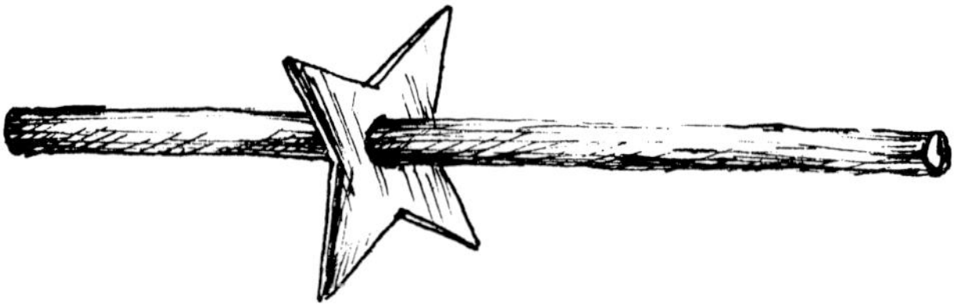

R. FORD, PAT. AUGUST 22, 1876

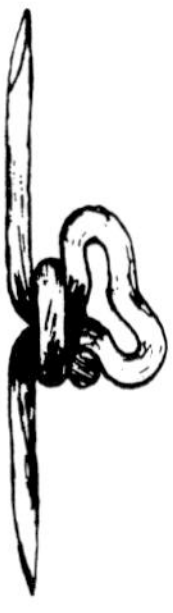

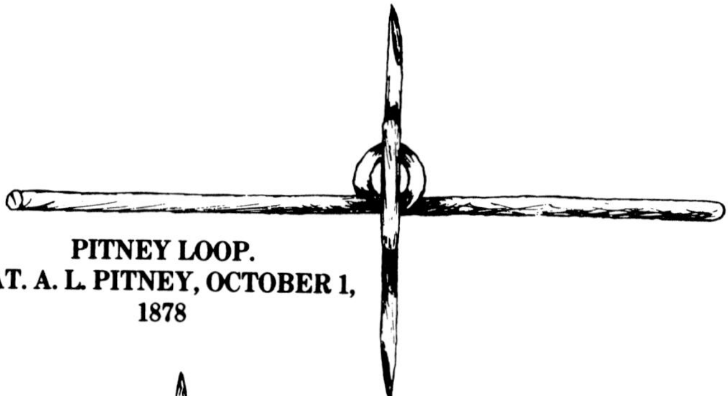

PITNEY LOOP.
PAT. A. L. PITNEY, OCTOBER 1, 1878

575

576

PARALLEL ELWOOD — PAT. A. ELWOOD, JANUARY 31, 1882

577
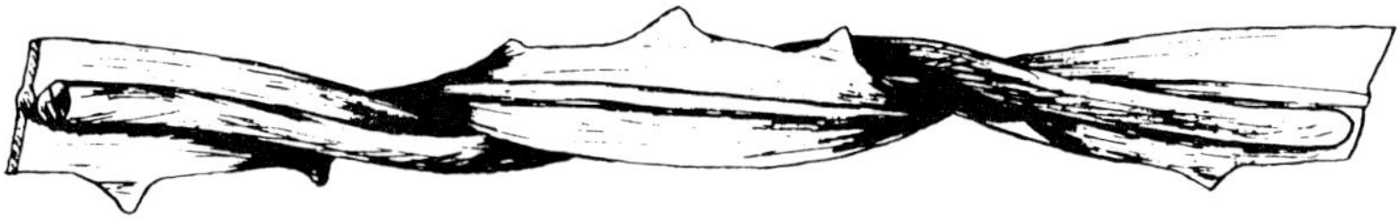

BUCKTHORN WITH SOLID LINE COPPER PAT. T. V. ALLIS, JULY 26, 1881

578
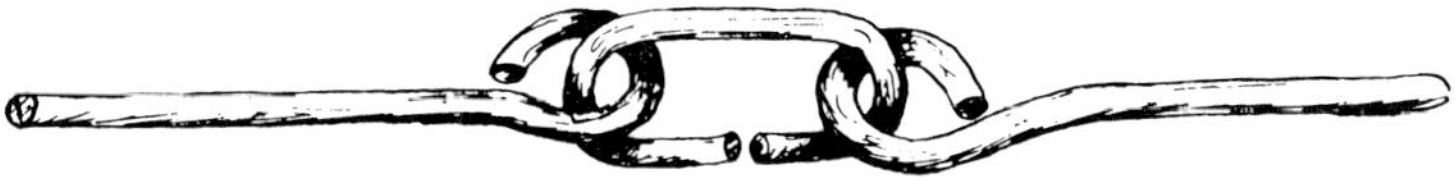

FENCE STAY LINK

579
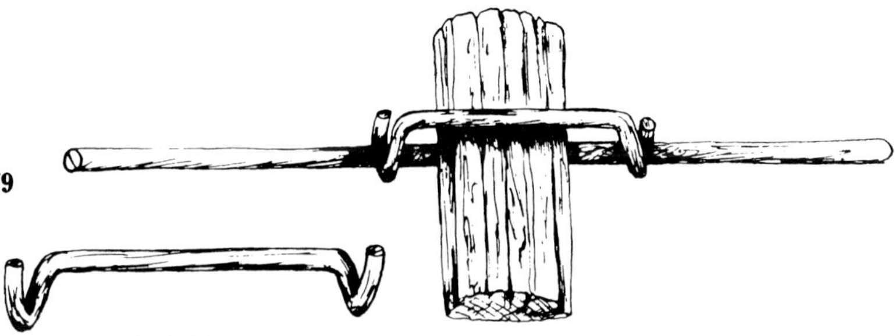

FENCE STAY CLIP FOR SNOW, SHEEP OR SMALL STOCK

580
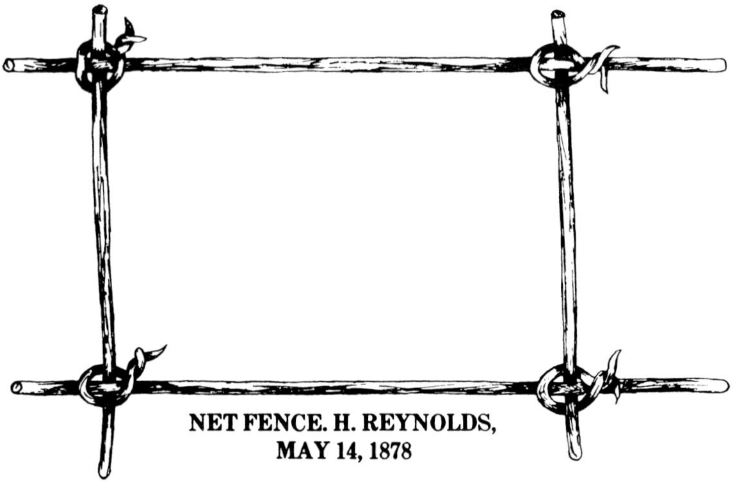

NET FENCE. H. REYNOLDS, MAY 14, 1878

581
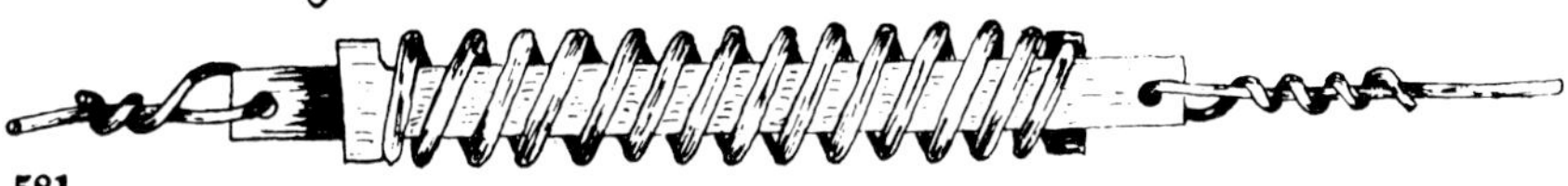

SPRING TYPE EXPANSION

582

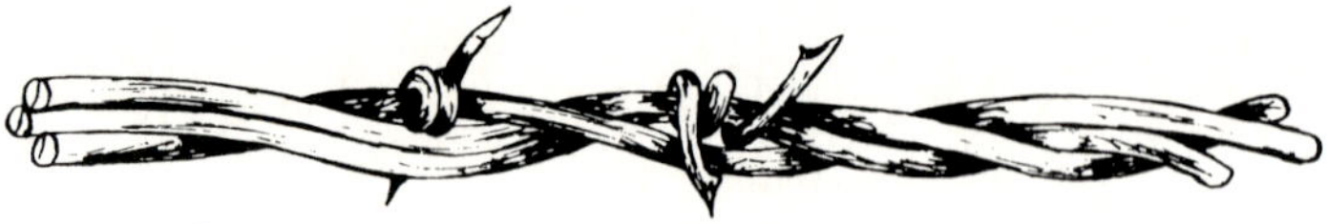

ROSE & GLIDEEN THREE LINE, J. F. GLIDDEN
PAT. NOVEMBER 24, 1874

583

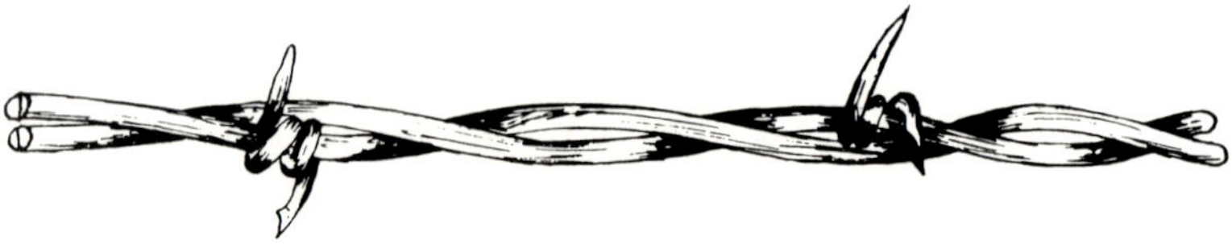

ROSE FLAT POINT ON BOTH LINES
H. M. ROSE, PAT. DECEMBER 25, 1877

584

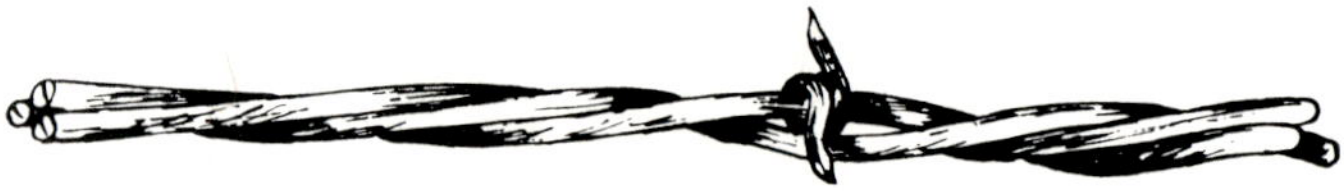

STOVER THREE LINE
D. C. STOVER, PAT. MAY 1, 1877

585

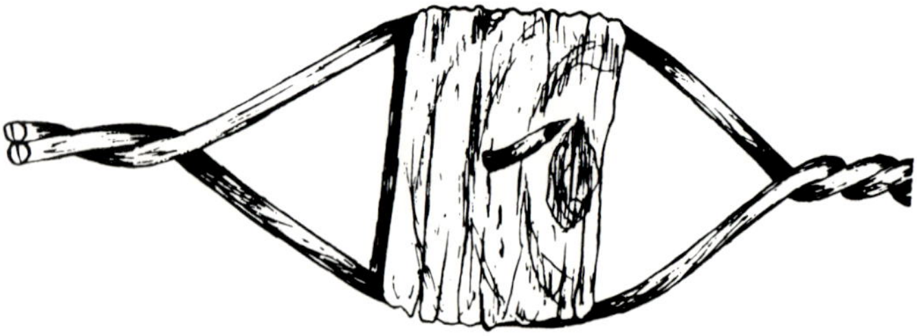

PIN BLOCK SCUTTS — PAT. FEBRUARY 10, 1880, H. B. SCUTT

586

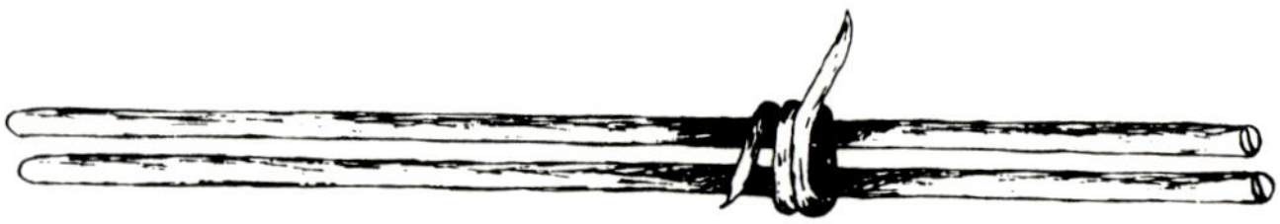

GLIDDEN TWO POINT PARALLEL
PAT. J. F. GLIDDEN, FEBRUARY 8, 1876

587

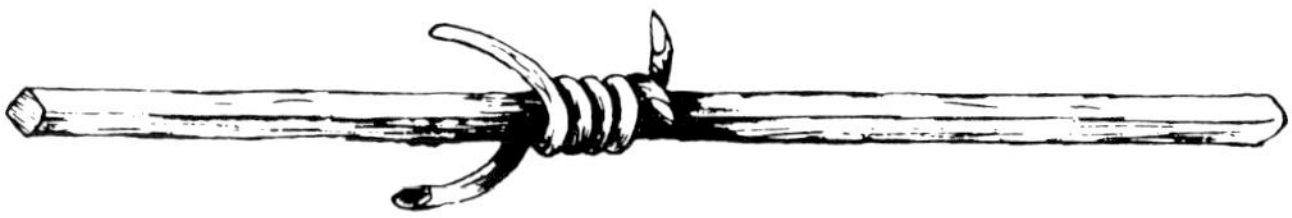

EARLY MERRILL ON SQUARE LINE
PAT. L. & J. MERRILL, SEPTEMBER 29, 1871

588

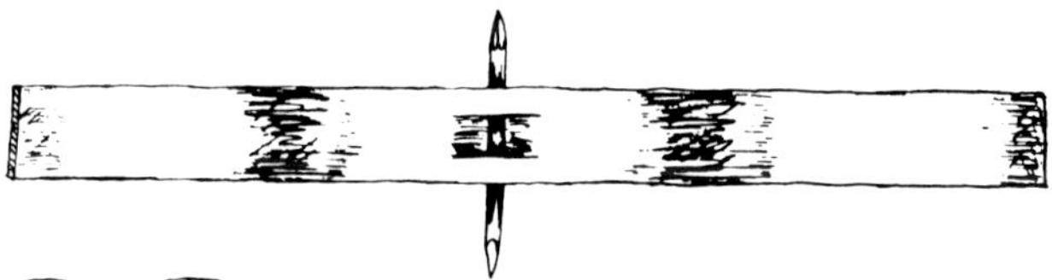

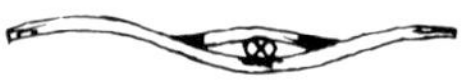

WILLIAM BURTIS CORRUGATED RIBBON
PAT. DECEMBER 30, 1884

589

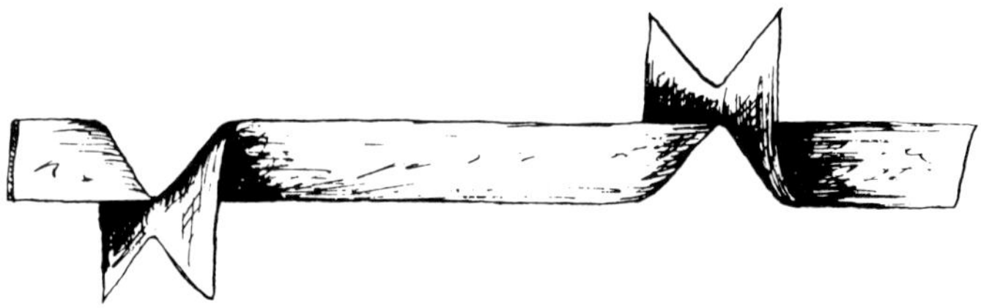

KIRCHHOFFER STRETCHER RIBBON
PAT. GEORGE KIRCHHOFFER, MARCH 1, 1881

590

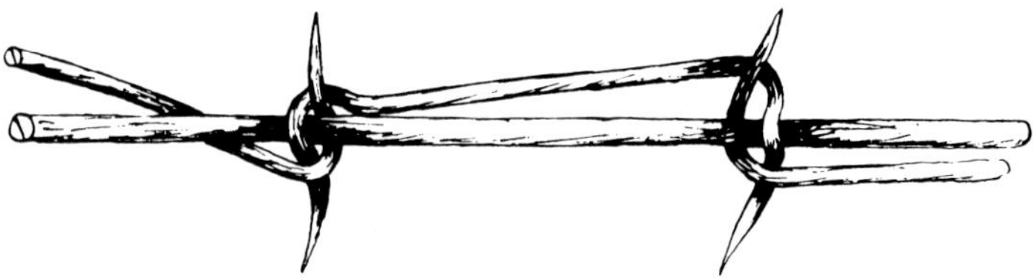

PETER P. HILL LINK
PAT. OCTOBER 3, 1876

591 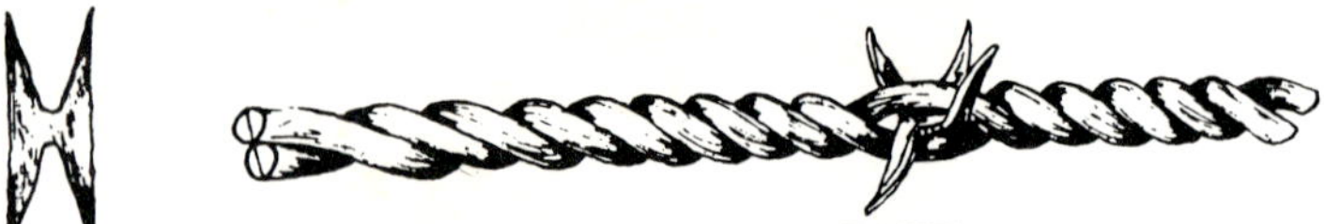

SWAN'S H. PLATE
PAT. FRANK SWAN, JUNE 10, 1879

592 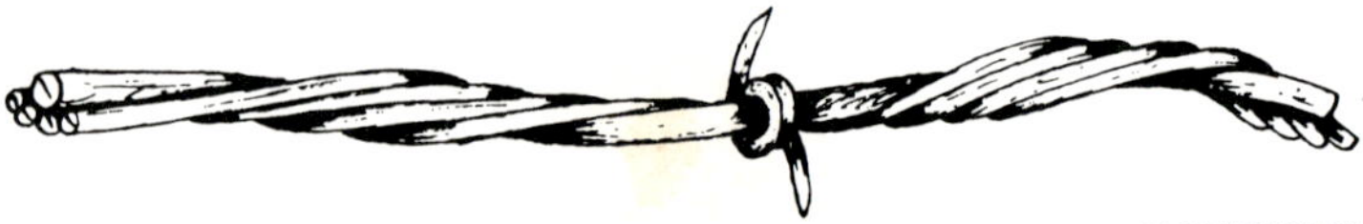

GLIDDEN FOUR SMALL, ONE LARGE LINE, BARB ON LARGE LINE
PAT. J. F. GLIDDEN, NOVEMBER 24, 1874

593 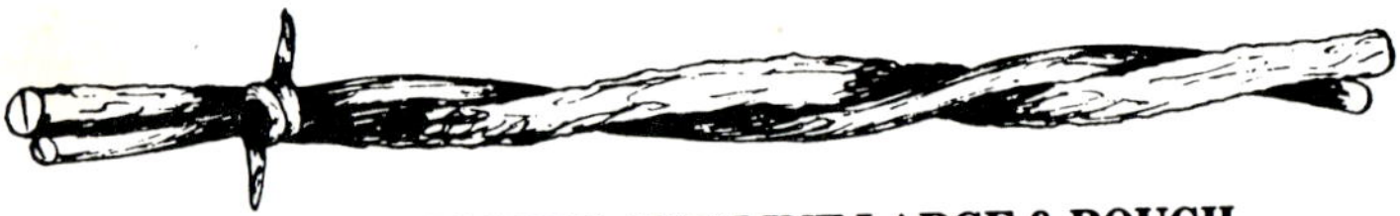

GLIDDEN TWO LINE, ONE LINE LARGE & ROUGH
PAT. J. F. GLIDDEN, NOVEMBER 24, 1874

594 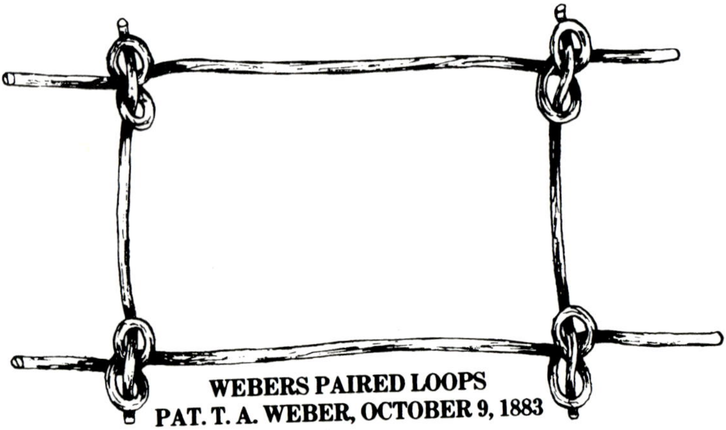

WEBERS PAIRED LOOPS
PAT. T. A. WEBER, OCTOBER 9, 1883

595 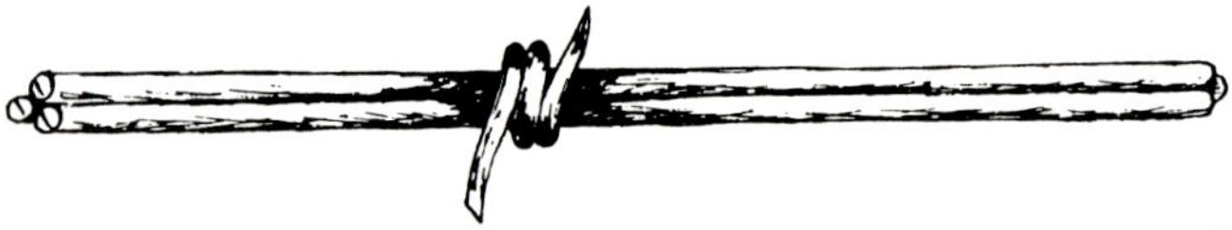

GLIDDEN THREE LINE PARALLEL, BARB AROUND THREE LINES,
FEB. 8, 1876

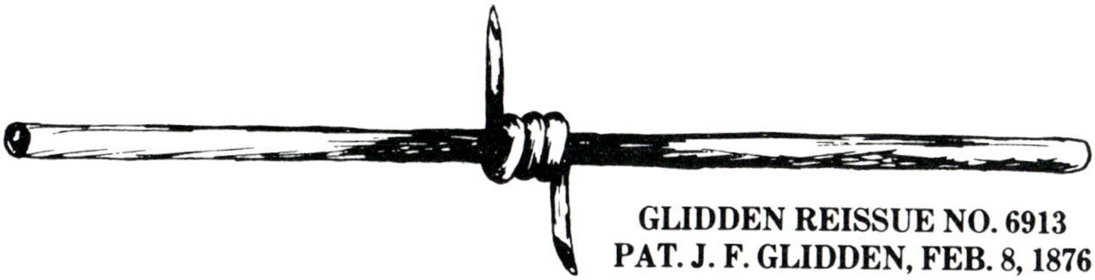

596

GLIDDEN REISSUE NO. 6913
PAT. J. F. GLIDDEN, FEB. 8, 1876

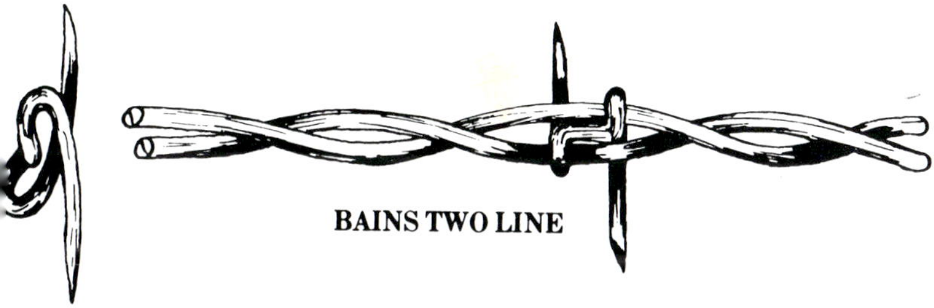

BAINS TWO LINE

597

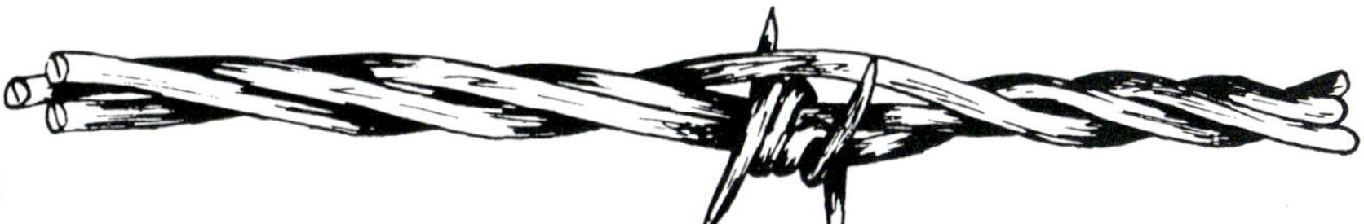

598

THREE LINE SHINN
PAT. M. C. SHINN, MARCH 1, 1881

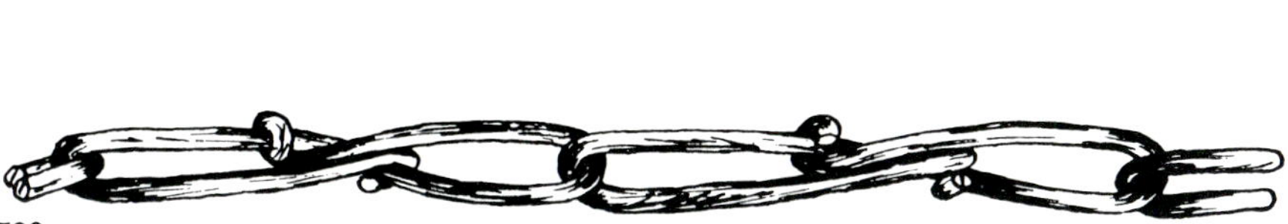

599

VERY SMALL CRANDALS LINK
PAT. NOV. 28, 1876

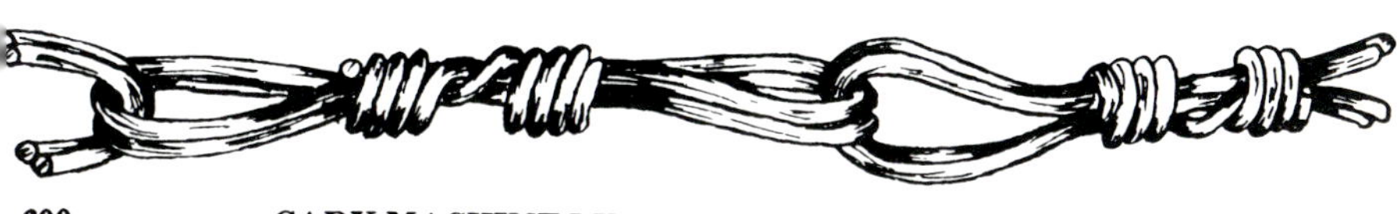

600

CADY MACHINE LINK DOUBLE SHORT LINK

601 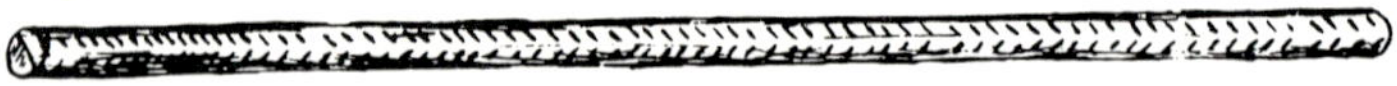

THREE CRIMPED ROWS ON LARGE SINGLE LINE. BARBLESS

602 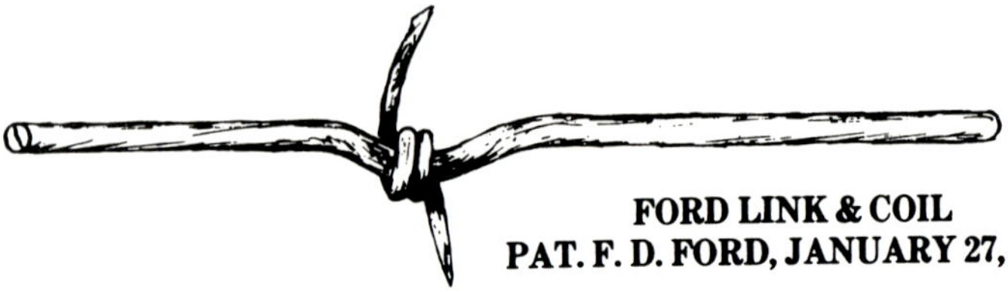

FORD LINK & COIL
PAT. F. D. FORD, JANUARY 27, 1885

603 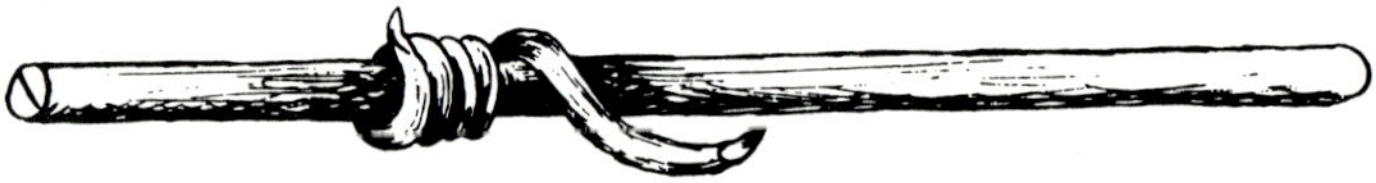

KING RANCH WIRE. MANU ON KING RANCH
1870 PERIOD FLATTENED LARGE ROUND LINE

604 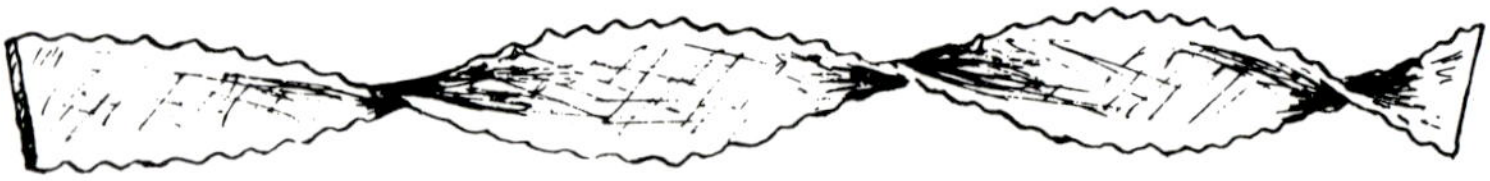

DOUBLE SAW TOOTH
PAT. T. V. ALLIS, JULY 26, 1881

605 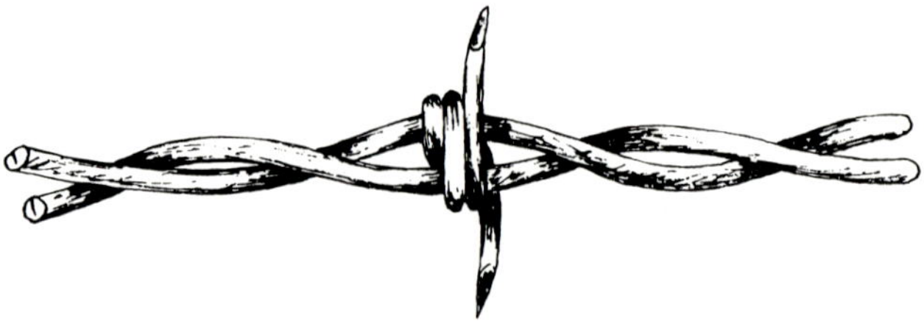

HAISHS TWO LINE
PAT. J. HAISH, AUGUST 31, 1885

606

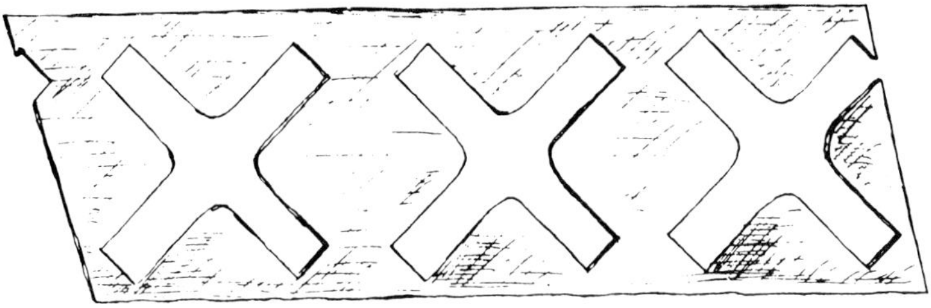

ORNAMENTAL PUNCH PRESS TEXAS RANCH

607

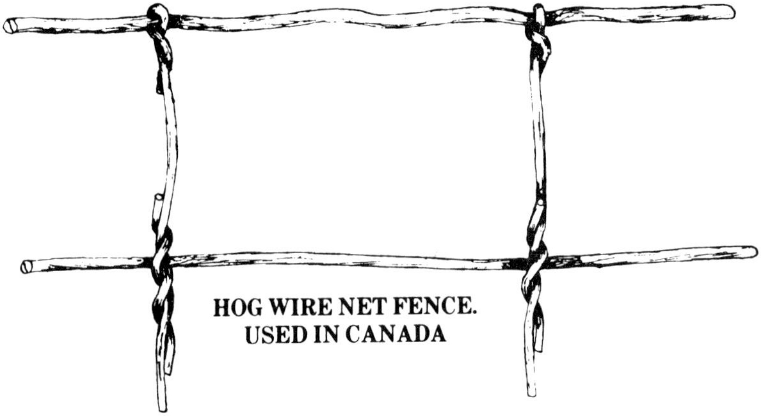

HOG WIRE NET FENCE.
USED IN CANADA

608

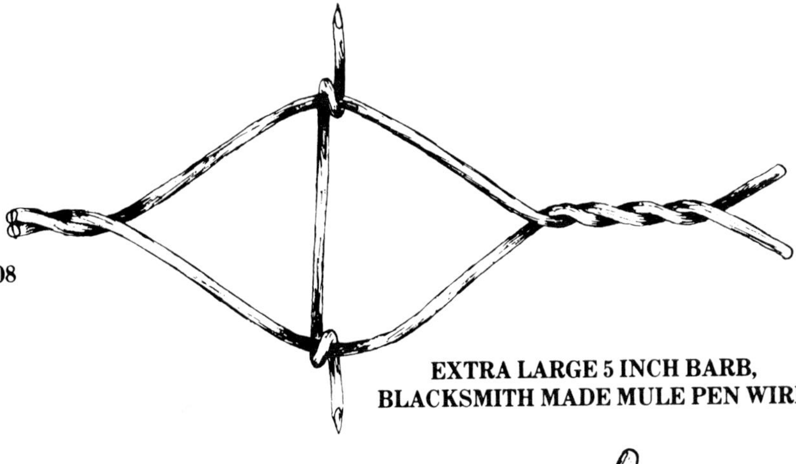

EXTRA LARGE 5 INCH BARB,
BLACKSMITH MADE MULE PEN WIRE

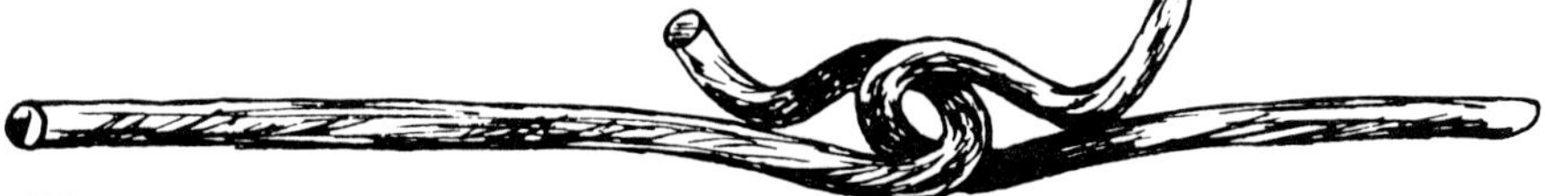

609

E. M. CRANDAL PORTABLE LARGE FOLDING LINK WIRE
PAT. MAR. 14, 1876

610 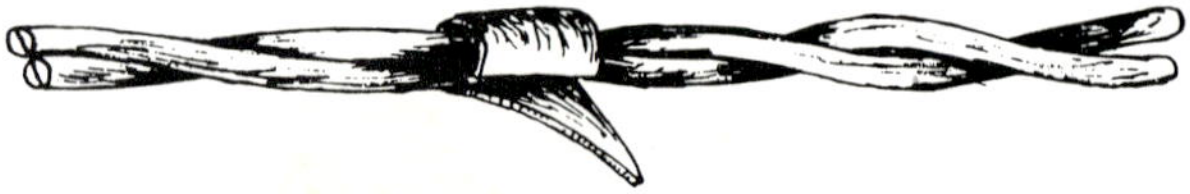

PUNCH PRESS BARB
PAT. JOHN C. MILLIGAN, NOVEMBER 28, 1882

611 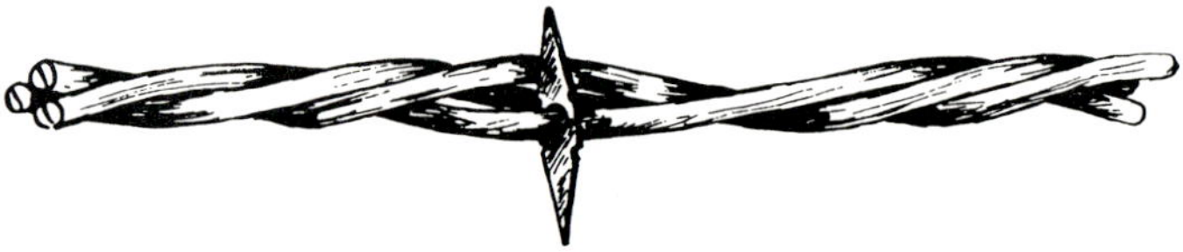

THREE LINE KELLY
PAT. M. KELLY, NOVEMBER 17, 1868

612 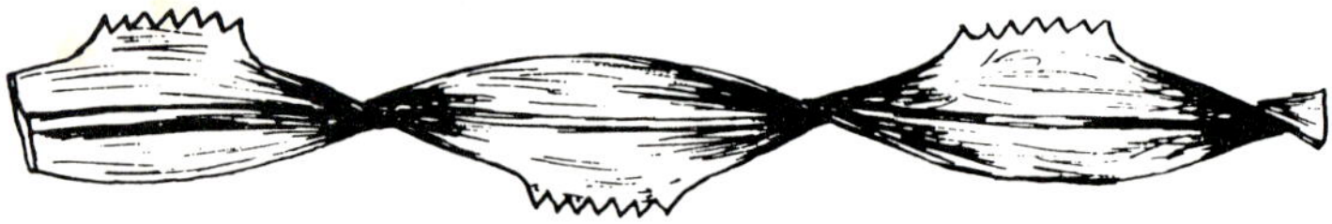

SPACED SAW TOOTH
PAT. T. V. ALLIS, NOV. 30, 1880

613 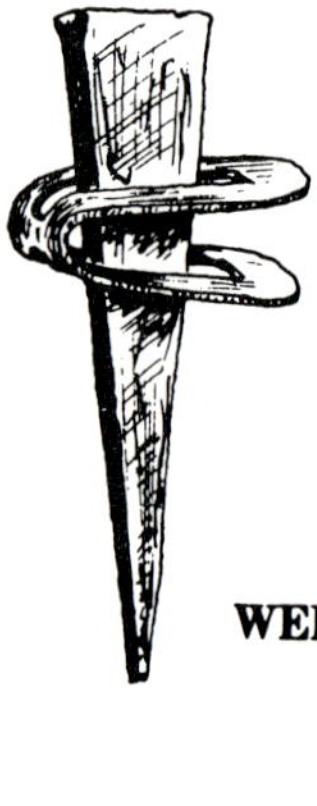

WEDGE & CLEVIS STAY & STAPLE

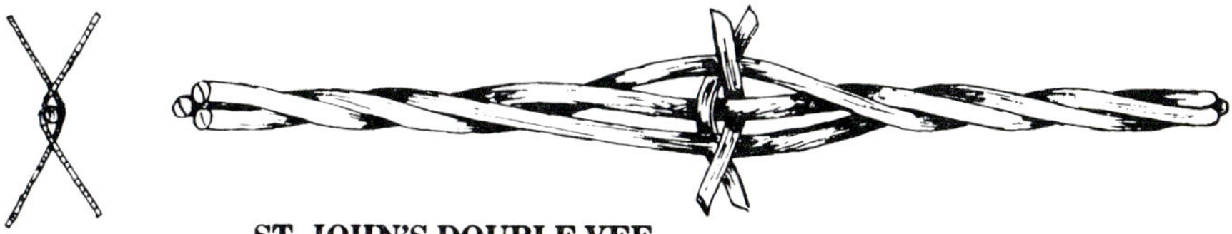

614

ST. JOHN'S DOUBLE VEE
PAT. S. H. & J. M. ST. JOHN, JULY 2, 1878

615

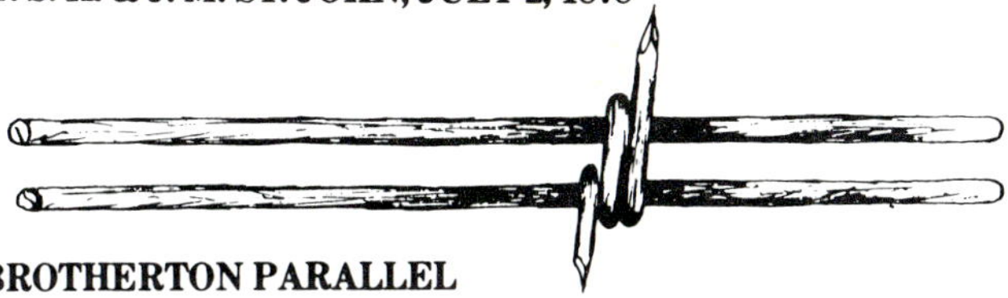

BROTHERTON PARALLEL
PAT. J. BROTHERTON, SEPT. 3, 1887

616

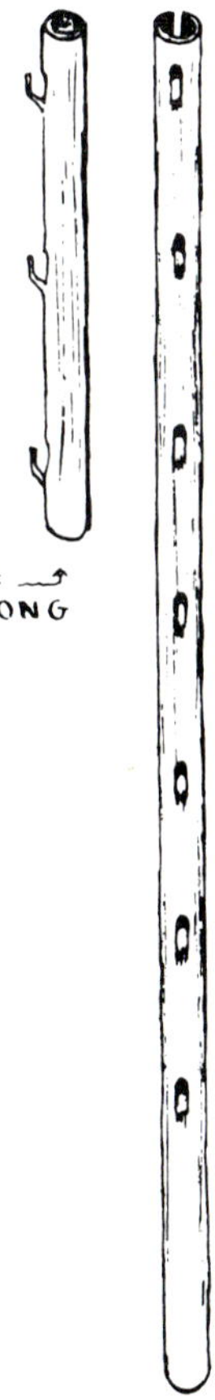

TUBE PERFORATED POST

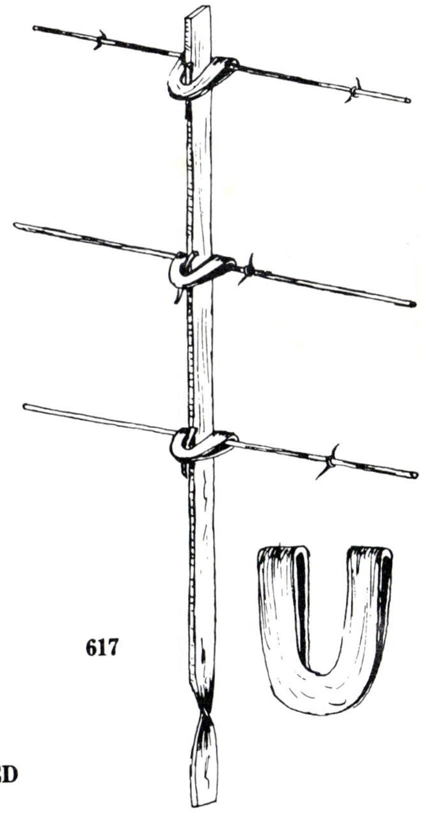

617

CLEVIS TYPE STEEL POST

618

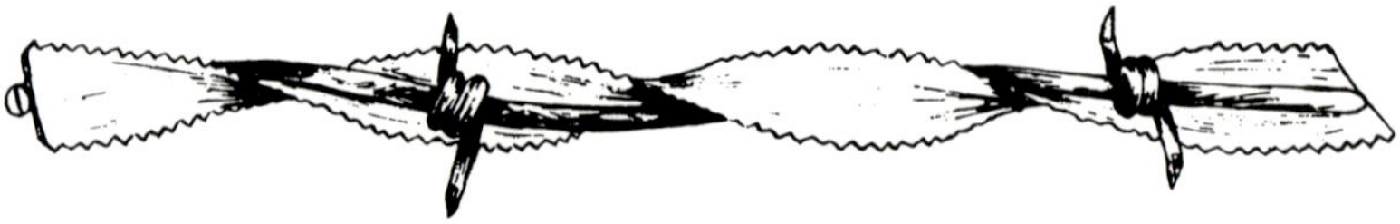

BRINK-MARTELLE DOUBLE SAW TOOTH
AUG. 11, 1885

619

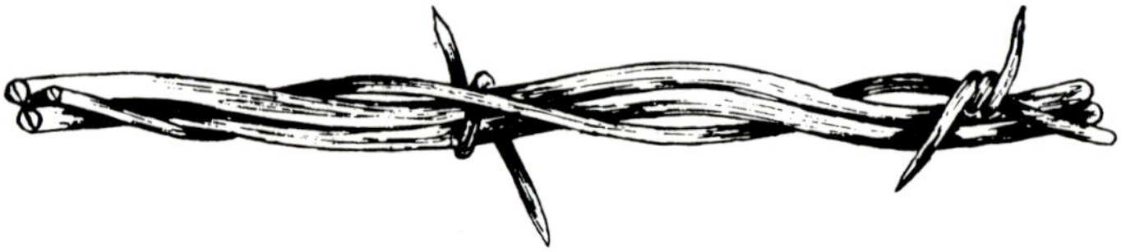

UPHAM & GLIDDEN
PAT. J. F. GLIDDEN, NOV. 24, 1874

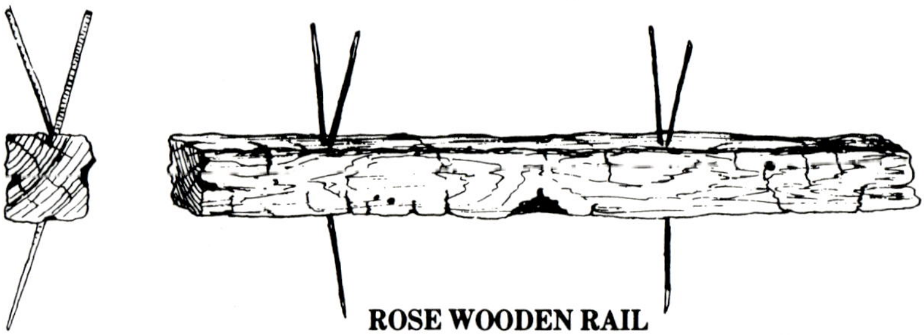

620

ROSE WOODEN RAIL
PAT. HENRY ROSE, 1873

621

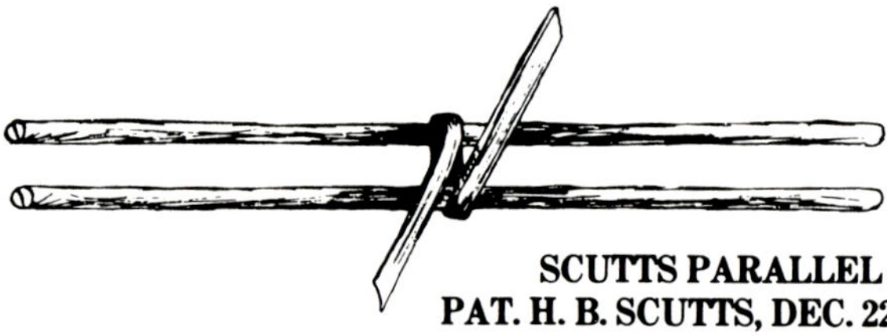

SCUTTS PARALLEL
PAT. H. B. SCUTTS, DEC. 22, 1885

622

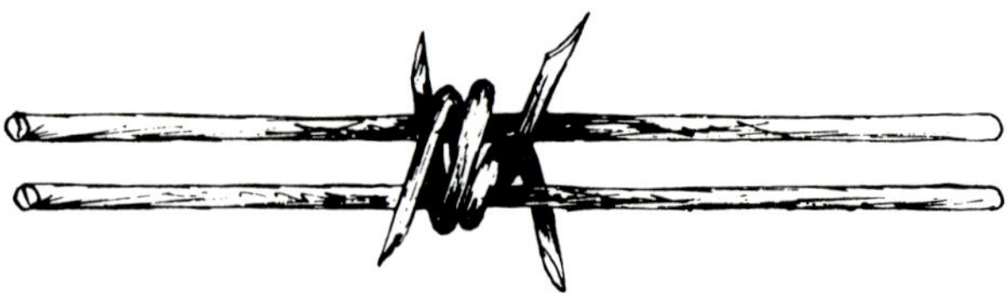

PARALLEL BURNELL
PAT. A. S. BURNETT, JUNE 19, 1877

623
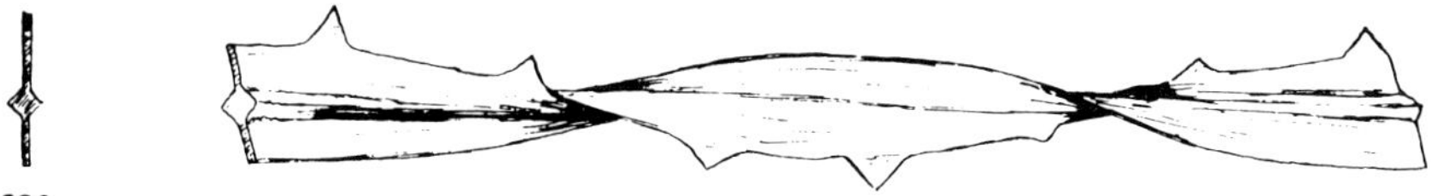

BUCKTHORN WITH DIAMOND CENTER WIRE
T. V. ALLIS, JULY 26, 1881

624
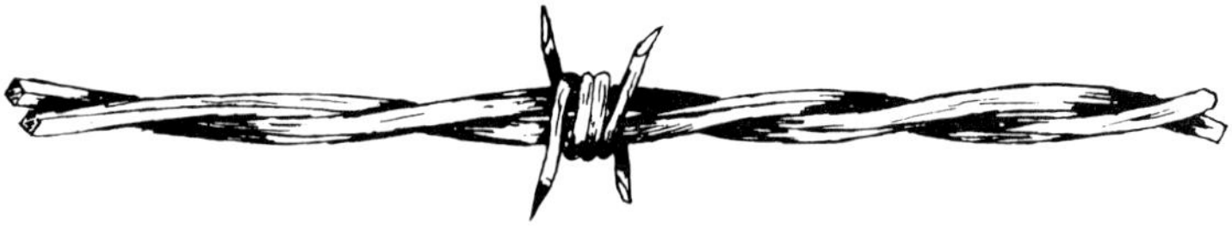

GUILLEAUME TWO LINE
GUILLEAUME PAT. MAY 9, 1893

625
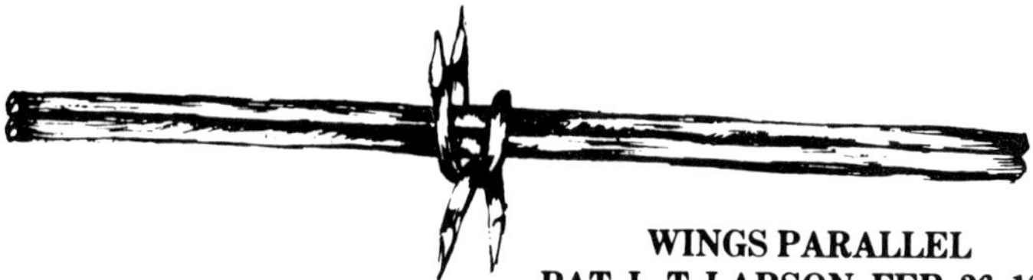

WINGS PARALLEL
PAT. L. T. LARSON, FEB. 26, 1878

626
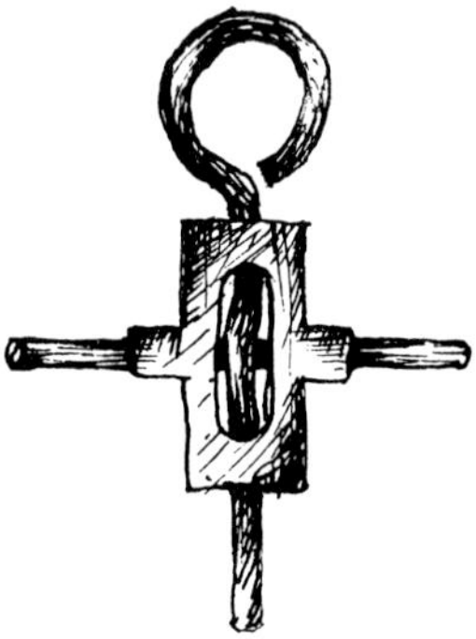

FENCE STAY WITH
4 CLIP FASTNER'S

627

THREE LINE SOLID COPPER YARD FENCE LINE.

628

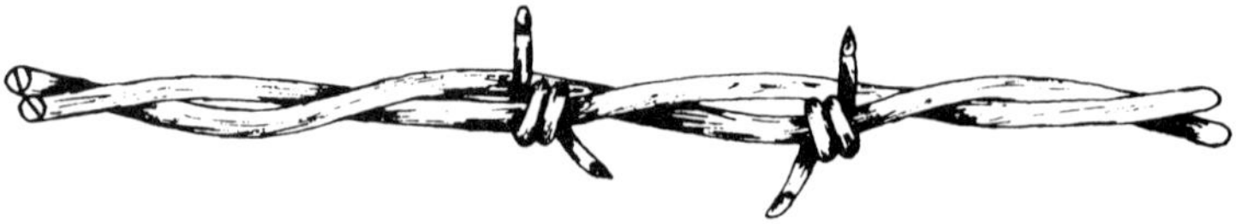

GLIDDEN BARB ON BOTH LINES.
PAT. J. F. GLIDDEN, NOV. 24, 1874

629

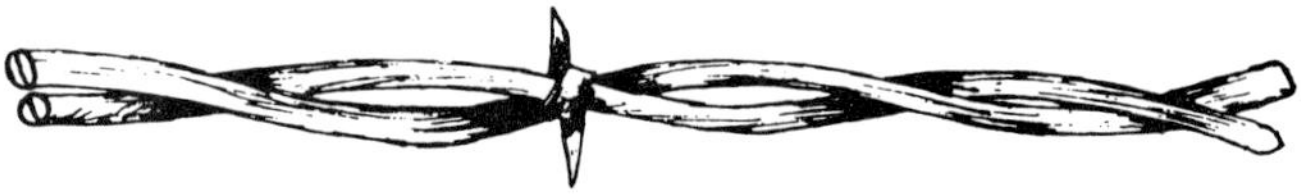

BAKER, ONE LINE OVAL, ONE LINE ROUND
PAT. G. C. BAKER, FEB. 27, 1883

630

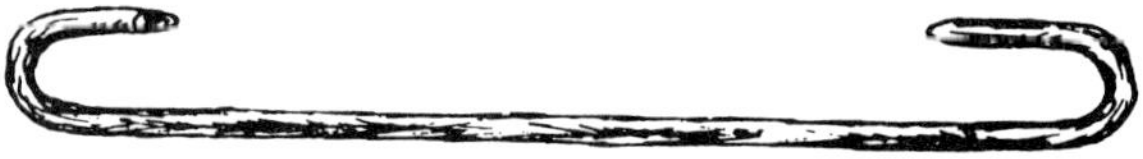

SECURITY WIRE FENCE STAY DIFF. LENGTHS
PAT. CHICAGO WIRE, FENCE & TOOL CO., 1898

631

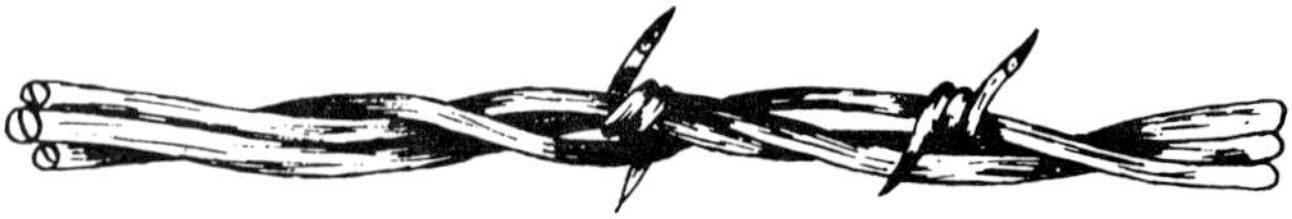

GLIDDEN & ROSE THREE LINE, TWO SMALL, ONE LARGE
PAT. J. F. GLIDDEN, NOV. 24, 1874

632

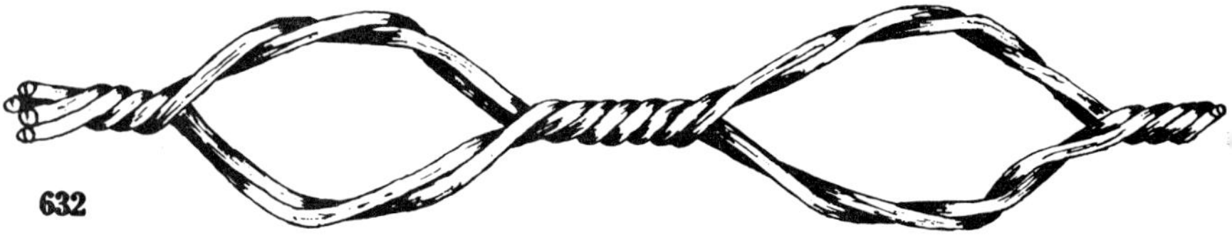

TWO LINE PICKET FENCE.

633 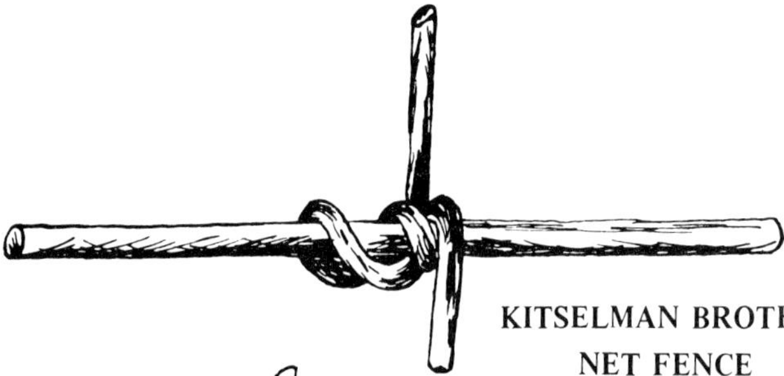

KITSELMAN BROTHER'S
NET FENCE

634 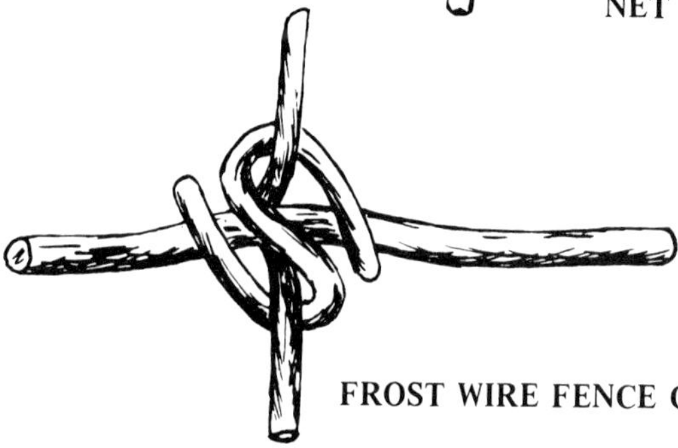

FROST WIRE FENCE CO. LARGE "S"

635 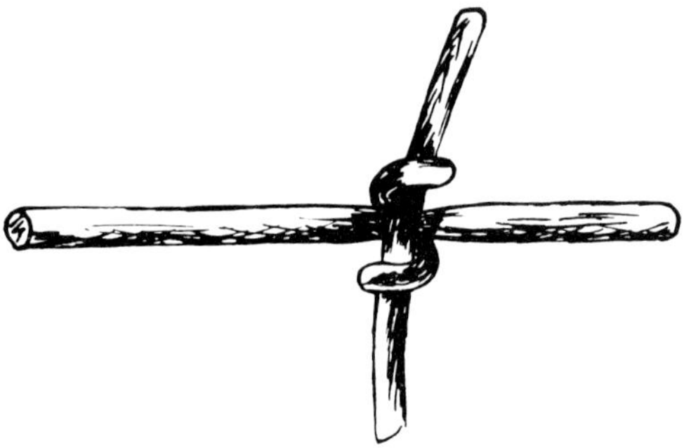

"S" NET WIRE SOLD BY SEARS, ROEBUCK & CO.

636

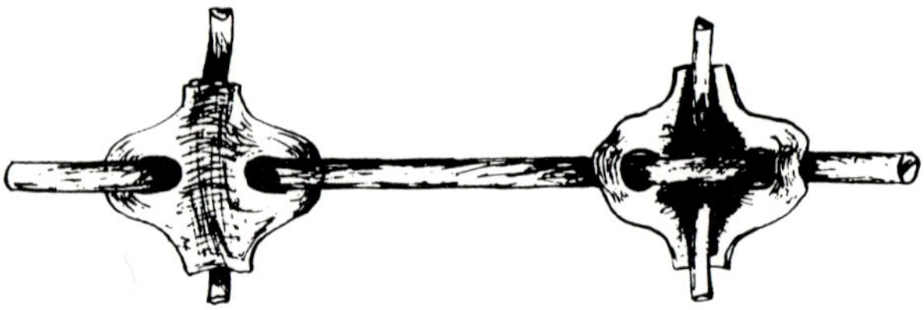

SHIELD PLATE FENCE TIE

637

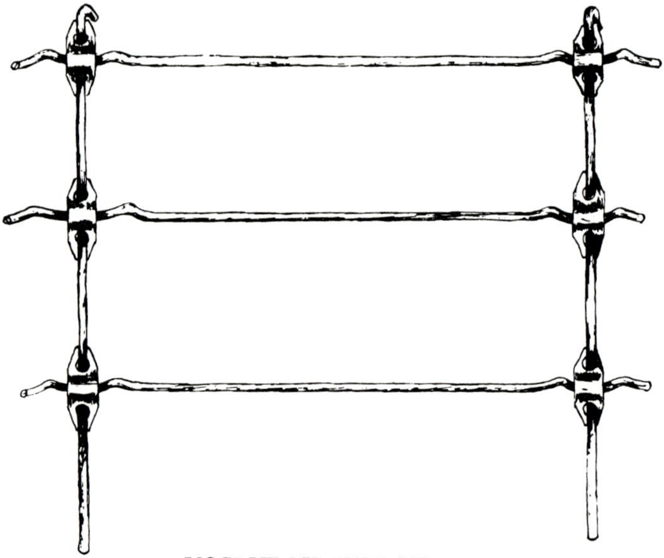

VOSBURGH TYPE NET TIE

638

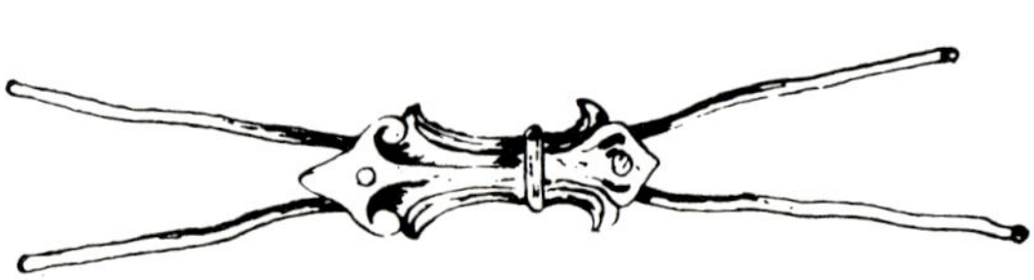

ORNAMENTAL FENCE TIE

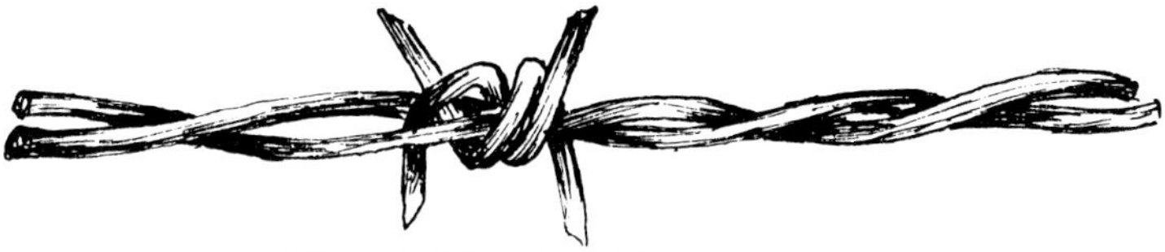

639 A. S. BURNELL'S HEAVY COPPER COATED 4 POINT
PAT. JUNE 19, 1877

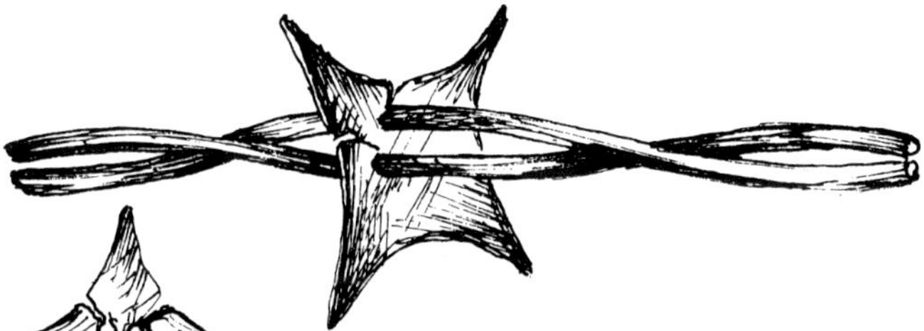

640

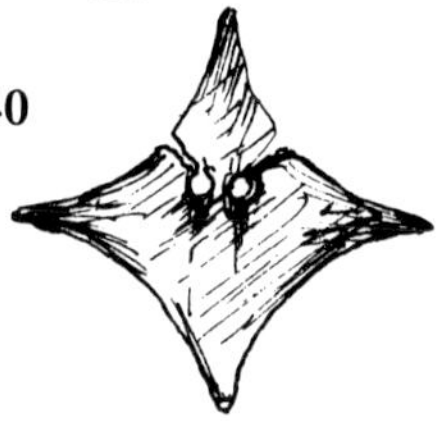

ORLANDO M. POND'S FOUR POINT
PAT. JAN. 2, 1883

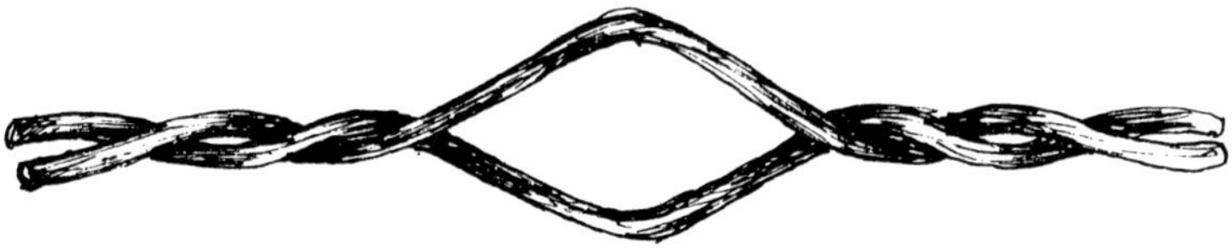

641 SOLID COPPER WIRES USED ON RANDOLF HEARST'S RANCH

642

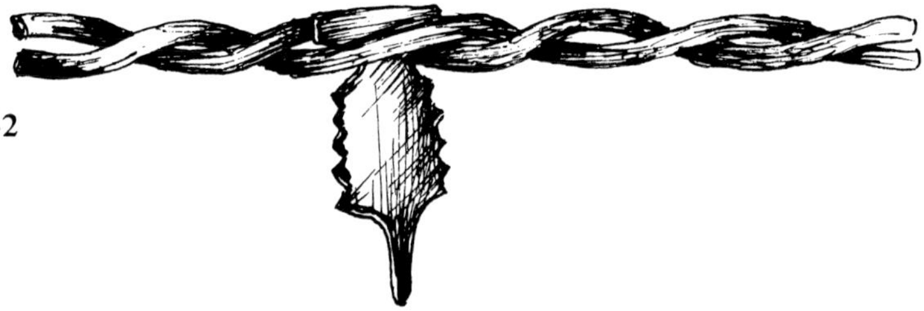

ARNOLD DEIFF'S LEAF PAT. JAN. 17, 1893

643 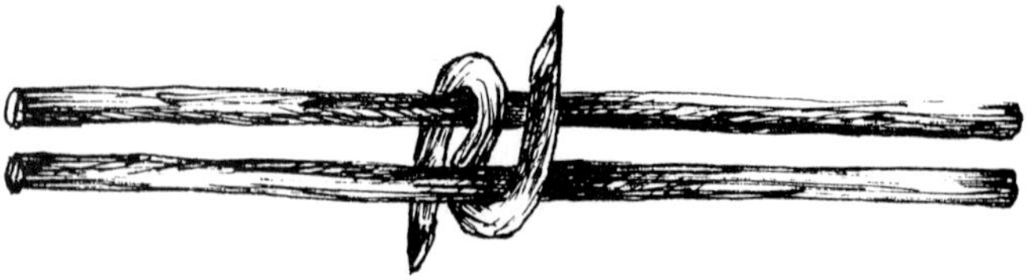

JACOB HAISH PARALLEL. LARGE LINES
PAT. AUG. 31, 1875

644 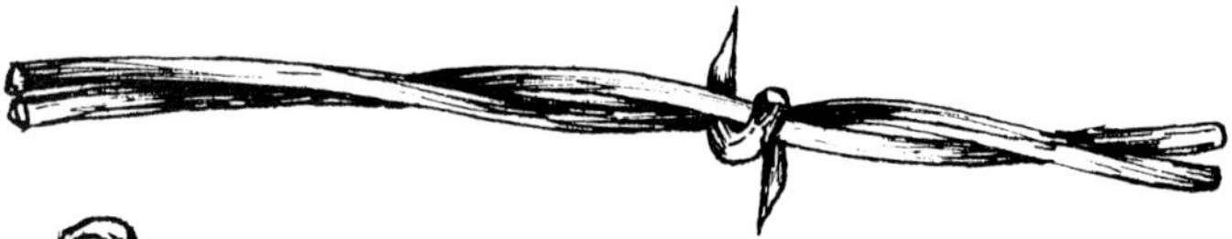

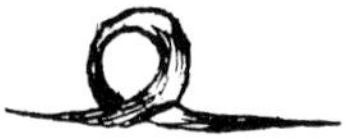

JACOB HAISH'S TRIPLE CRIMP BARB
PAT. JULY 25, 1882

645 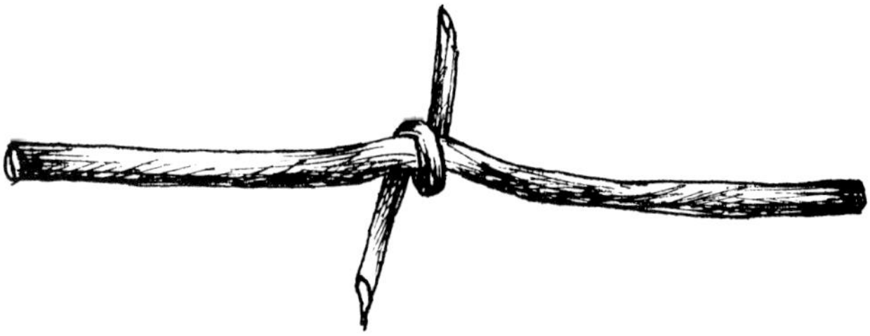

DOBB'S AND BOOTH KINK LINE
PAT. DEC. 14, 1875

646 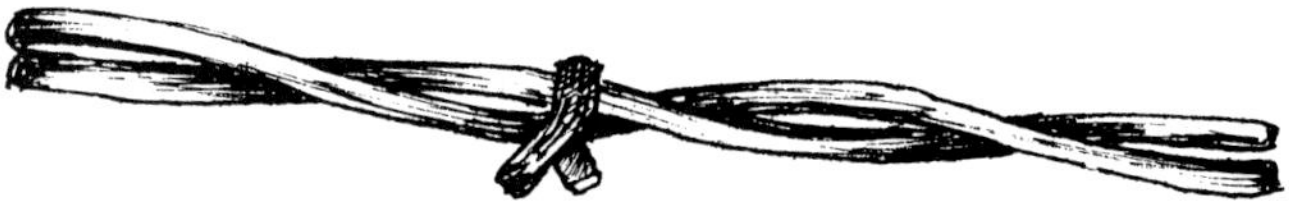

TWO LINE "HUMANE" BARB. BLUNT CUT

647 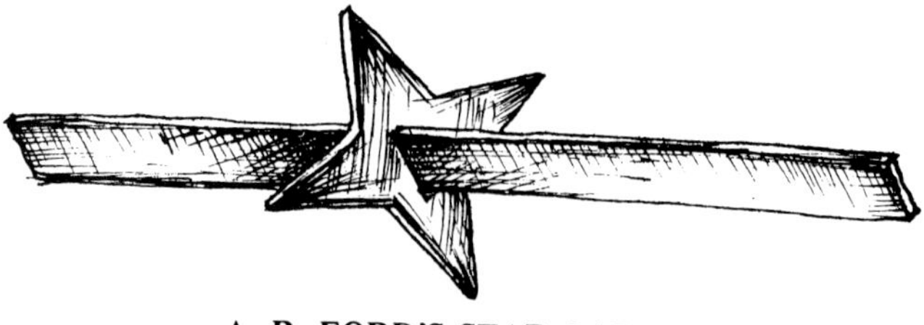

A. R. FORD'S STAR BARB RIBBON
PAT. AUG. 22, 1876

648 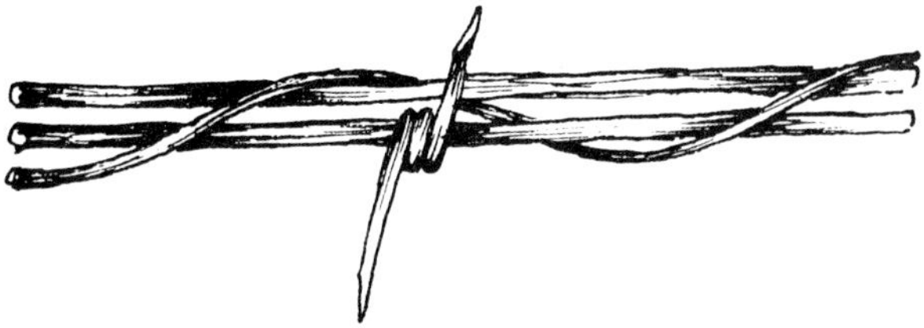

OX PEN WIRE PAT. BY JOSEPH GLIDDEN FEB. 8, 1876

649 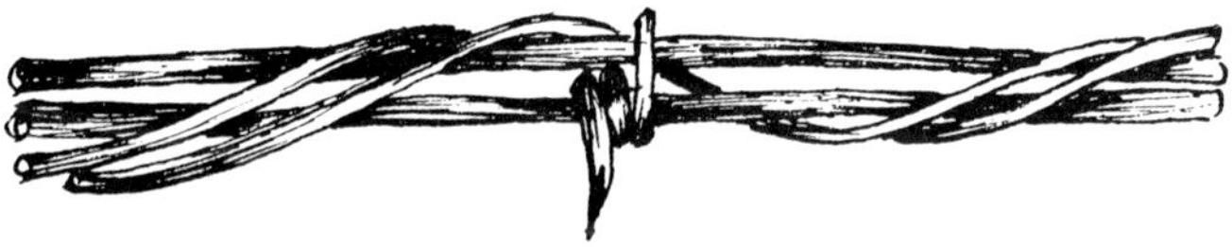

OX PEN WIRE PAT. BY JOSEPH GLIDDEN FEB. 8, 1876

650 

OX PEN WIRE PAT. BY JOSEPH GLIDDEN FEB. 8, 1876

651 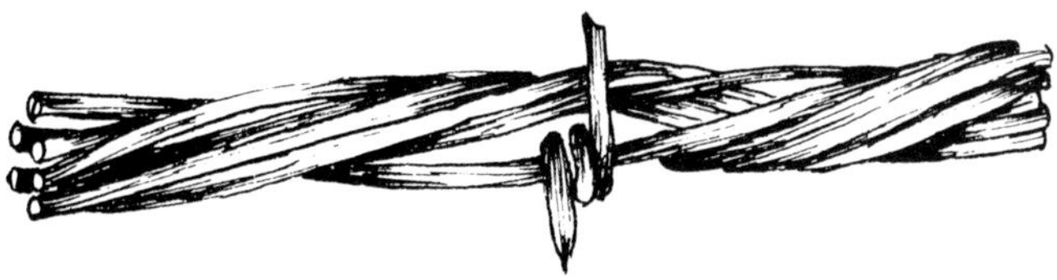

OX PEN WIRE PAT. BY JOSEPH GLIDDEN FEB. 8, 1876

652 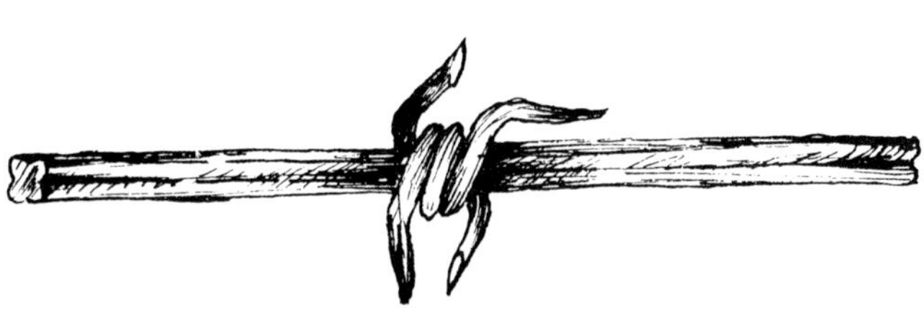

DODGE & WASHBURN'S BARB ON 4 CONCAVED LINE
PAT. JAN. 24, 1882

653 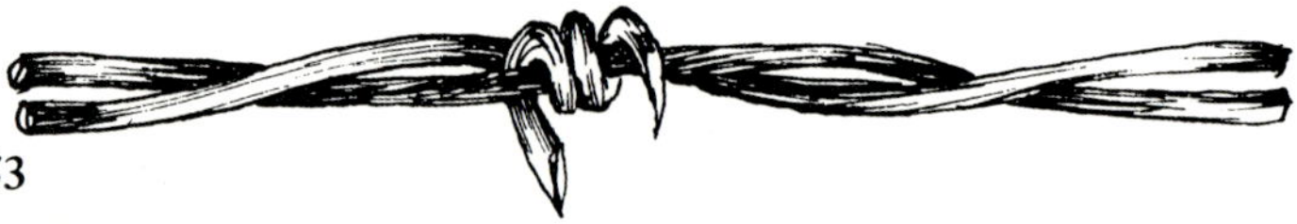

GLIDDEN "WRAP AND A HALF" PAT. FEB. 8, 1876

654 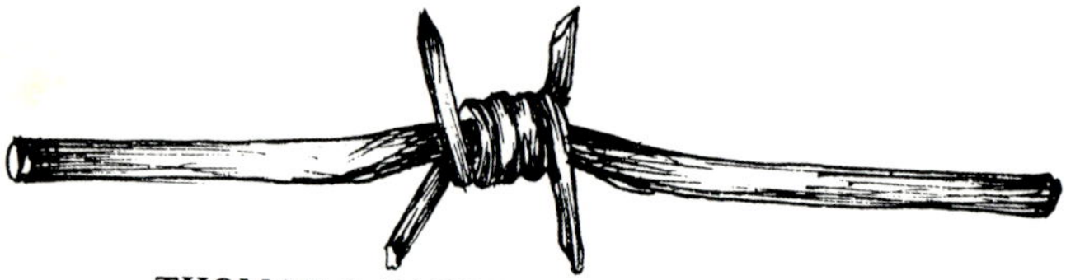

THOMAS & PATRICK HOWE "WACO COILS"
PAT. JUNE 8, 1880

655 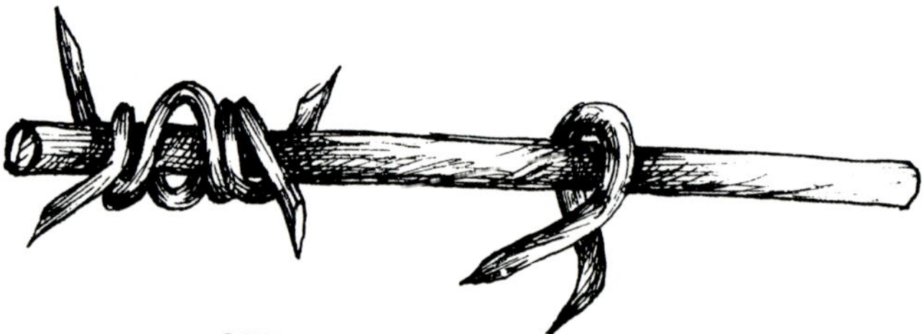

MERRILL'S WITH STAPLE BARB
PAT. DEC. 26, 1876

656 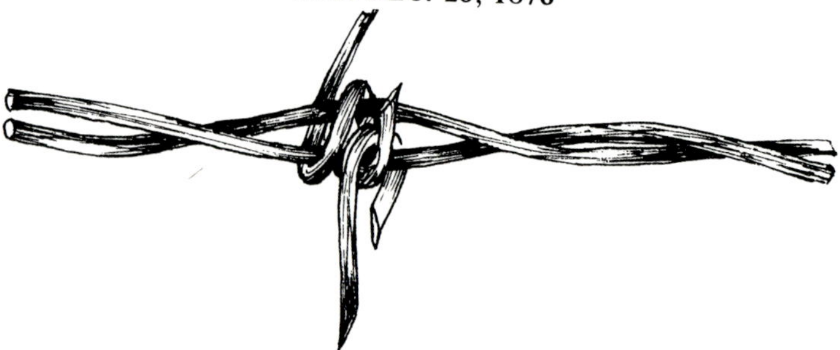

FRANK BILLING'S ONE BETWEEN LONG BARB
PAT. JUNE 25, 1878

657 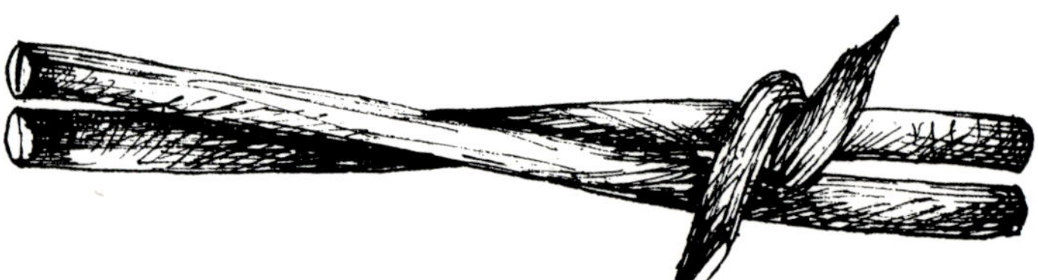

BIG JOHN WITH TWO LARGE TWISTED LINES
PAT. 1875

658 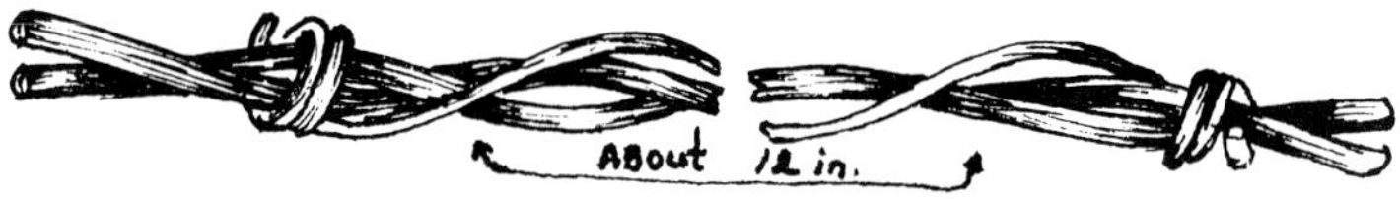

TWO LARGE LINES WITH SMALL WRAPPED AROUND

659 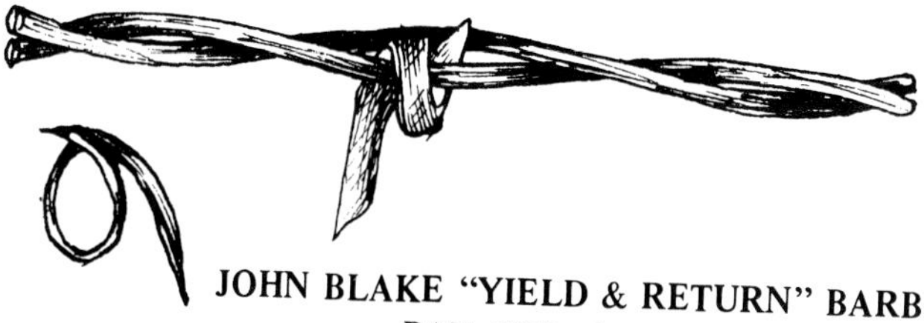

JOHN BLAKE "YIELD & RETURN" BARB
PAT. FEB. 17, 1891

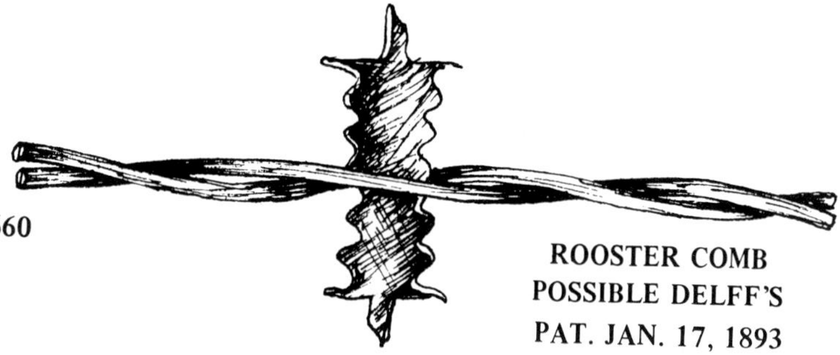
660

ROOSTER COMB
POSSIBLE DELFF'S
PAT. JAN. 17, 1893

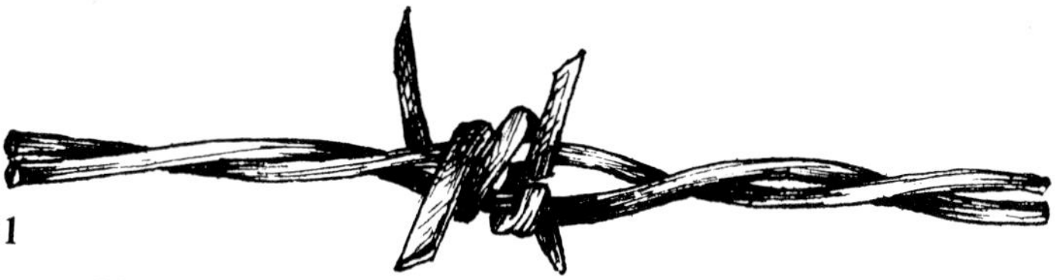
661

JOHN CURTIS FLAT BARB PAT. MARCH 28, 1893

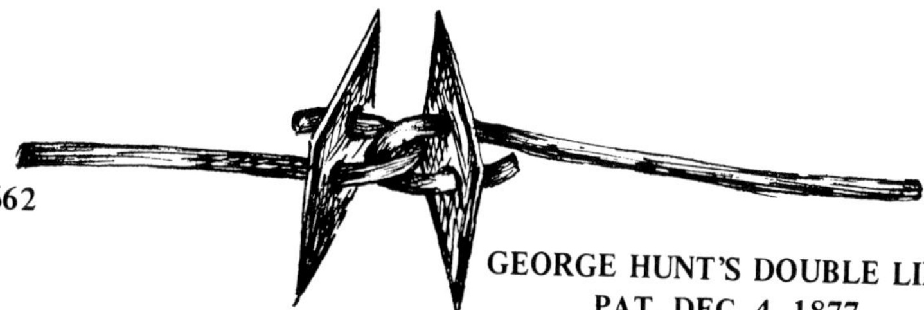
662

GEORGE HUNT'S DOUBLE LINK
PAT. DEC. 4, 1877

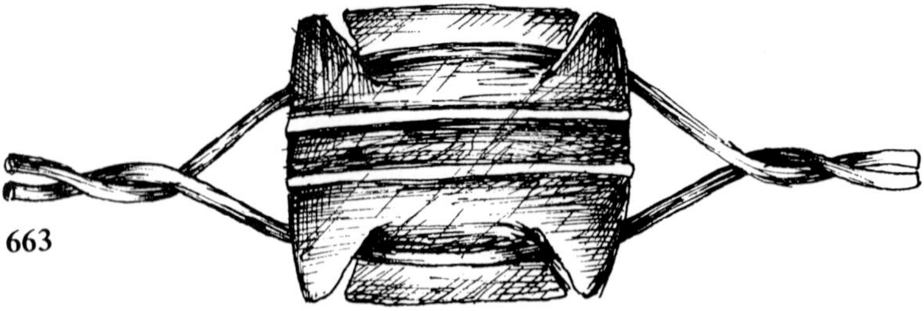

663

JOHN STUBBE'S PLATE WITH RAISED LINES
PAT. OCT. 23, 1883

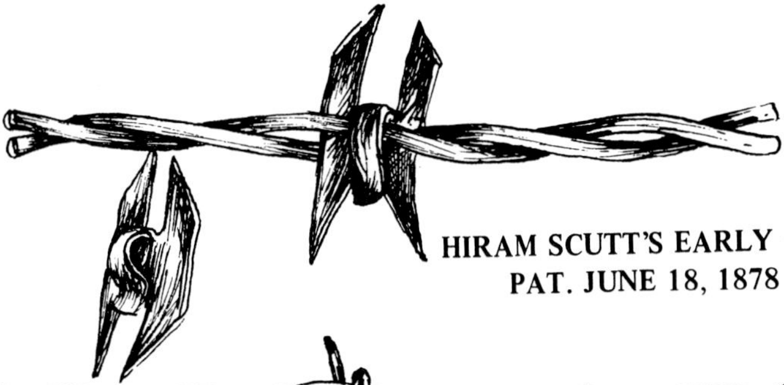

664

HIRAM SCUTT'S EARLY CLIP
PAT. JUNE 18, 1878

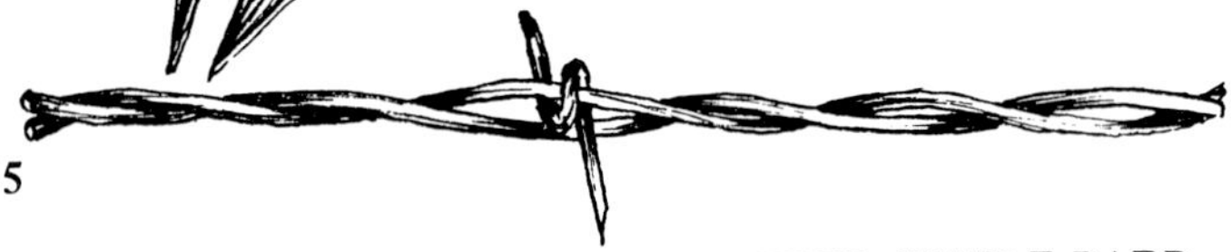

665

UPHAM'S VERY SMALL HAMMERED STAPLE BARB

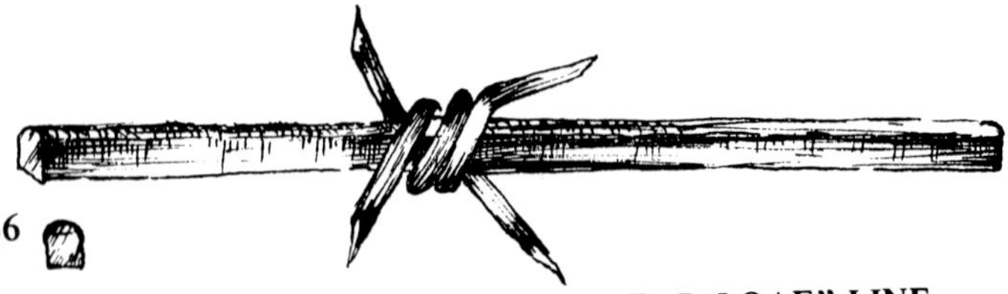

666

DODGE & WASHBURN "BREAD LOAF" LINE
PAT. JAN. 24, 1882

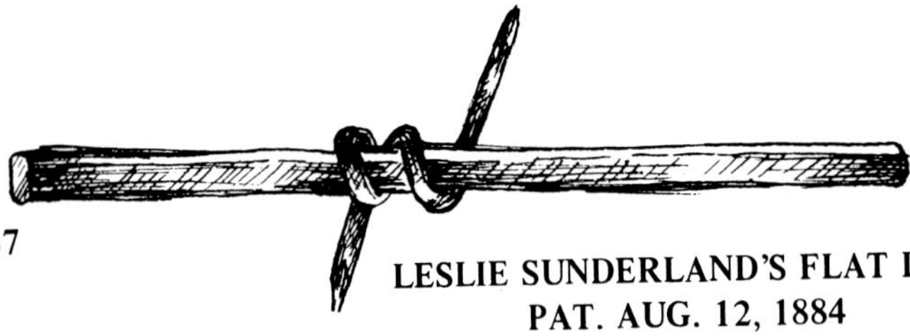

667

LESLIE SUNDERLAND'S FLAT LINE
PAT. AUG. 12, 1884

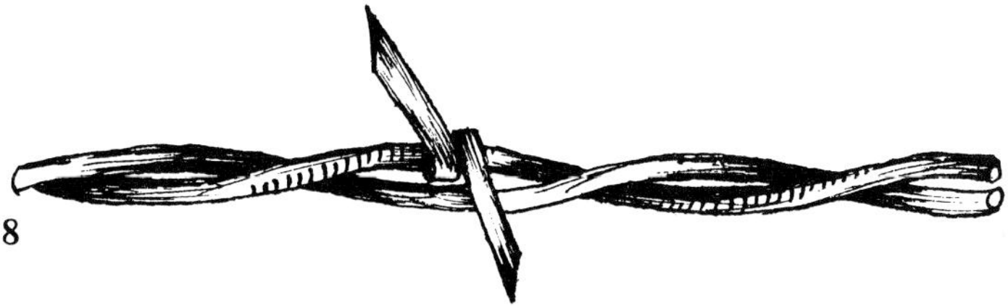

668

GEORGE BAKER NOTCHED LINE
PAT. FEB. 27, 1883

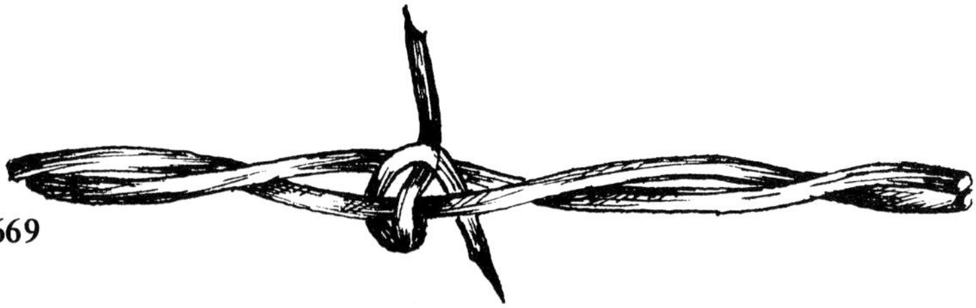

669

ANDREW UPHAM'S COILED BARB
PAT. JULY 2, 1878

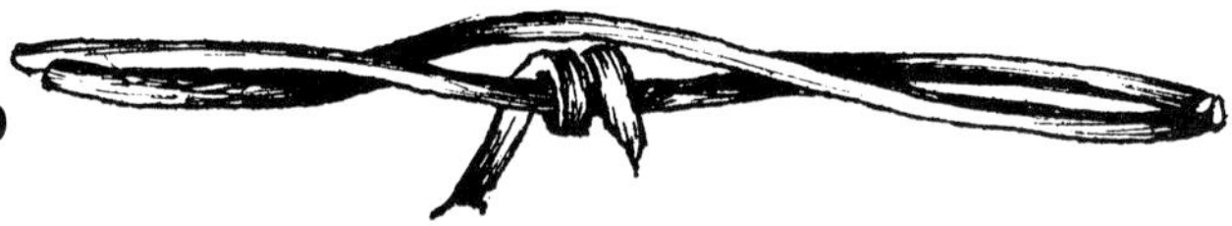

670

GEORGE BAKER'S "WRAP AND A HALF"
PAT. FEB. 27, 1883

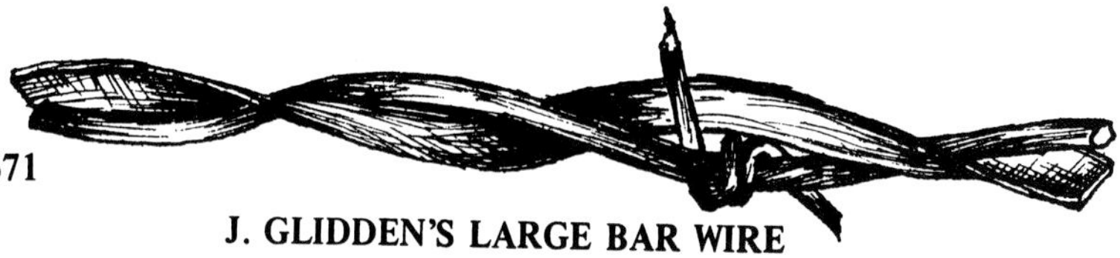

671

J. GLIDDEN'S LARGE BAR WIRE
PAT. FEB. 8, 1876

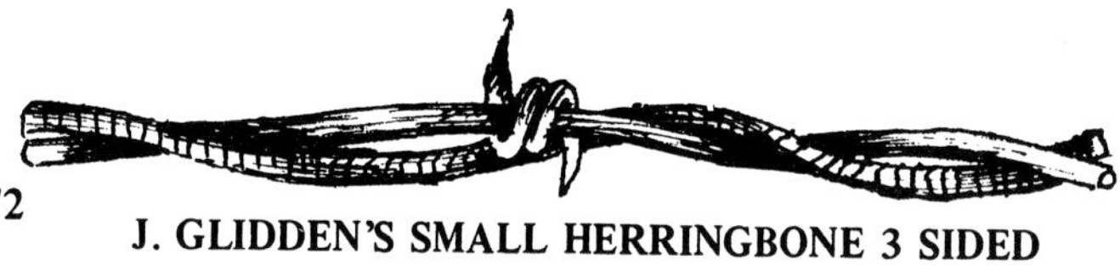

672

J. GLIDDEN'S SMALL HERRINGBONE 3 SIDED
PAT. FEB. 8, 1876

673
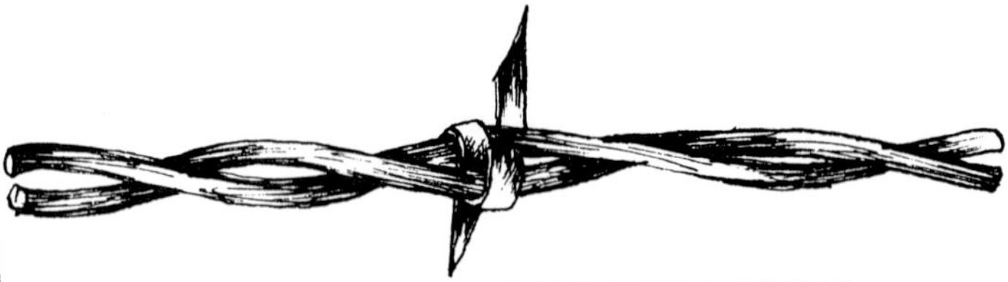
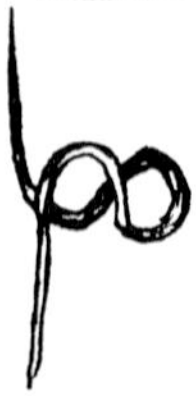

JOHN NADELHOFFER'S "GULL WING"
PAT. NOV. 4, 1884

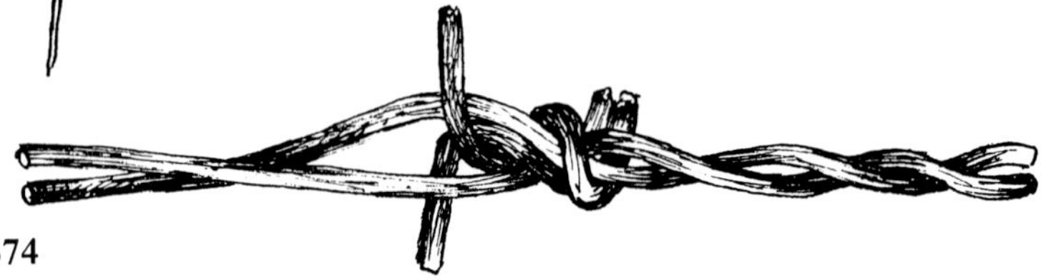
674

CANADIAN SPLIT WRAP
HOME MACHINE MADE IN THE 1880's

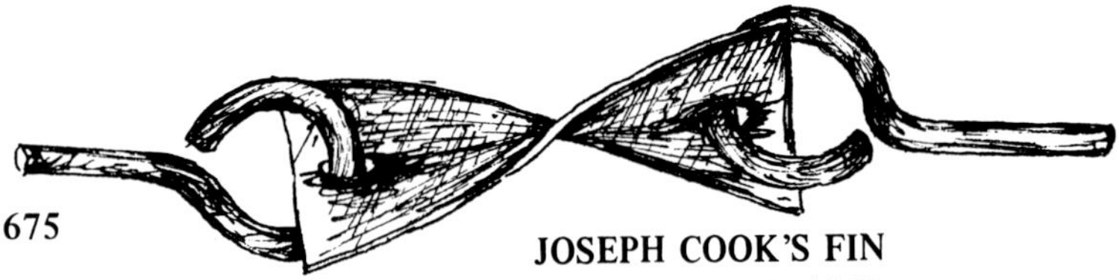
675

JOSEPH COOK'S FIN
PAT. SEPT. 26, 1882

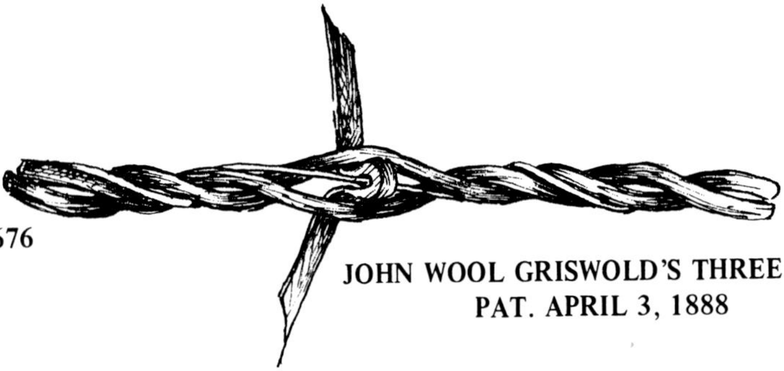
676

JOHN WOOL GRISWOLD'S THREE LINE
PAT. APRIL 3, 1888

677
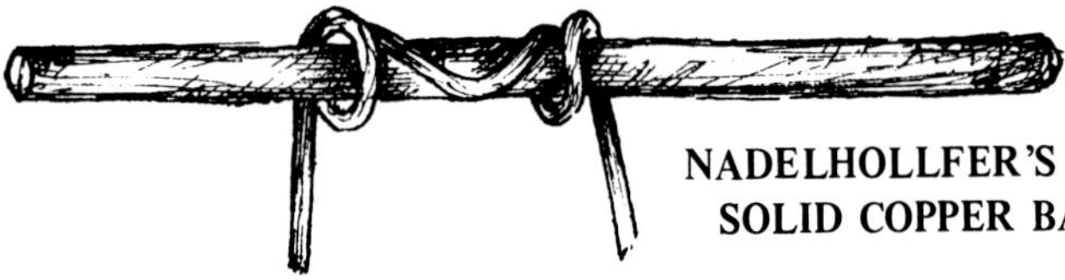

NADELHOLLFER'S TYPE
SOLID COPPER BARB

678 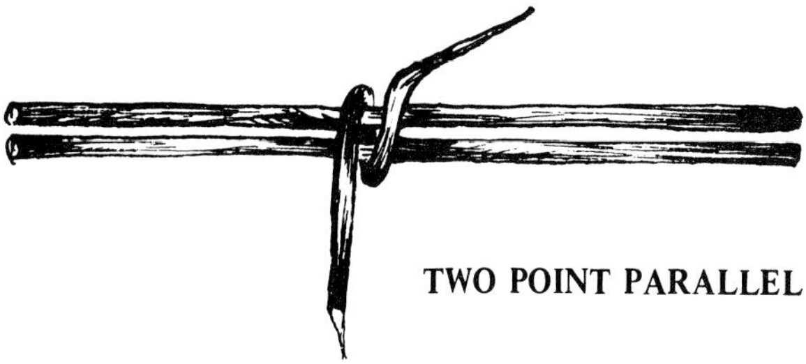

TWO POINT PARALLEL

679 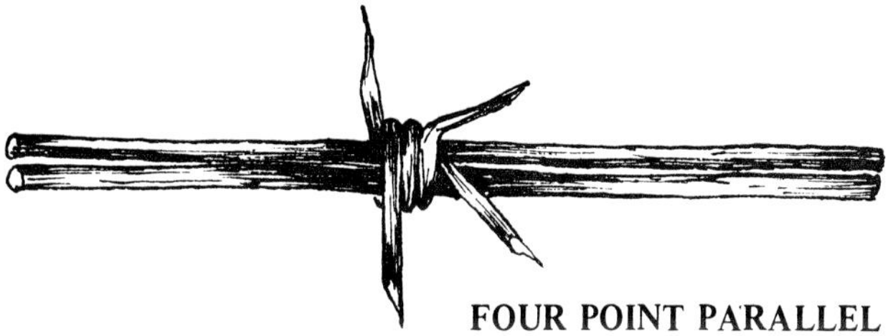

FOUR POINT PARALLEL

680 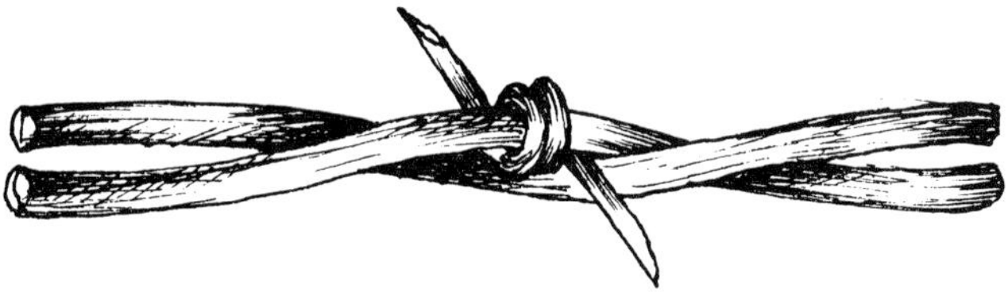

LARGE LINE WIRES ABOVE WIRES MADE BY DOUBLE H STEEL & WIRE CO., ROCKY FORD, COLO.

681 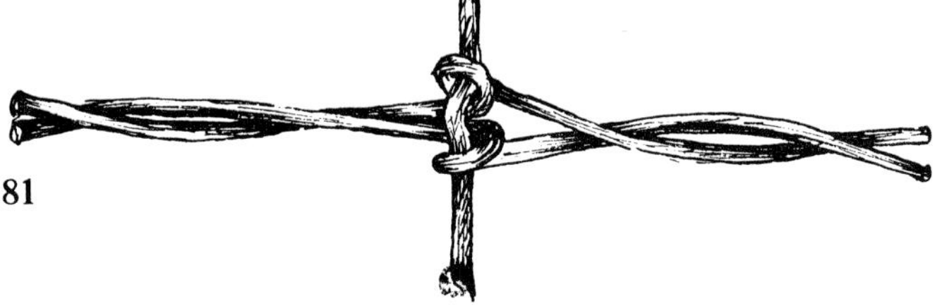

A. J. UPHAM'S LOOPED LINE WIRES
PAT. MARCH 3, 1885

682 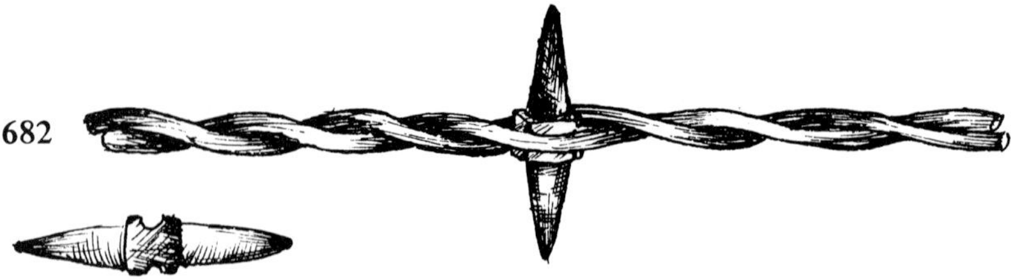

CHARLES KELLEY'S CAST BARB. SOLID BRASS & IRON
PAT. NOV. 20, 1877

683 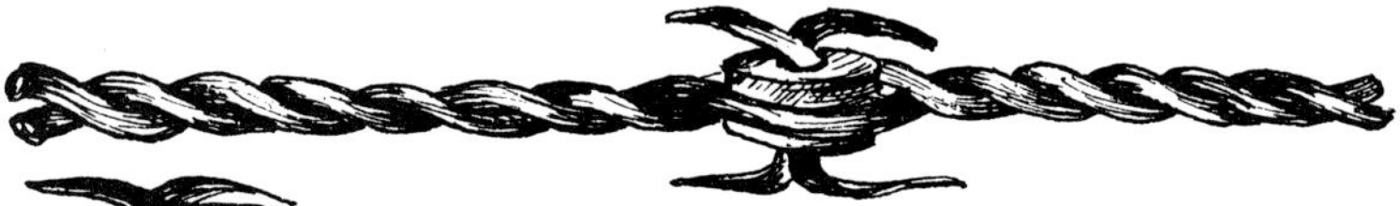

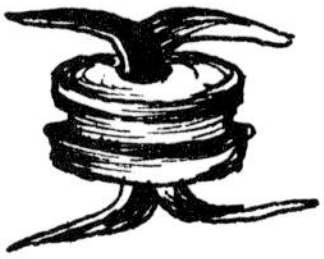

DANIEL FUNCHEON ROTATING BARB
PAT. MARCH 7, 1893

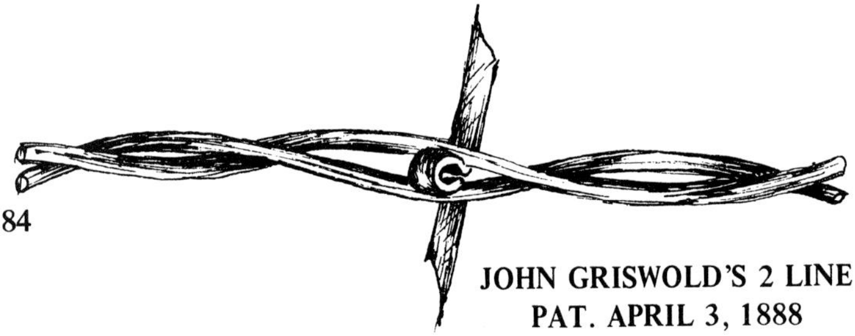
684

JOHN GRISWOLD'S 2 LINE
PAT. APRIL 3, 1888

685 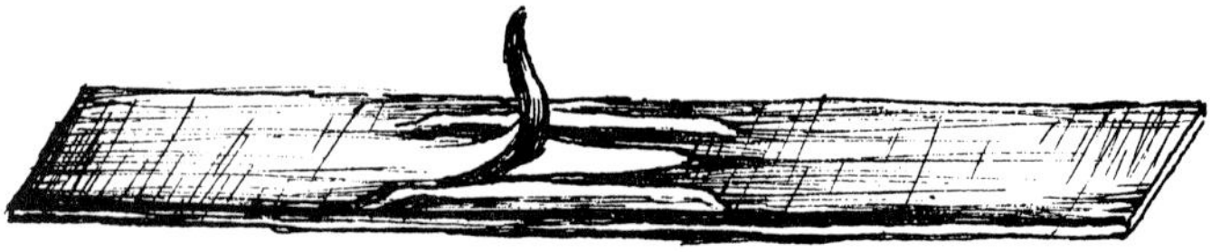

L. M. WOODCOCK'S RAIL PAT. OCT. 17, 1876

686

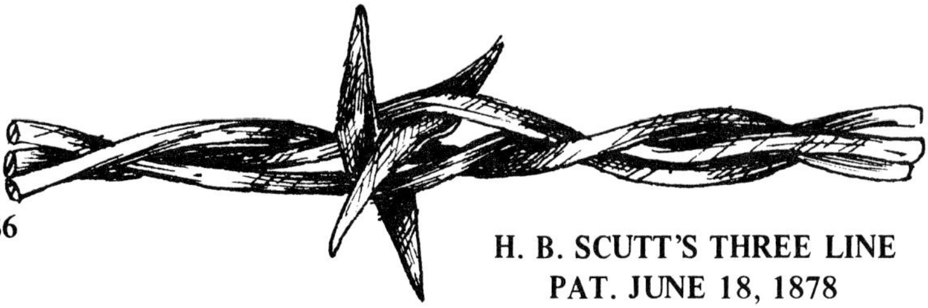

H. B. SCUTT'S THREE LINE
PAT. JUNE 18, 1878

687

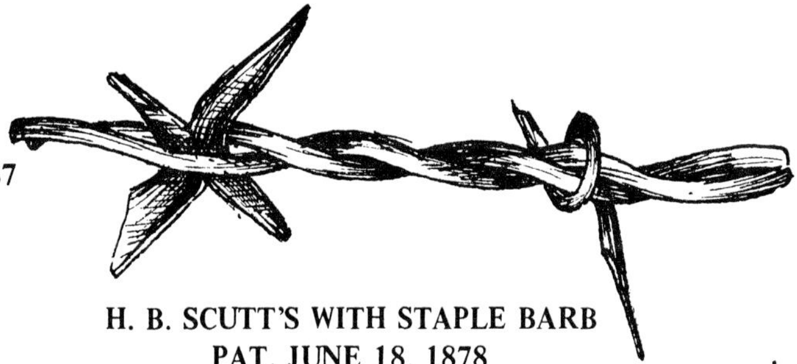

H. B. SCUTT'S WITH STAPLE BARB
PAT. JUNE 18, 1878

688

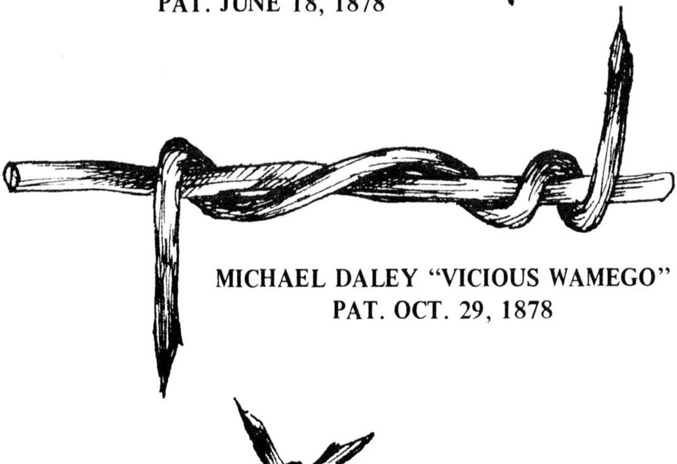

MICHAEL DALEY "VICIOUS WAMEGO"
PAT. OCT. 29, 1878

689

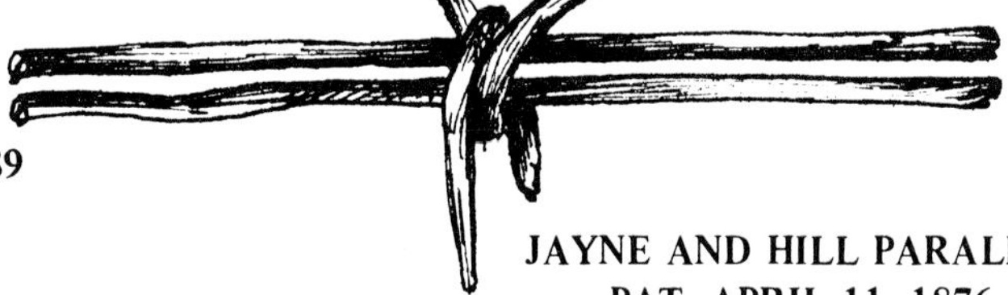

JAYNE AND HILL PARALLEL
PAT. APRIL 11, 1876

690 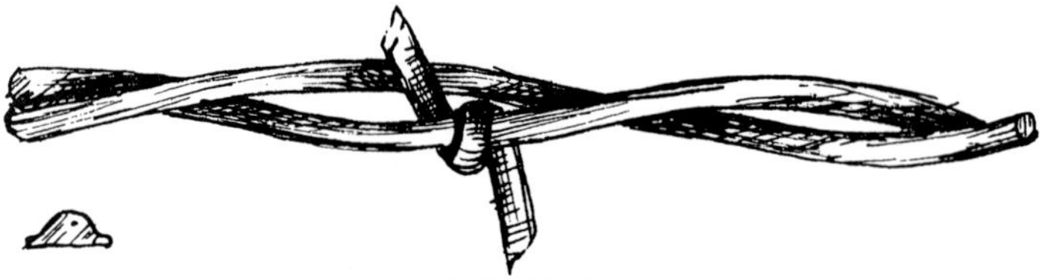

G. BAKER CHANNELED BARB
PAT. FEB. 27, 1883

691 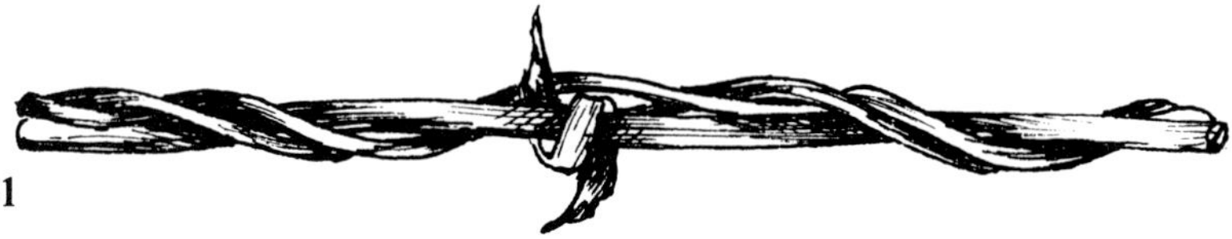

G. BAKER'S LARGE AND TWO SMALL
PAT. FEB. 27, 1883

692 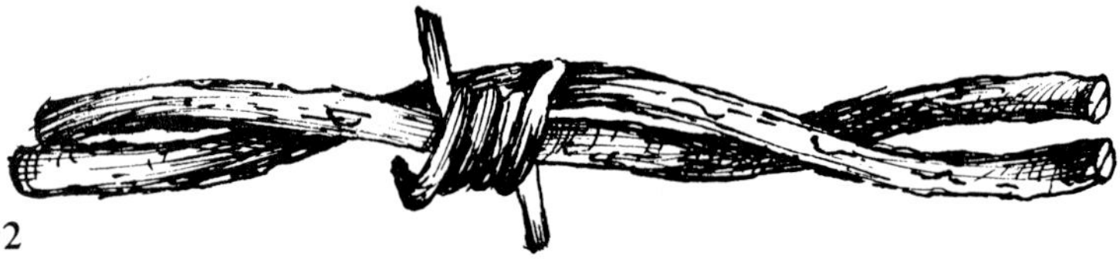

DODGE & WASHBURN LARGE LINES
PAT. JAN. 24, 1882

693 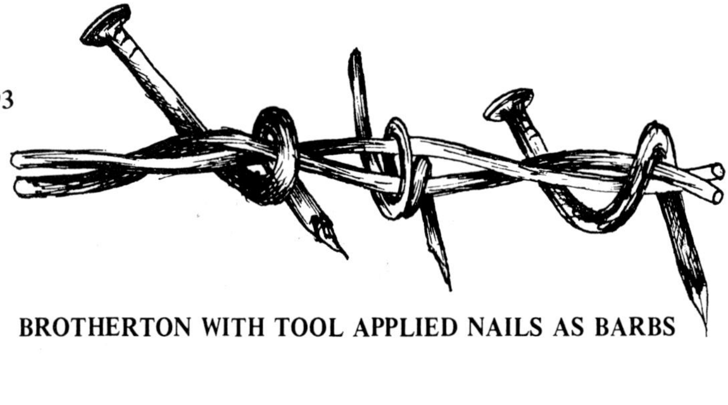

BROTHERTON WITH TOOL APPLIED NAILS AS BARBS

694

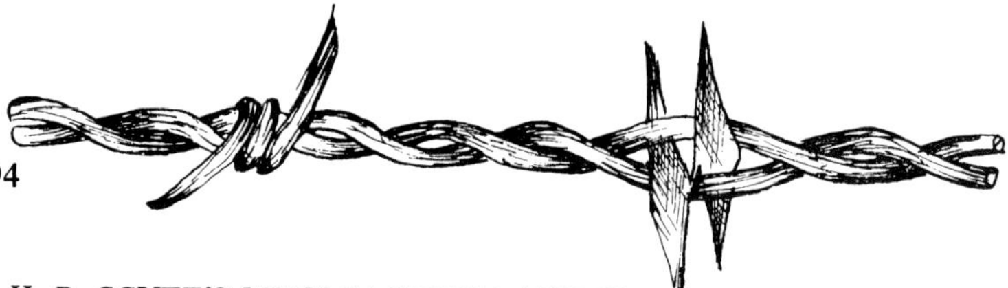

H. B. SCUTT'S WITH FACTORY APPLIED
GLIDDEN TYPE BARB

695

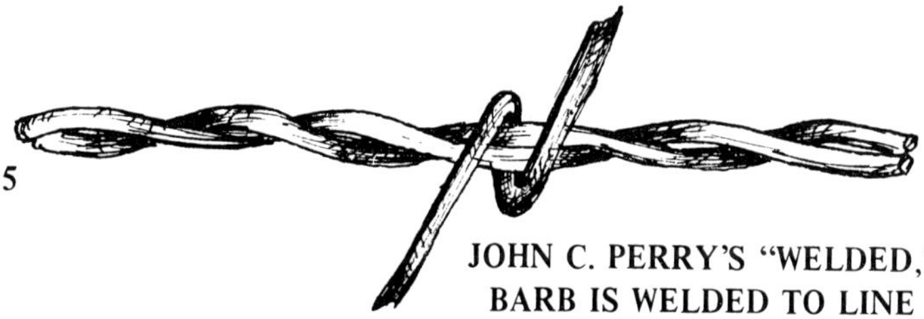

JOHN C. PERRY'S "WELDED,"
BARB IS WELDED TO LINE
PAT. AUG. 24, 1897

696

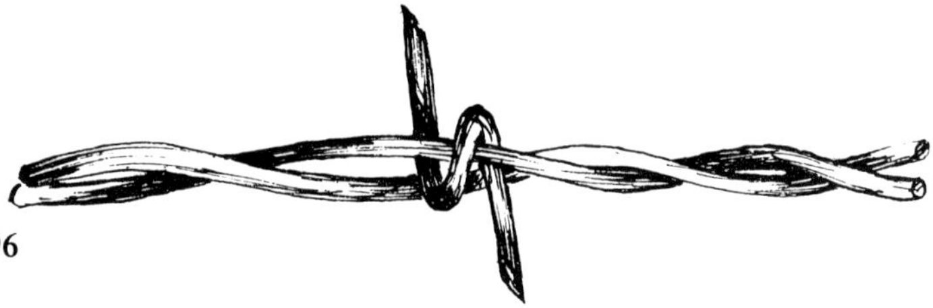

PERRY'S WELDED BARB AROUND ONE LINE
PAT. AUG. 24, 1897

697

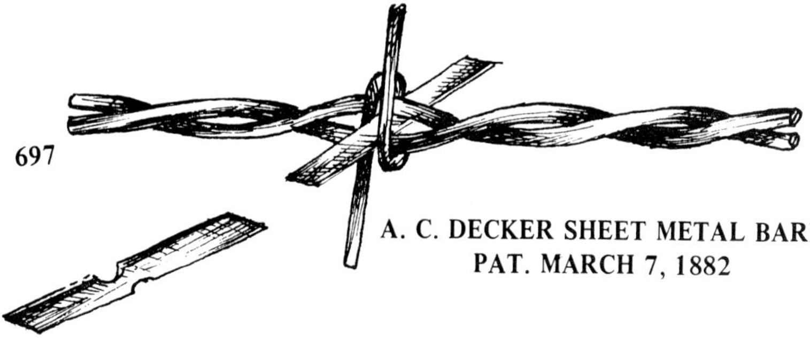

A. C. DECKER SHEET METAL BARB
PAT. MARCH 7, 1882

698

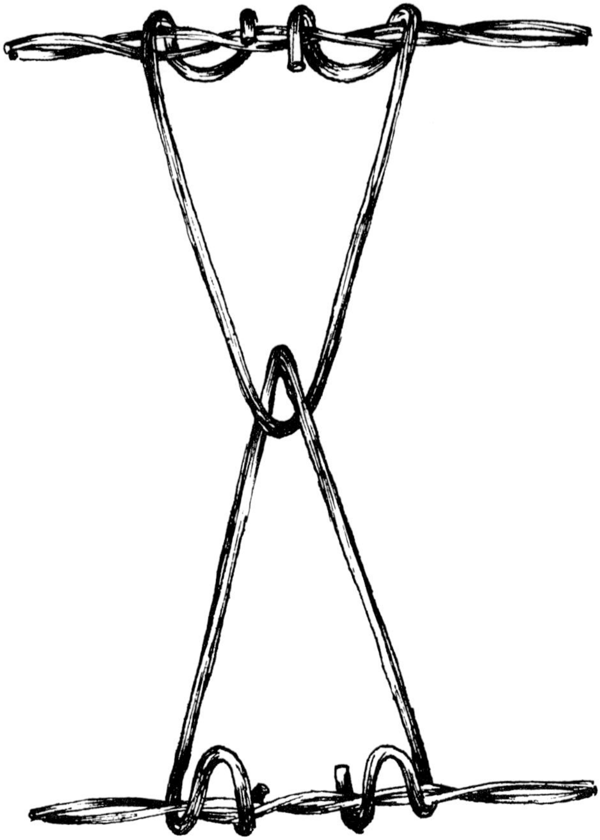

TWO PIECE STAY THAT WILL GIVE

699

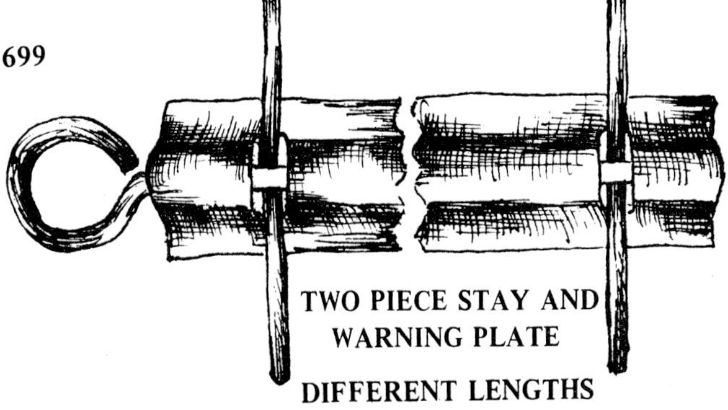

TWO PIECE STAY AND WARNING PLATE

DIFFERENT LENGTHS

700 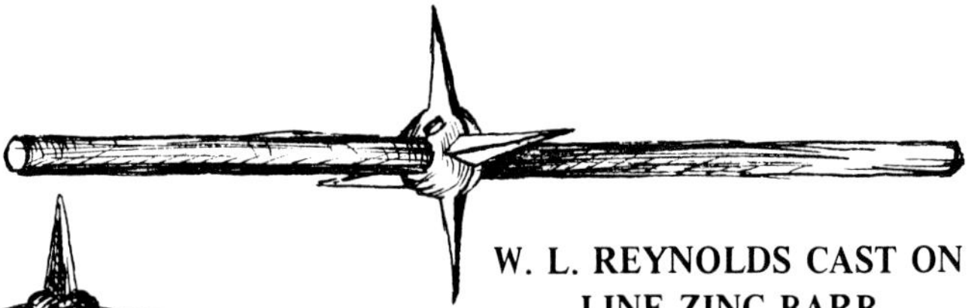

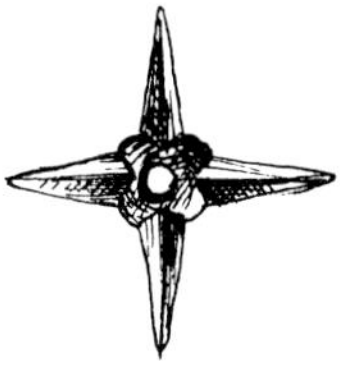

W. L. REYNOLDS CAST ON
LINE ZINC BARB
PAT. FEB. 6, 1877

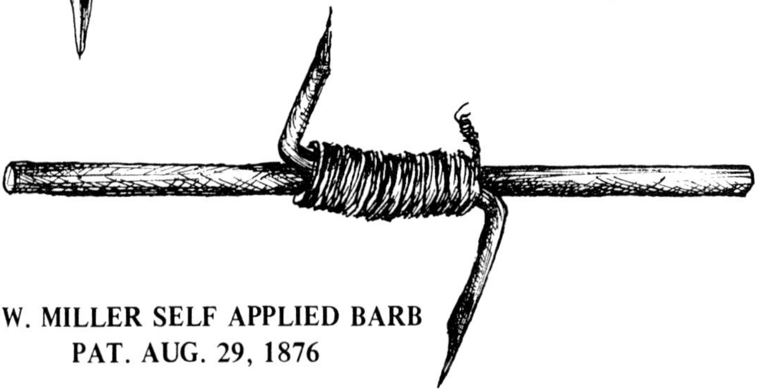

701

C. W. MILLER SELF APPLIED BARB
PAT. AUG. 29, 1876

702 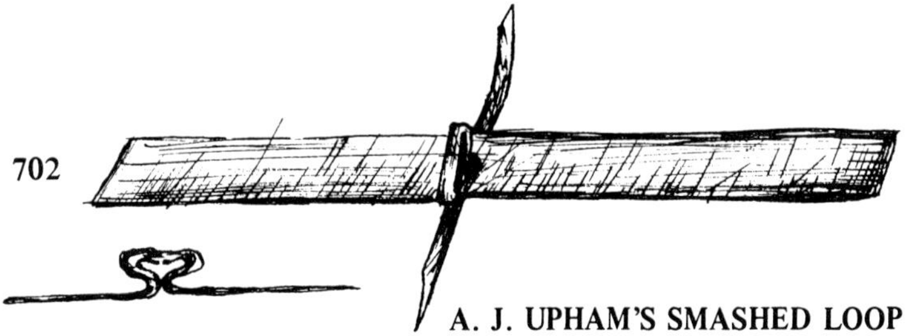

A. J. UPHAM'S SMASHED LOOP
PAT. JUNE 26, 1883

703 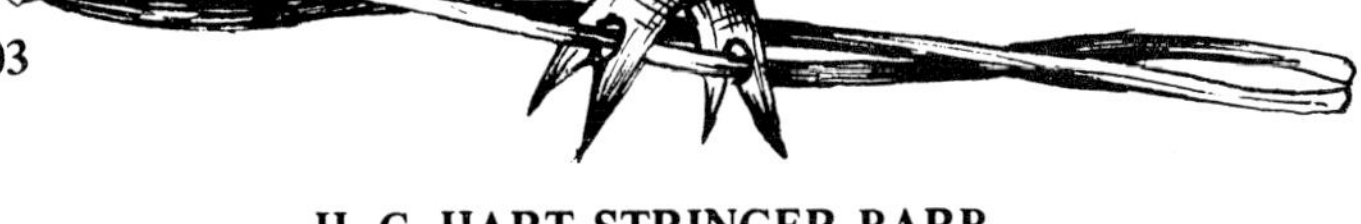

H. C. HART STRINGER BARB
PAT. FEB. 17, 1885

704

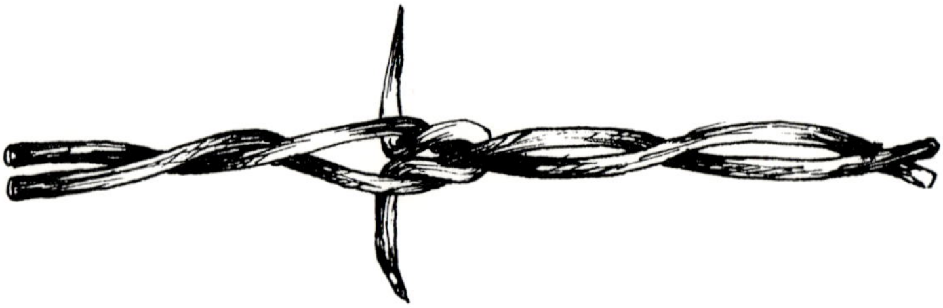

BAKER & BESTOR LOOPED CENTER BARB
PAT. SEPT. 17, 1878

705

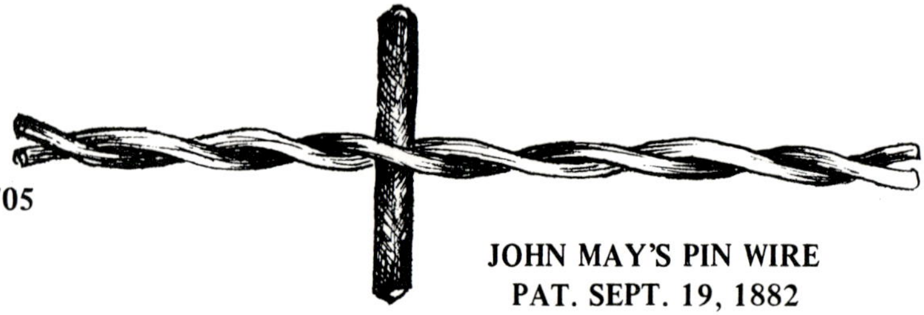

JOHN MAY'S PIN WIRE
PAT. SEPT. 19, 1882

706

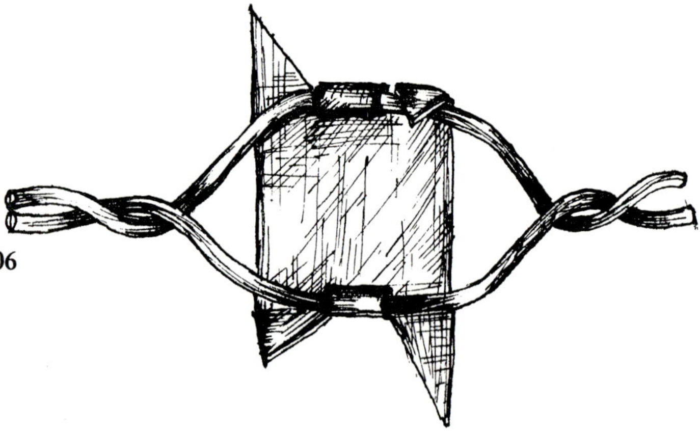

GEORGE ELSEY FOUR POINT PLATE BARB
PAT. JULY 18, 1882

707

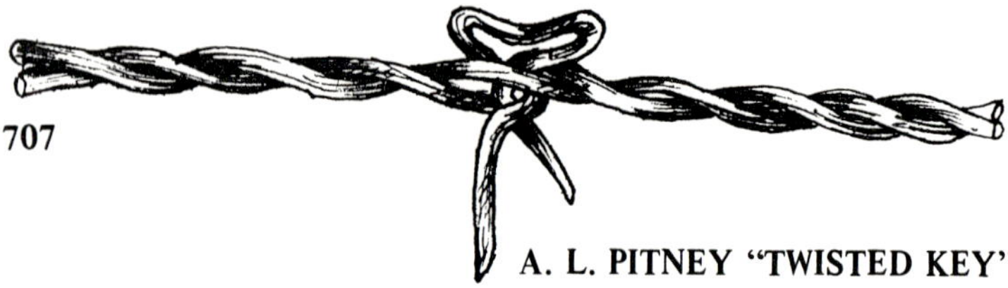

A. L. PITNEY "TWISTED KEY"
PAT. OCT. 1, 1878

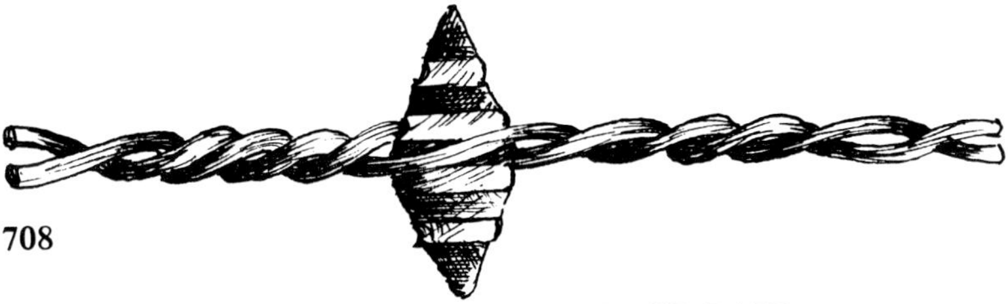

708

J. B. OLIVER TWO POINT RIBBED BARB
PAT. OCT. 2, 1883

709

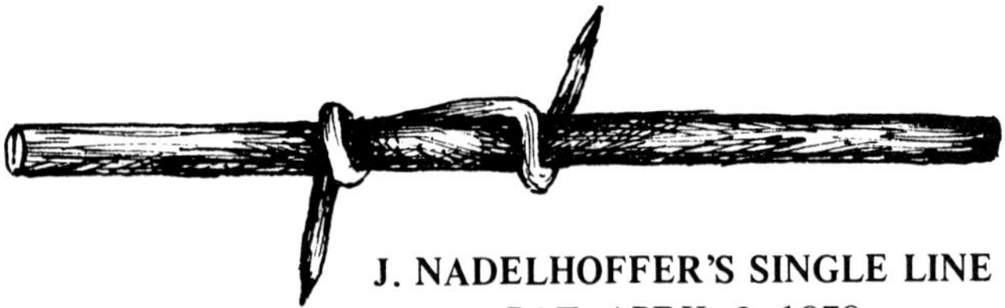

J. NADELHOFFER'S SINGLE LINE
PAT. APRIL 2, 1878

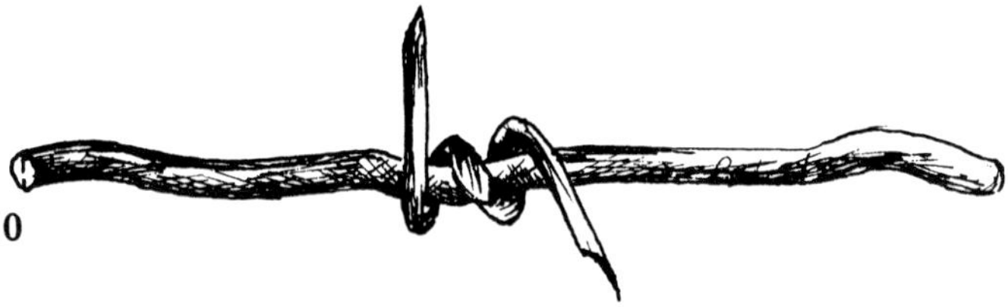

710

SUNDERLAND'S HAMMERED BARB
PAT. AUG. 12, 1884

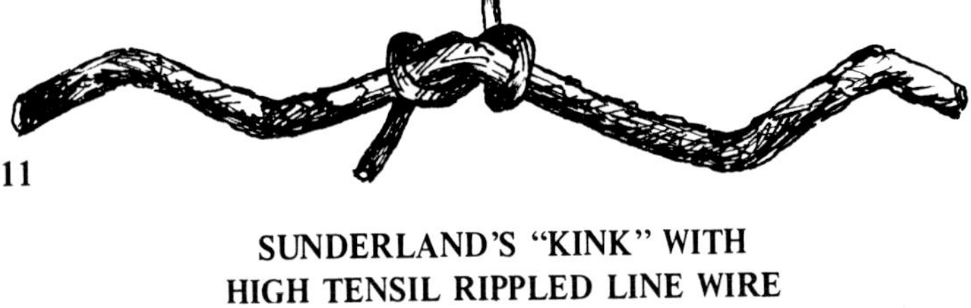

711

SUNDERLAND'S "KINK" WITH
HIGH TENSIL RIPPLED LINE WIRE

712

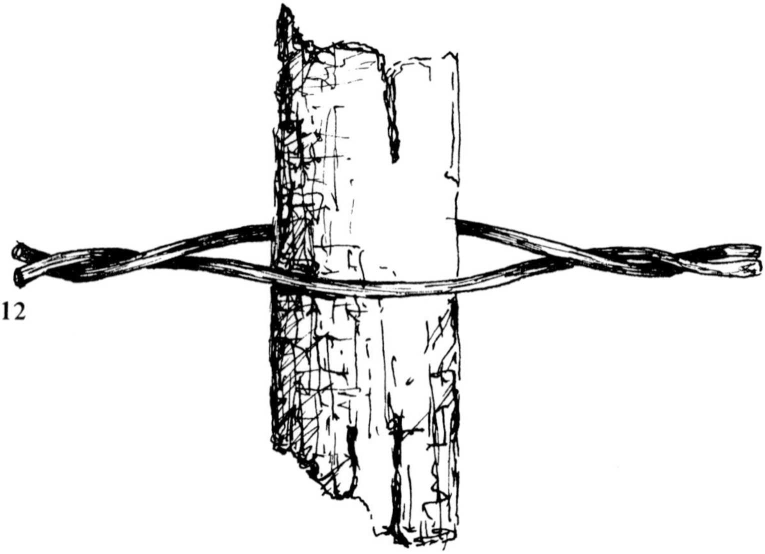

G. FLETCHER SR. BRUSHWOOD FENCE
PAT. NOV. 17, 1868

713

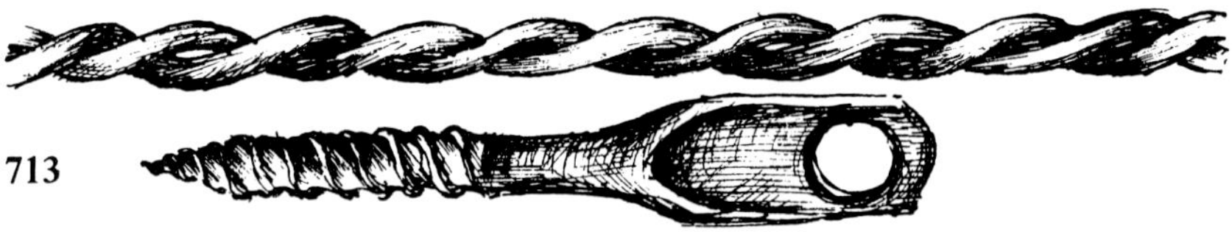

KING RANCH HIGH TENSIL
CORRAL WIRE & STAPLE

714

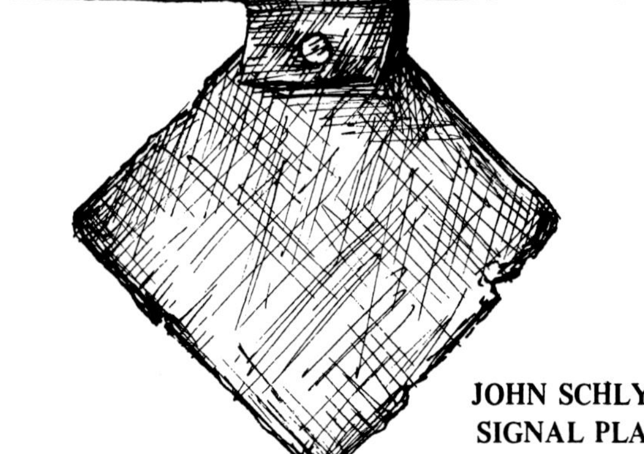

JOHN SCHLYER
SIGNAL PLATE
PAT. JUNE 25, 1889

715

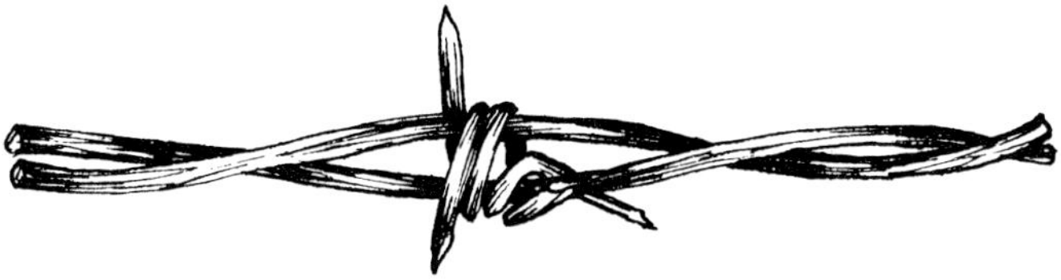

ROSS WITH BARB PAINTED RED
MODERN MFG.

716

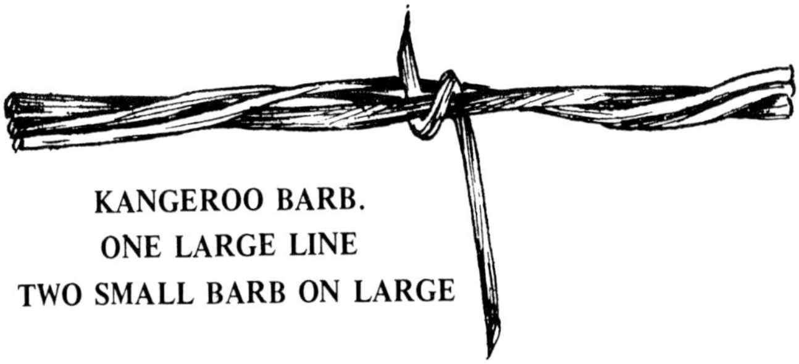

KANGEROO BARB.
ONE LARGE LINE
TWO SMALL BARB ON LARGE

717

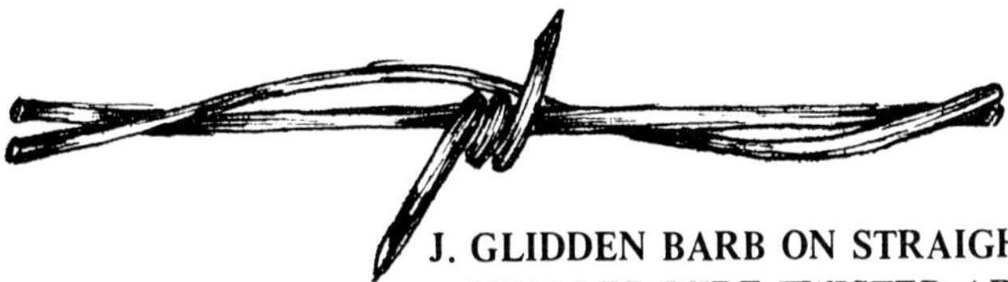

J. GLIDDEN BARB ON STRAIGHT LINE.
SMALLER WIRE TWISTED AROUND

718

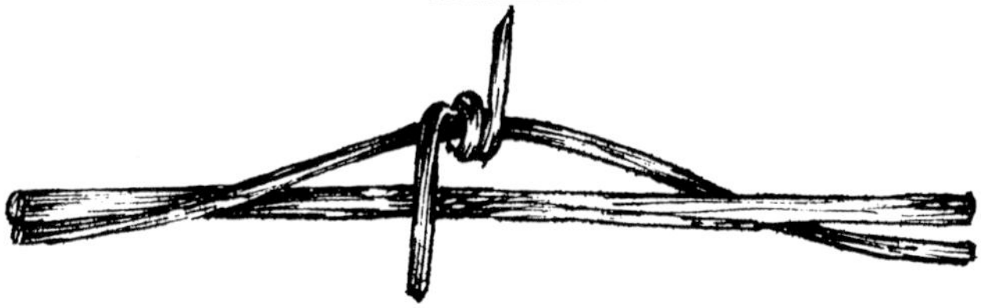

GLIDDEN BARB ON SMALL LINE.
LARGE LINE STRAIGHT
PAT. FEB. 8, 1876

719

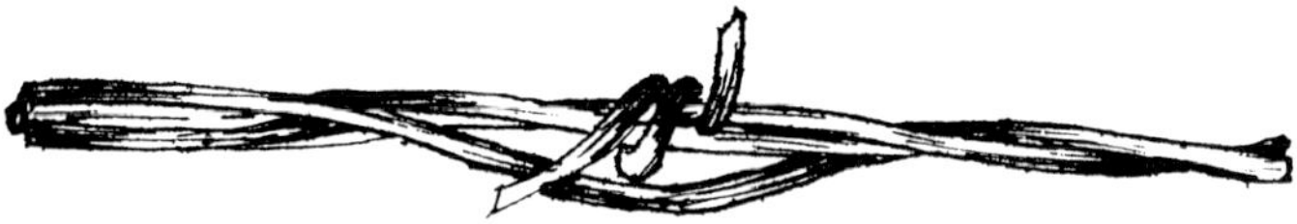

J. BROTHERTON'S 3 LINE. BARB BETWEEN TWO.
PAT. SEPT. 3, 1878

720 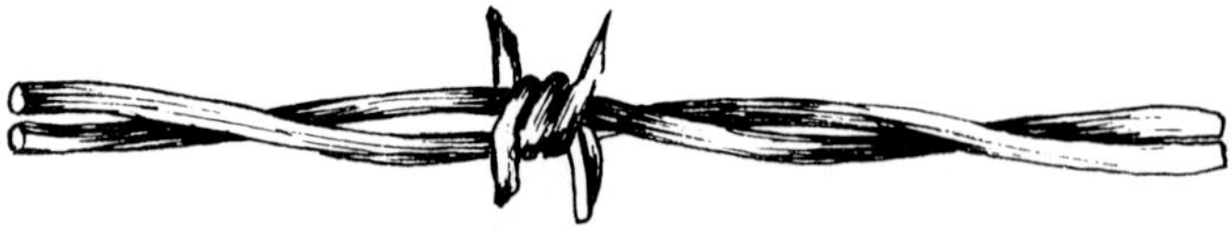

VERY SMALL LINE WIRES &
DODGE & WASHBURN HALF ROUND BARB
PAT. JAN. 24, 1882

721 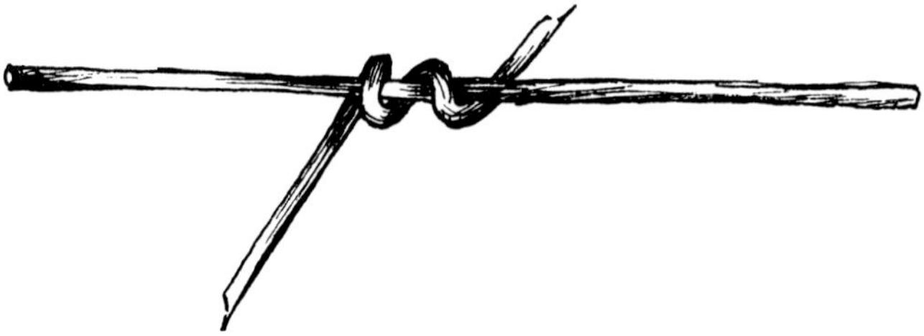

SINGLE LINE MILITARY ENTANGLEMENT WIRE
MODERN MFG.

722 HIGH TENSIL WATERGAP WIRE.
WILL GIVE & TAKE

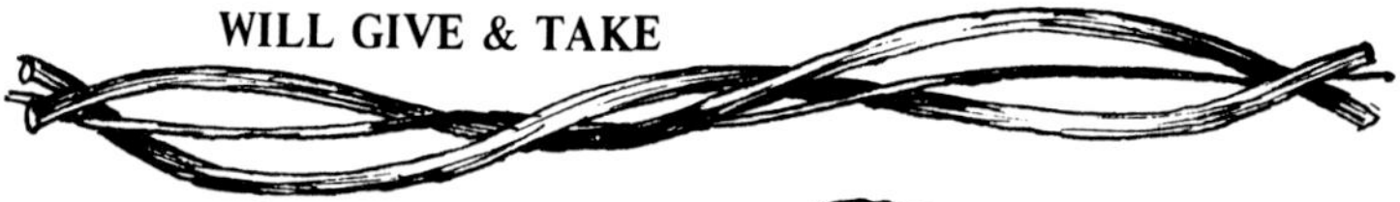

723 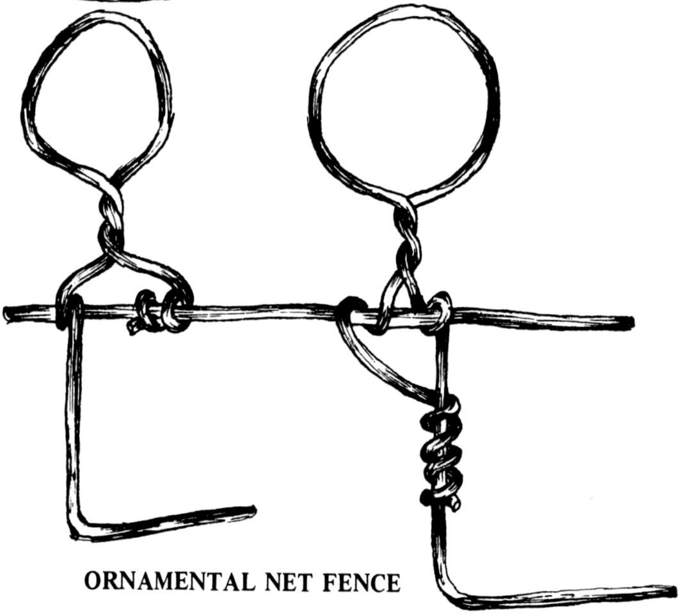

ORNAMENTAL NET FENCE

724

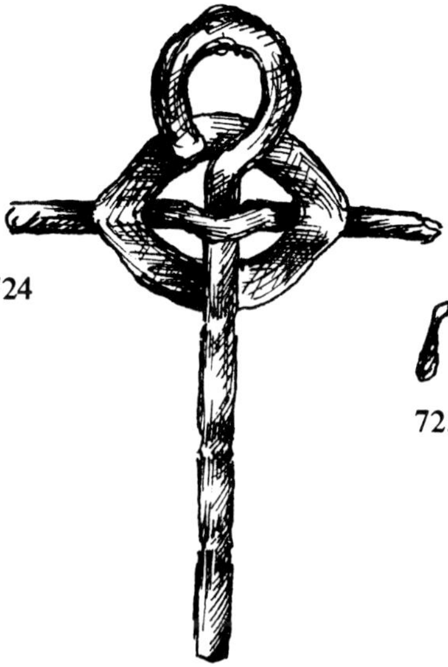

RIPPLE STAY & WASHER

725

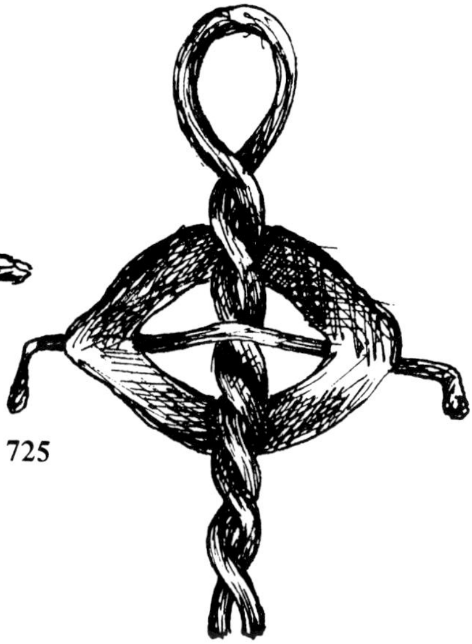

TWISTED STAY & WASHER

726

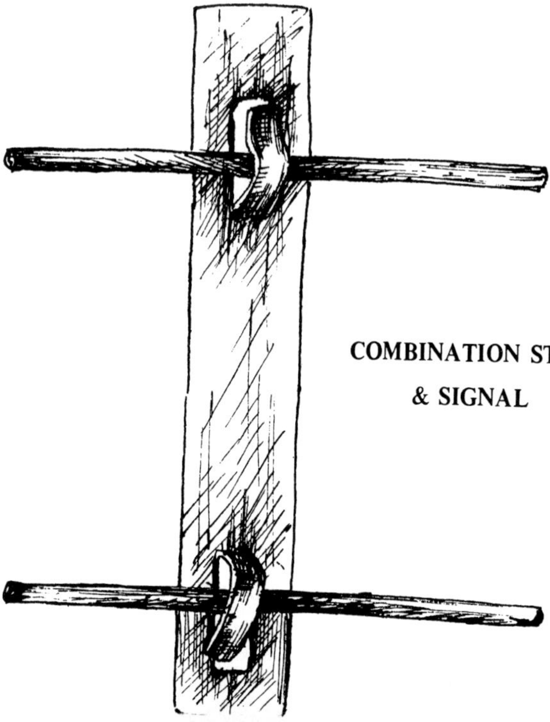

COMBINATION STAY & SIGNAL

727

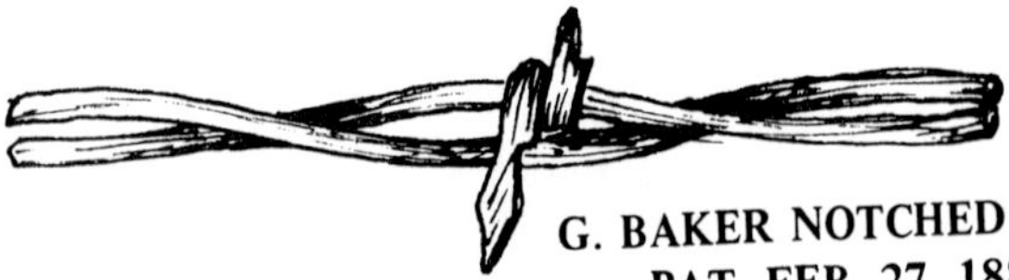

G. BAKER NOTCHED BARB
PAT. FEB. 27, 1883

728

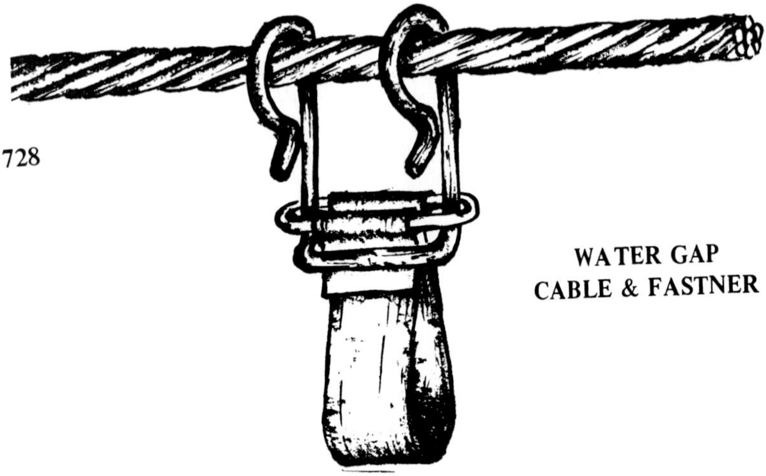

WATER GAP
CABLE & FASTNER

729

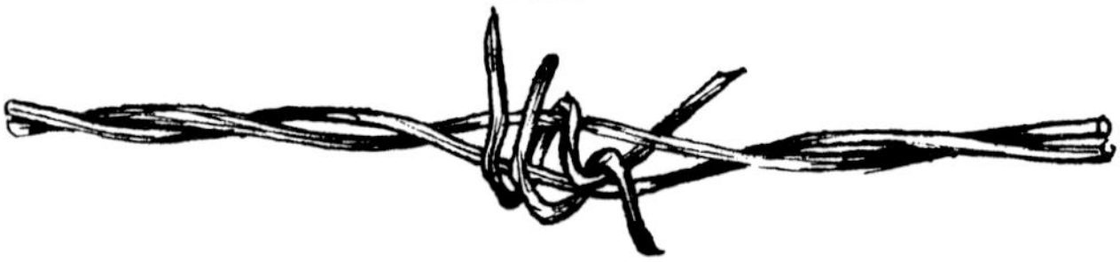

VERY SMALL HIGH TENSIL WIRE.
MODERN MFG.

730

J. GLIDDEN. STRAIGHT LARGE LINE
TWO SMALLER TWISTED

731

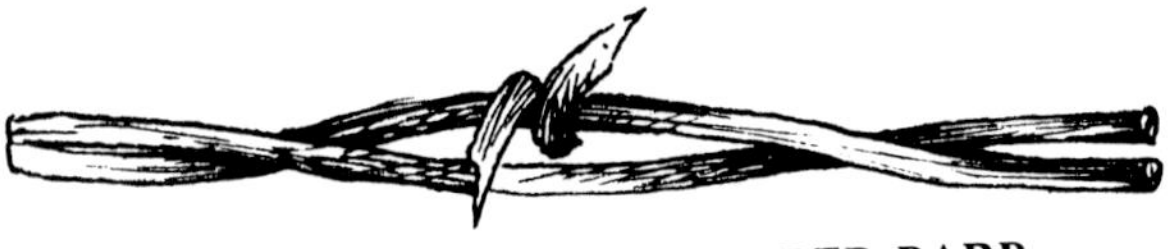

HALF ROUND PAINTED RED BARB.
MODERN MFG.

732

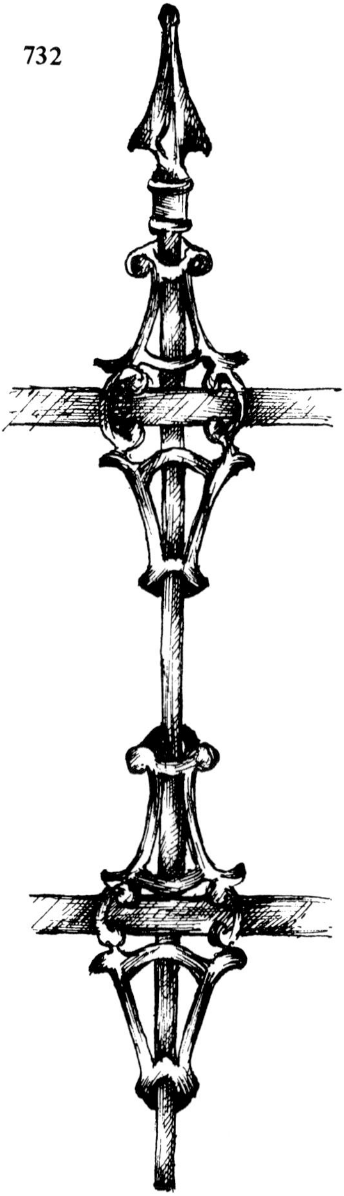

SECTION ORNAMENTAL FENCE

733

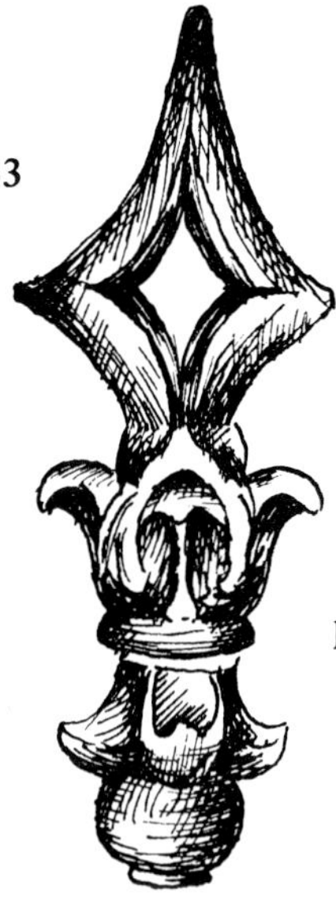

CAST IRON FENCE TOP

734

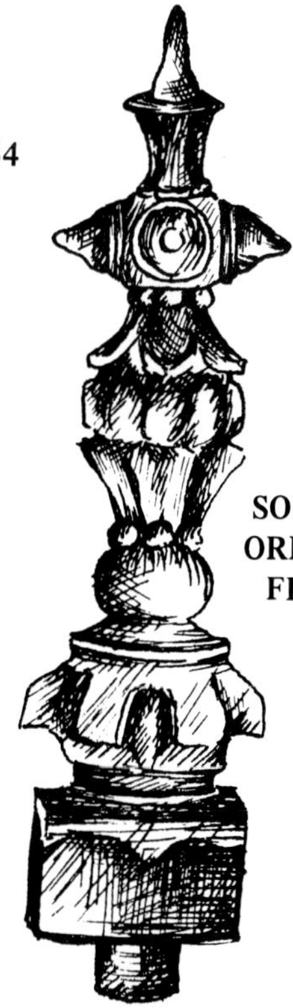

SOLID BRASS ORNAMENTAL FENCE TOP

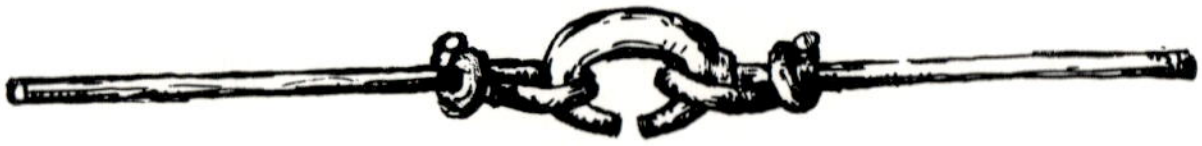

PLANTER WIRE, SOFT CAST
735 CENTER LINK

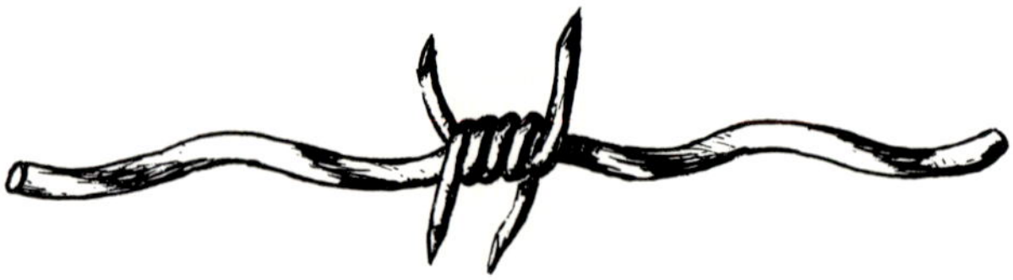

736 AUSTRALIAN 4 PT. SPIRAL LINE

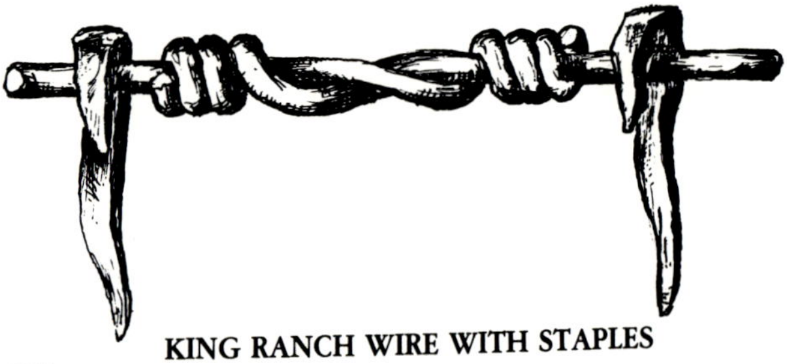

KING RANCH WIRE WITH STAPLES
737

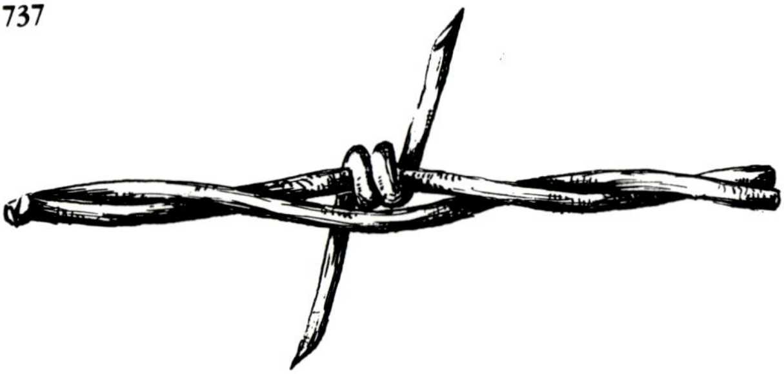

738 GLIDDEN VICIOUS
PAT. J.F. GLIDDEN NOV. 24, 1874

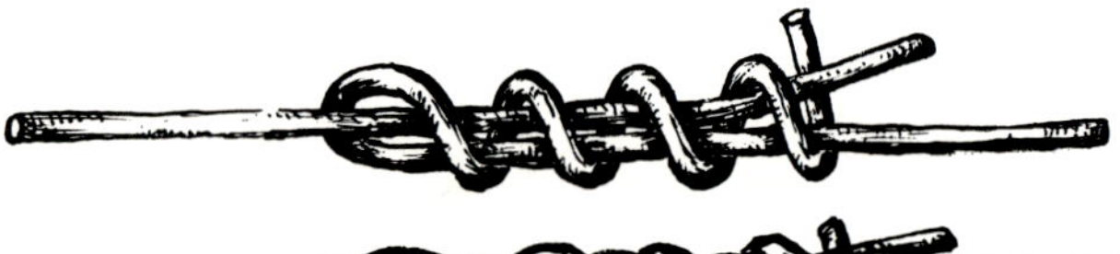

CANADIAN SPLICE FRONT
739 & SIDE VIEW

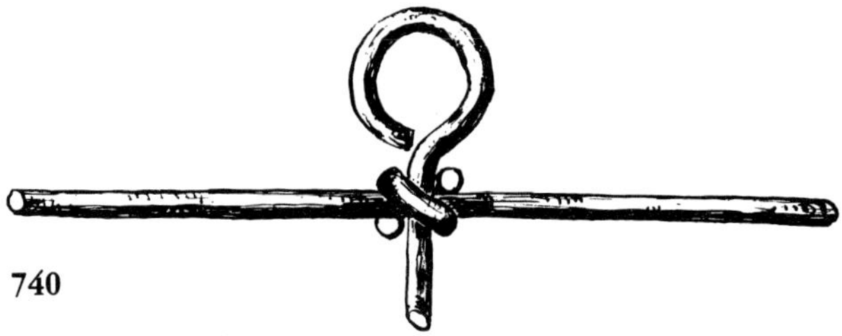

740

STAY ON HIGH TENSIL LINE

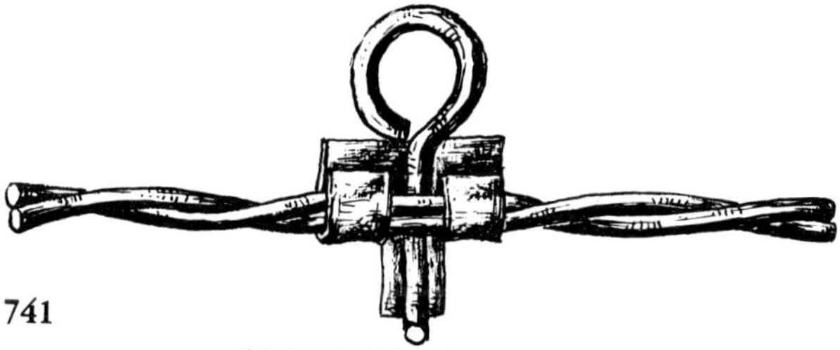

741

STAY WITH CRIMP HOLDER

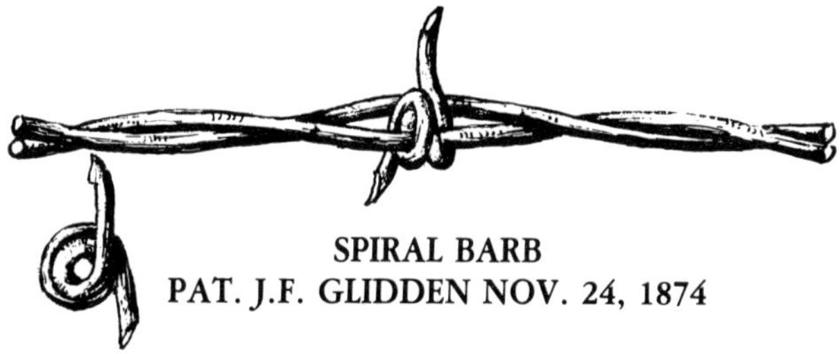

SPIRAL BARB
PAT. J.F. GLIDDEN NOV. 24, 1874

742

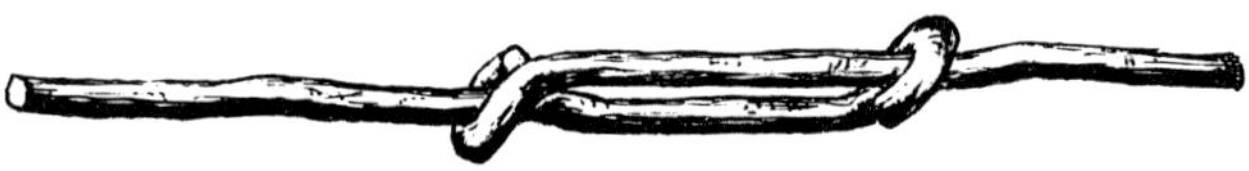

743 BROWN HOOK
PAT. R.G. BROWN SEPT. 19, 1876

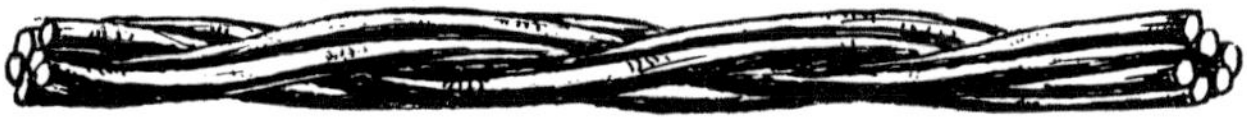

744 6 STRAND HORSE PEN WIRE
ALL WIRE CO'S. MADE THIS

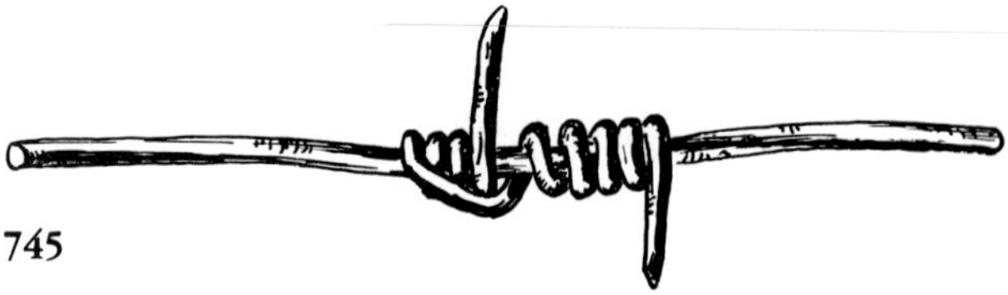

745

CANADIAN SINGLE LINE

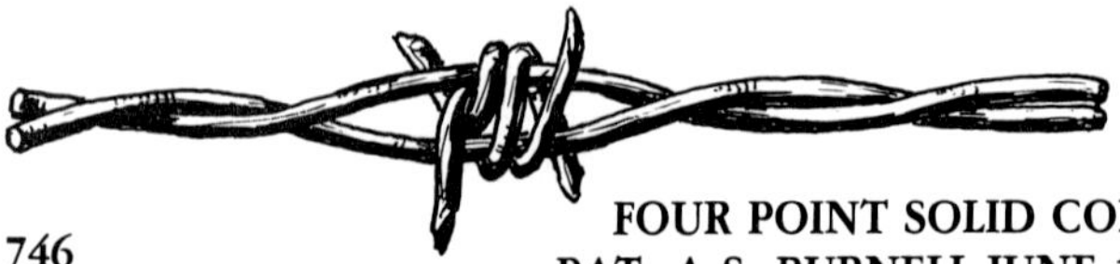

746

FOUR POINT SOLID COPPER
PAT. A.S. BURNELL JUNE 19, 1877

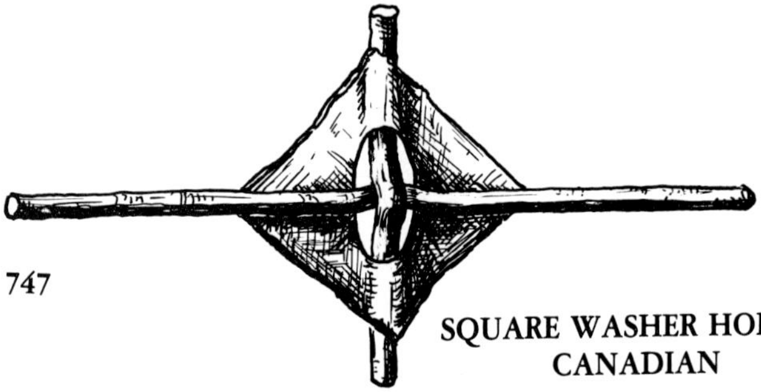

747

SQUARE WASHER HOLDER
CANADIAN

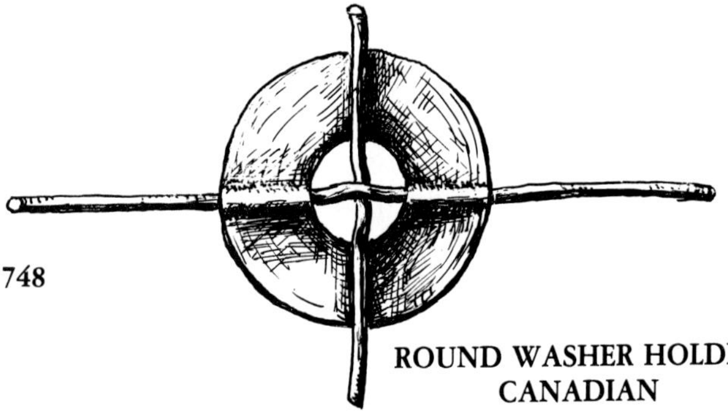

748

ROUND WASHER HOLDER
CANADIAN

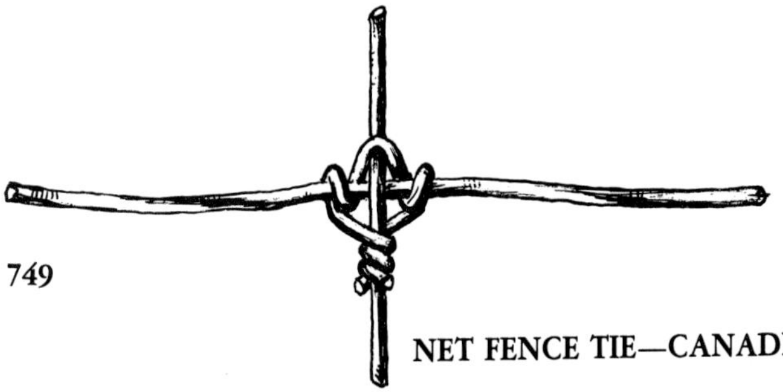

749

NET FENCE TIE—CANADIAN

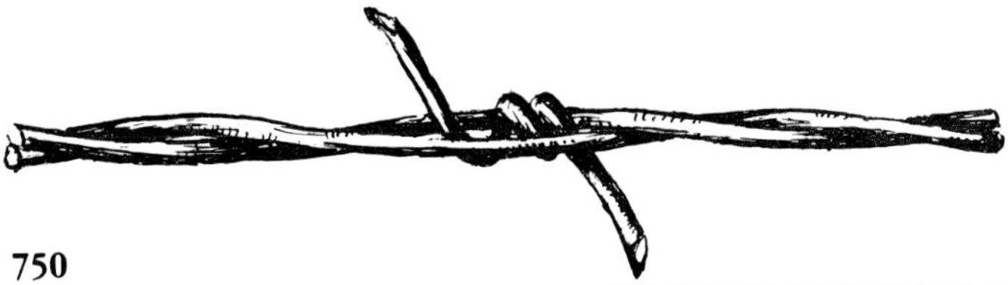

750

GLIDDEN LOOSE WRAP
PAT. J.F. GLIDDEN NOV. 24, 1874

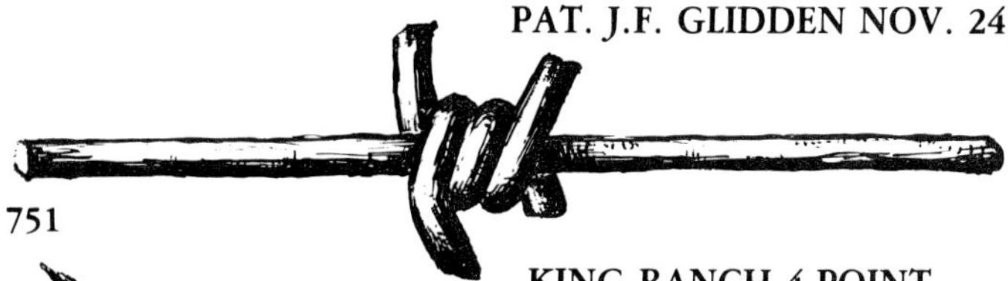

751

KING RANCH 4 POINT

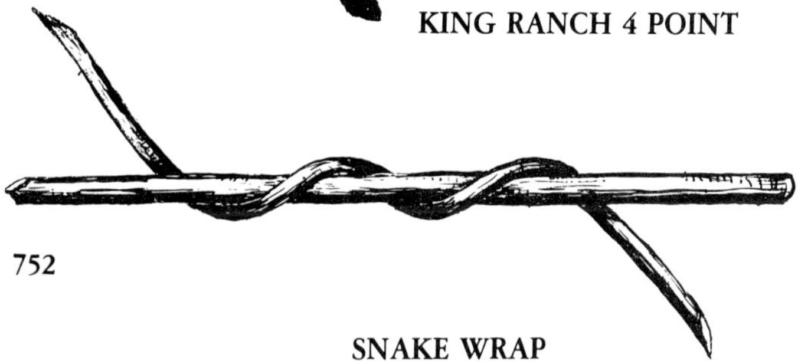

752

SNAKE WRAP

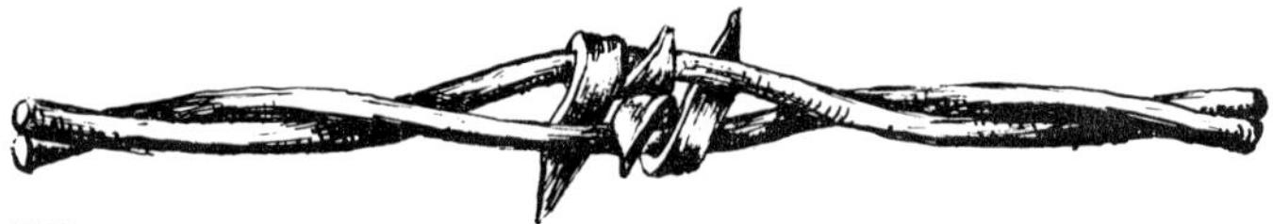

753

BARBS ON BOTH LINES
G.C. BAKER FEB. 27, 1883

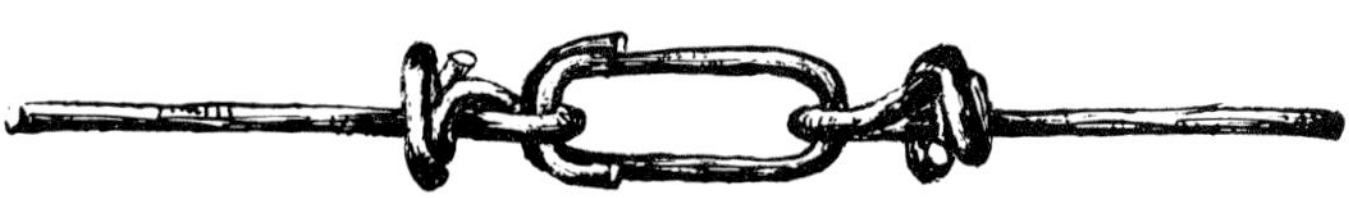

754

PLANTER TRIP WIRE

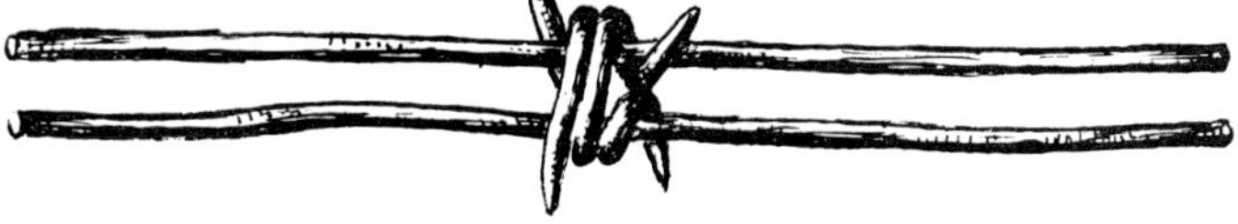

755

PARALLEL LINE
PAT. A.S. BURNELL JUNE 19, 1887

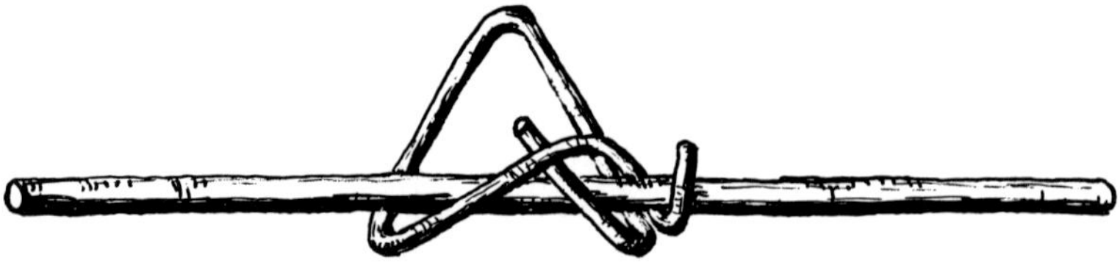

756

WATER GAP LOCK

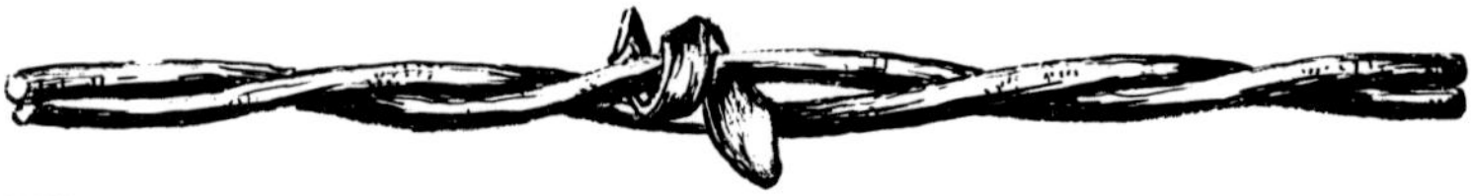

757

SPOON PAINT BARB PAT. G.C. BAKER
FEB. 27, 1883

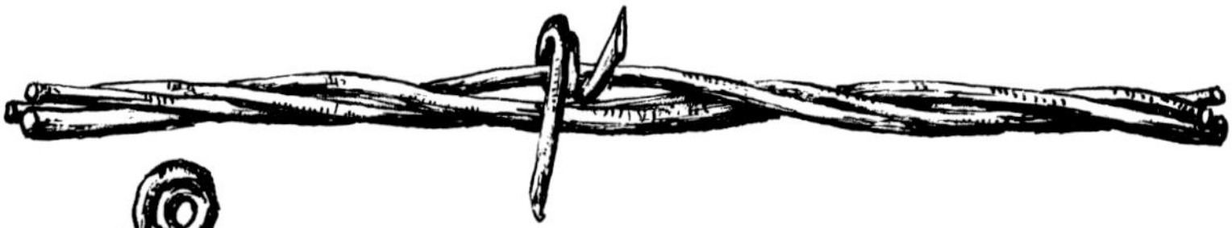

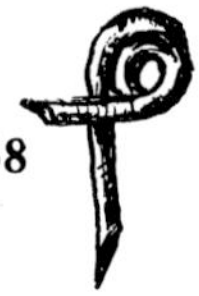

758 THREE LINE SPIRAL BARB PAT. J.F. GLIDDEN
NOV. 24, 1874

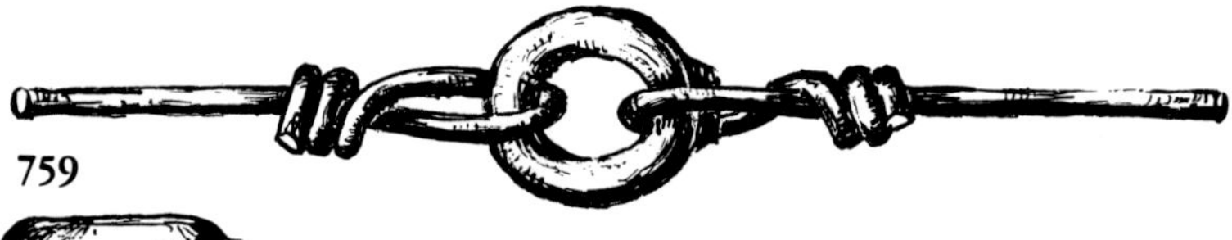

759

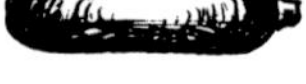

SOLID RING PLANTER WIRE

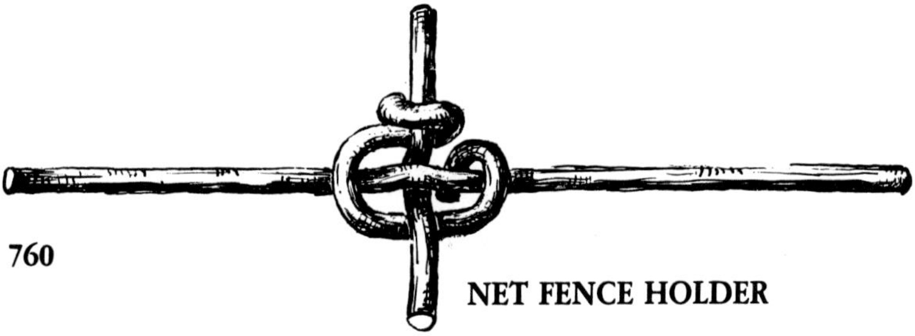

760

NET FENCE HOLDER

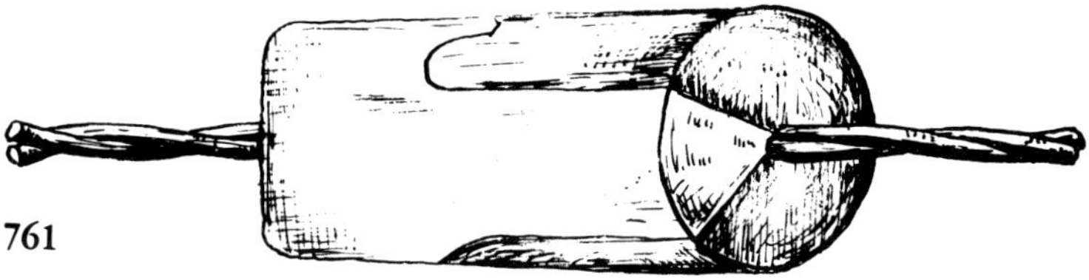

761

WEDGE FENCE SPLICER

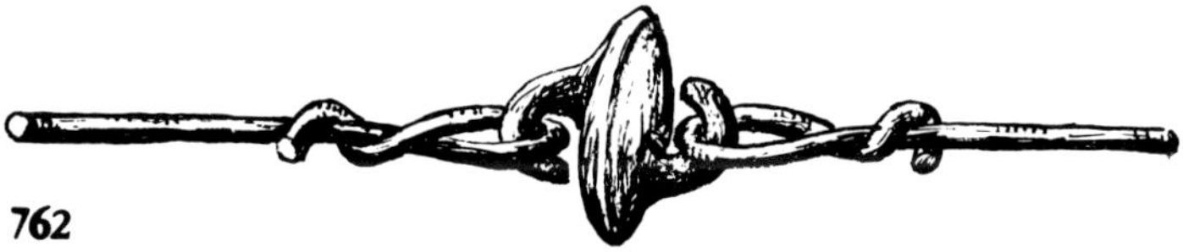

762

OFFSET PLANTER WIRE

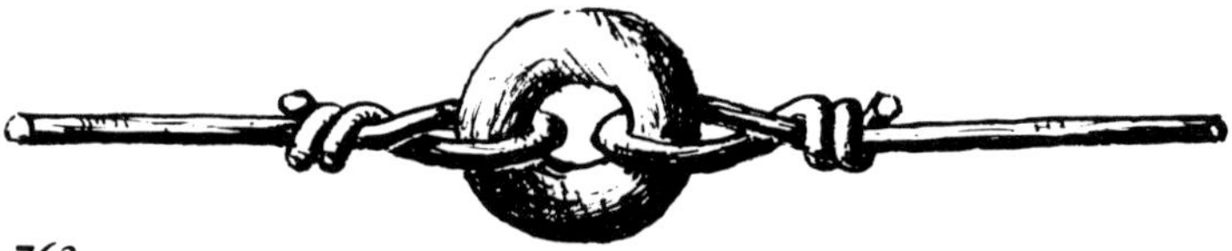

763

PLANTER RING—SMALL

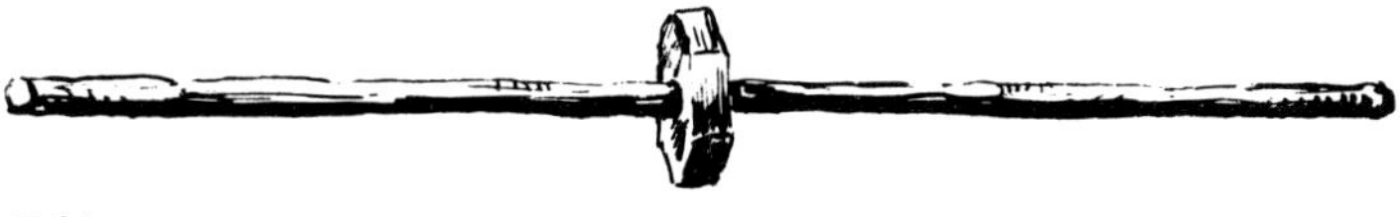

764

SQUARE PLANTER TRIP

765 SWISS DOUBLE TRIANGLE

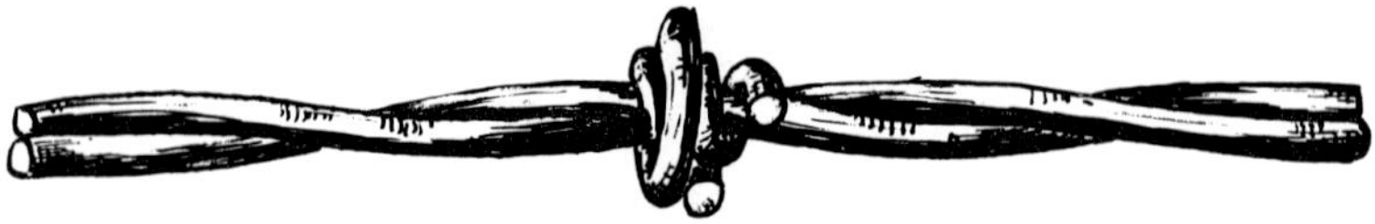

766 CANADIAN BLUNT CUT

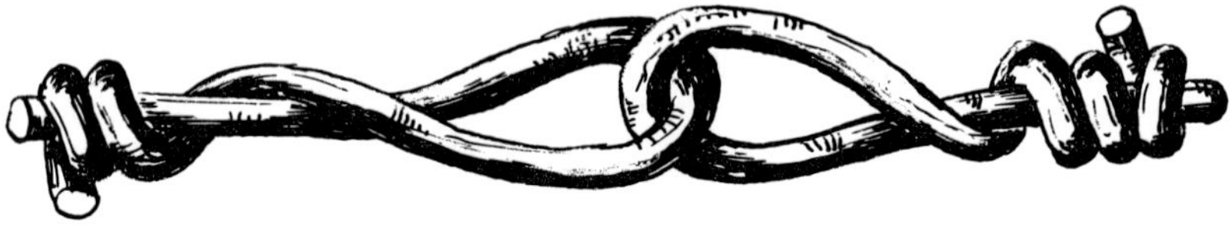

767 HEAVY SINGLE LINE SPLICE

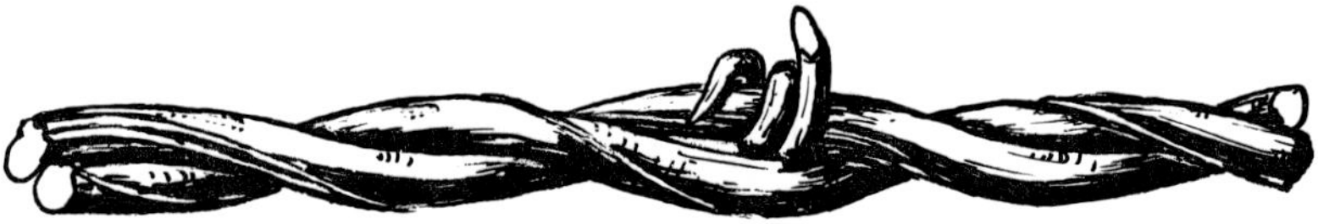

768 RAISED RIDGE WIRE
J.F. GLIDDEN NOV. 24, 1874

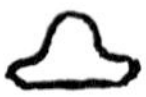

769 HEAVY HORSE PEN ROD WIRE

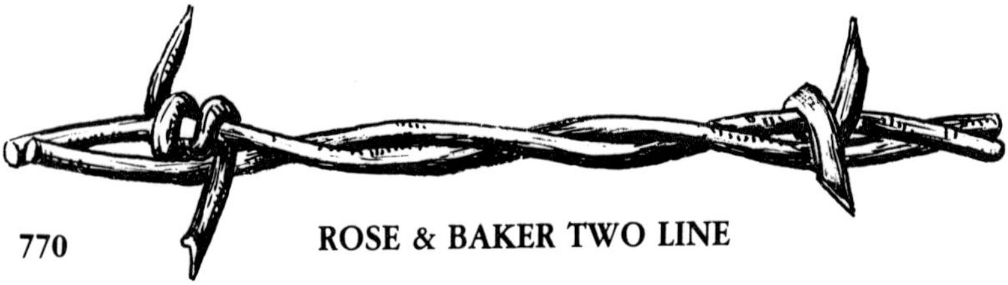

770 ROSE & BAKER TWO LINE

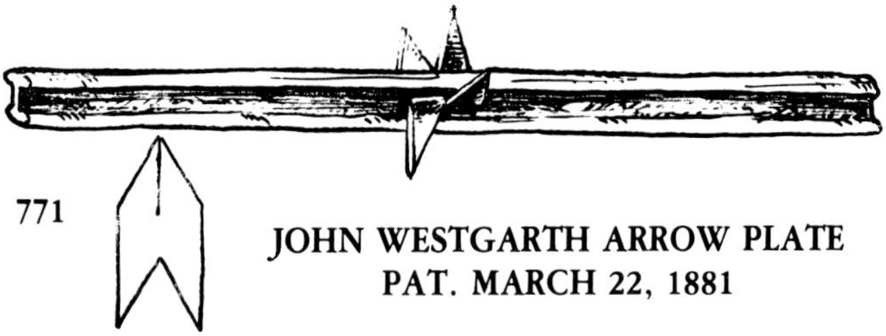

771 JOHN WESTGARTH ARROW PLATE
PAT. MARCH 22, 1881

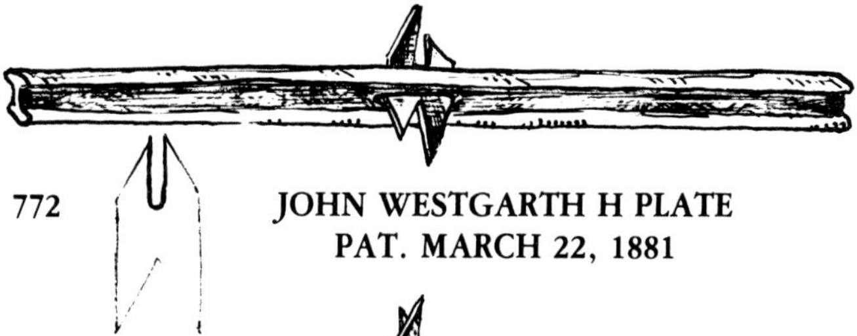

772 JOHN WESTGARTH H PLATE
PAT. MARCH 22, 1881

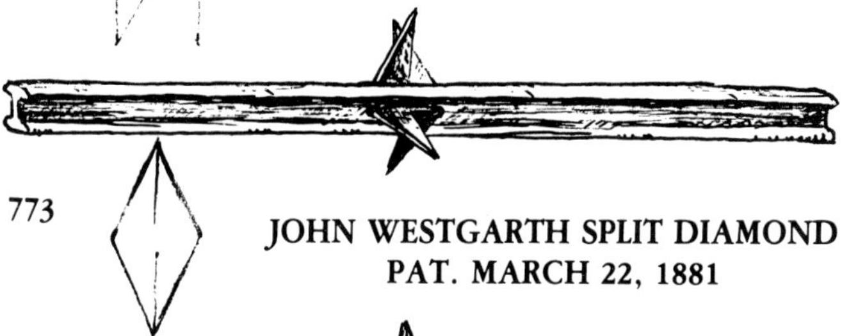

773 JOHN WESTGARTH SPLIT DIAMOND
PAT. MARCH 22, 1881

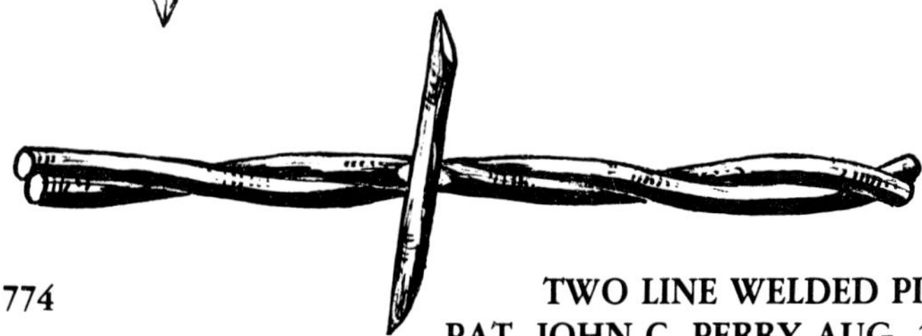

774 TWO LINE WELDED PIN
PAT. JOHN C. PERRY AUG. 24, 1897

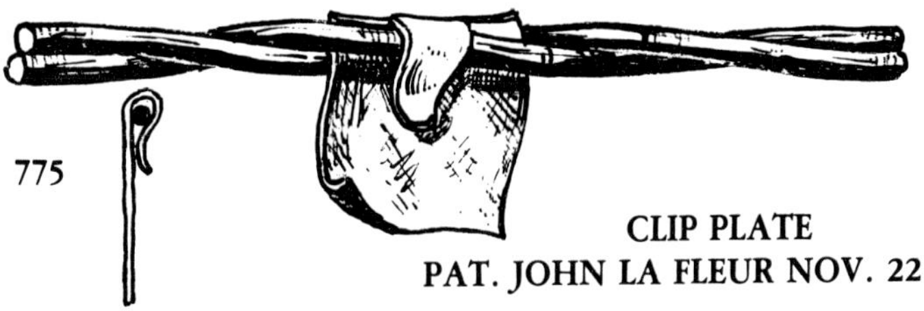

775 CLIP PLATE
PAT. JOHN LA FLEUR NOV. 22, 1881

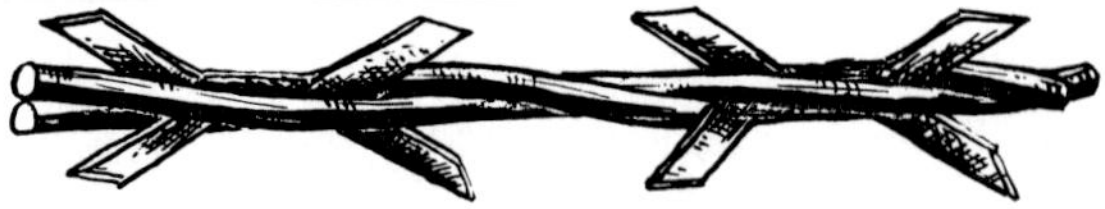

776

MILITARY WIRE FROM FRANCE

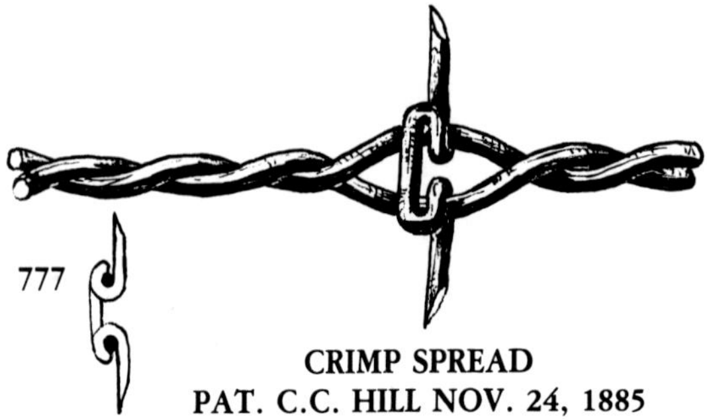

777

CRIMP SPREAD
PAT. C.C. HILL NOV. 24, 1885

778

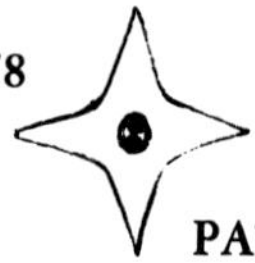

SHELLABERGER'S STAR
PAT. KIRK SHELLABERGER FEB. 13, 1883

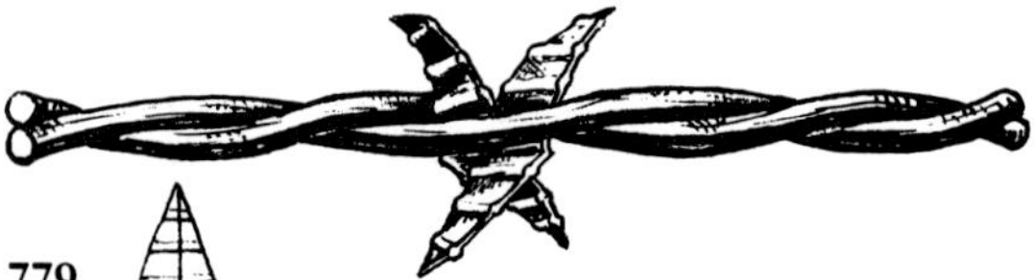

779

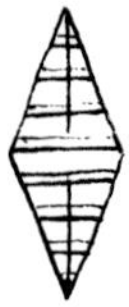

OLIVER SPLIT DIAMOND
PAT. J.B. OLIVER OCT. 2, 1883

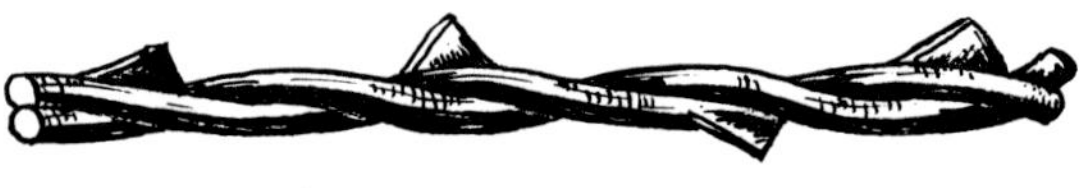

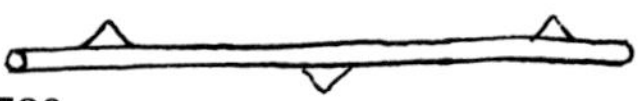

780

CARPENTERS UNDULATOR
PAT. JAMES CARPENTER JUNE 6, 1882

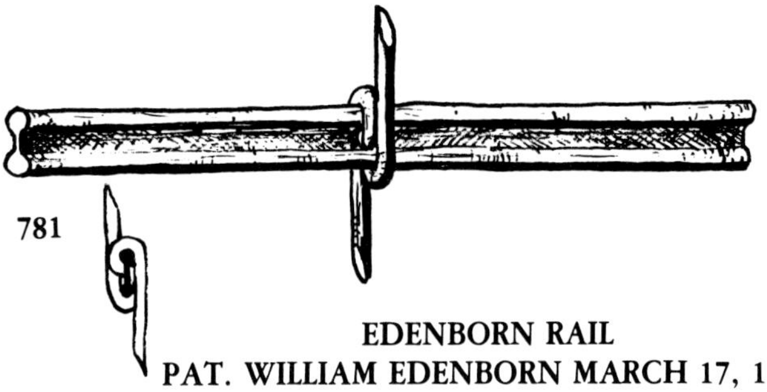

781

EDENBORN RAIL
PAT. WILLIAM EDENBORN MARCH 17, 1885

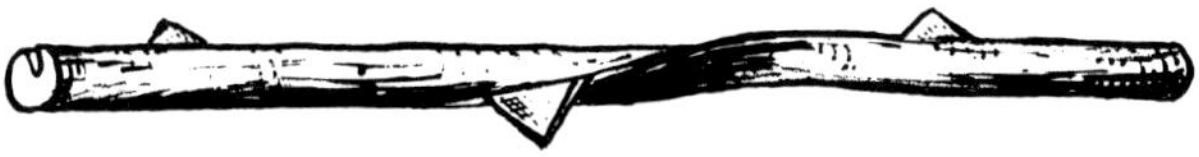

782

UPHAM GROOVED RAIL
PAT. ANDREW UPHAM APR. 5, 1881

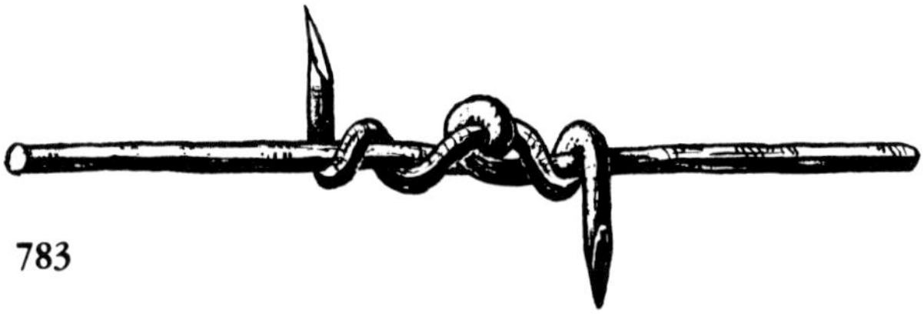

783

LOOP & TWIST
PAT. CHARLES LOCHES MAY 8, 1883

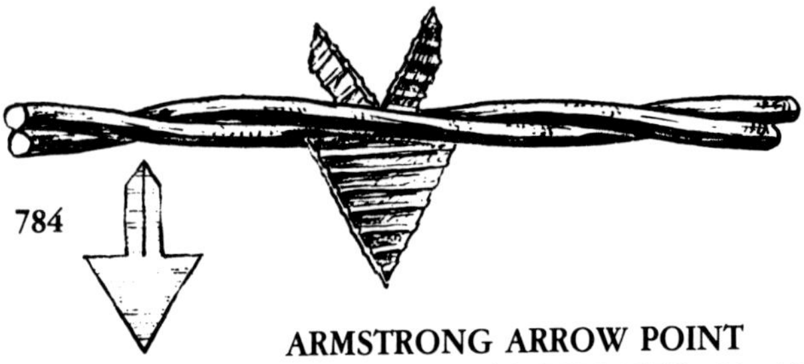

784

ARMSTRONG ARROW POINT
PAT. FRANK ARMSTRONG DEC. 31, 1875

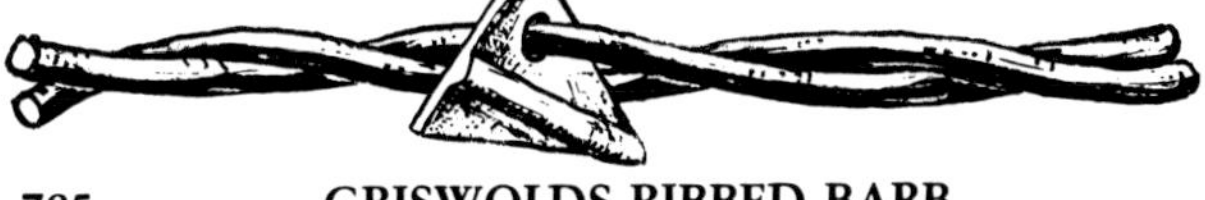

785

GRISWOLDS RIBBED BARB
PAT. J.W. GRISWOLD JAN. 31, 1888

786

NEUBAUERS STAR
PAT. CARL E. NEUBAUER NOV. 28, 1905

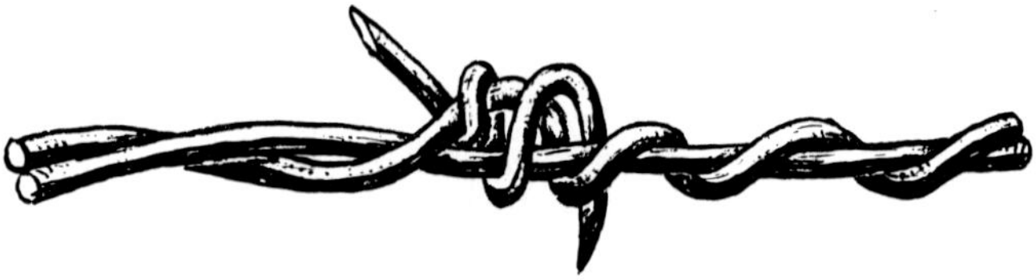

787

MOONSHINE WIRE

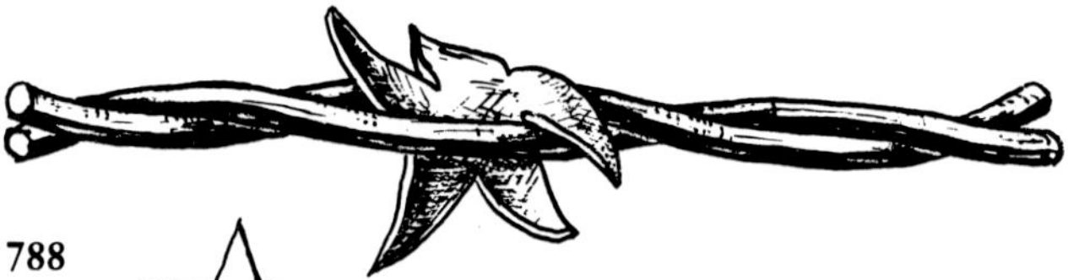

788

STOVER STAR
PAT. D.C. STOVER NOV. 22, 1881

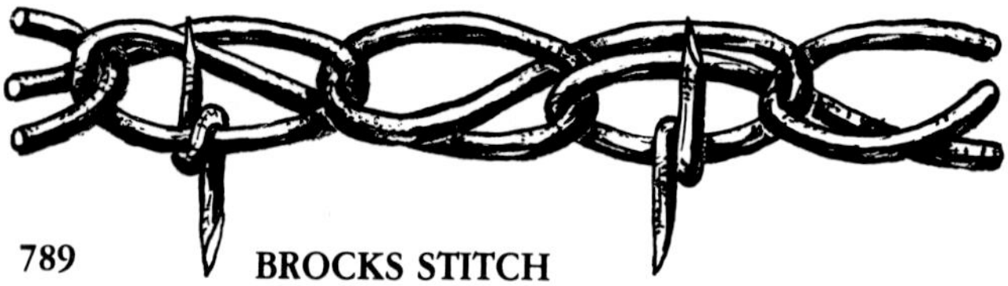

789

BROCKS STITCH
PAT. W.E. BROCK FEB. 12, 1884

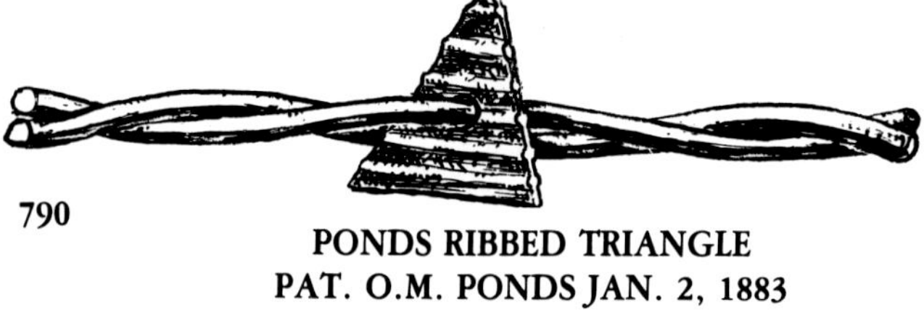

790

PONDS RIBBED TRIANGLE
PAT. O.M. PONDS JAN. 2, 1883

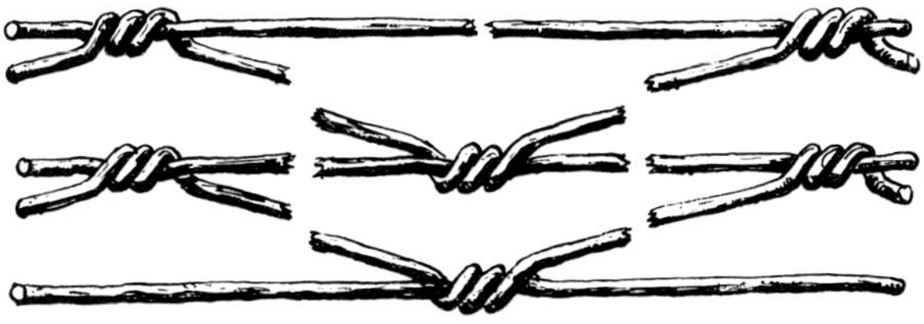

791 MILES BARBLES PARALLEL
PAT. PURCHES MILES MAY 22, 1883

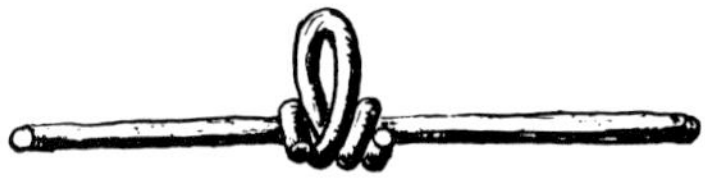

792 McNELLYS SINGLE LOOP
PAT. M.F. McNELLY MAY 14, 1895

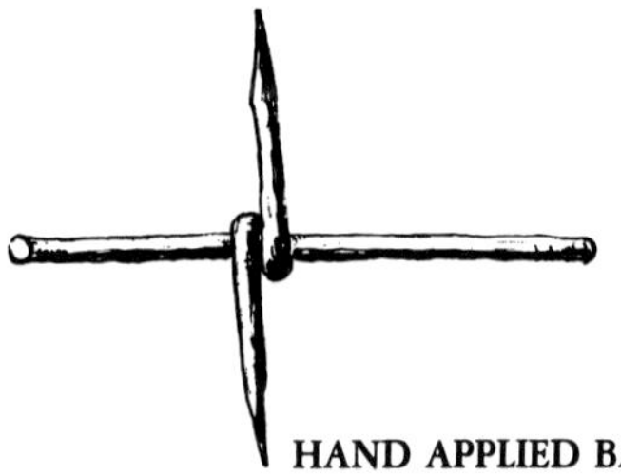

793

HAND APPLIED BARB

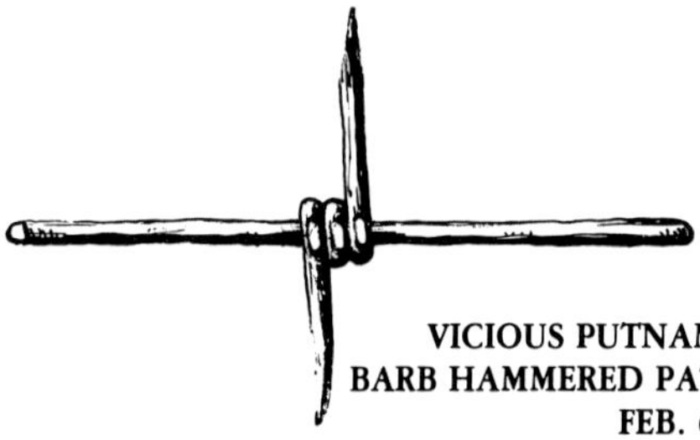

794 VICIOUS PUTNAM, BOTH SIDES OF
BARB HAMMERED PAT. HENRY W. PUTNAM
FEB. 6, 1877

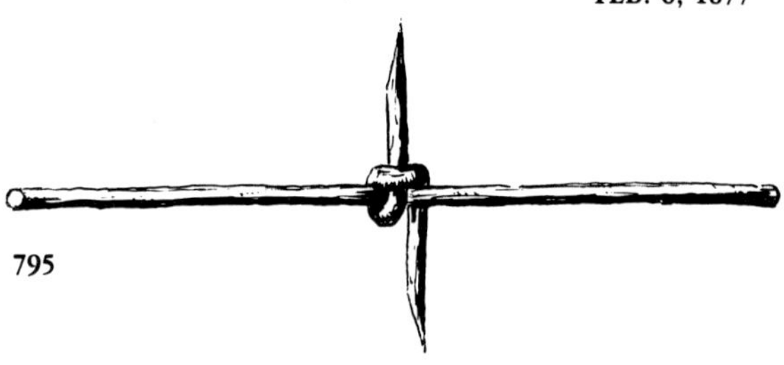

795

KITTLESONS SINGLE LINE
HALF HITCH PAT. O.O. KITTLESON APRIL 3, 1887

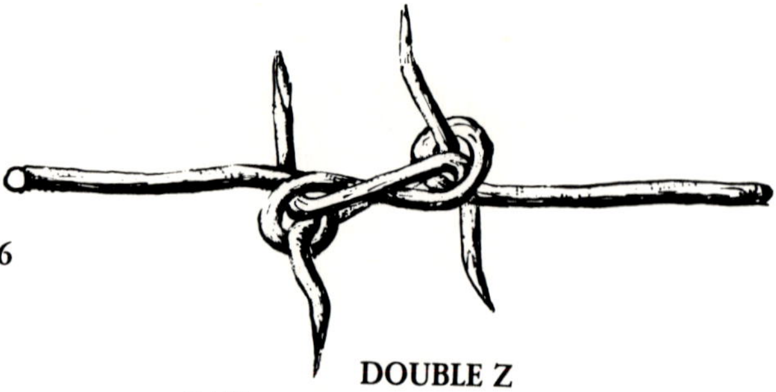

796

DOUBLE Z
PAT. A.M. MUNSON APR. 15, 1879

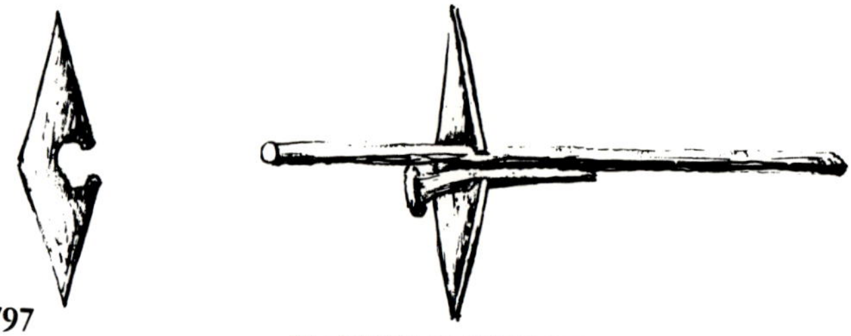

797

HARSHA'S WEDGE
PAT. M.S. HARSHA OCT. 3, 1876

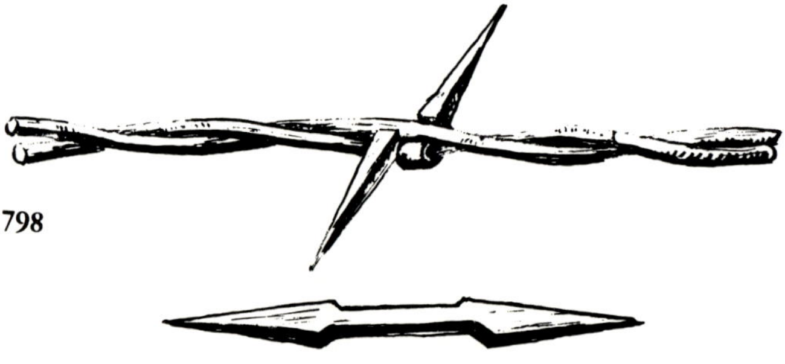

798

HAISH'S SPEAR
PAT. JACOB HAISH NOV. 24, 1891

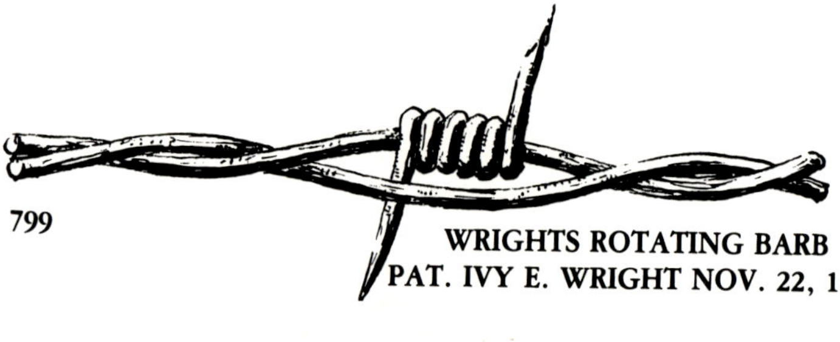

799

WRIGHTS ROTATING BARB
PAT. IVY E. WRIGHT NOV. 22, 1881

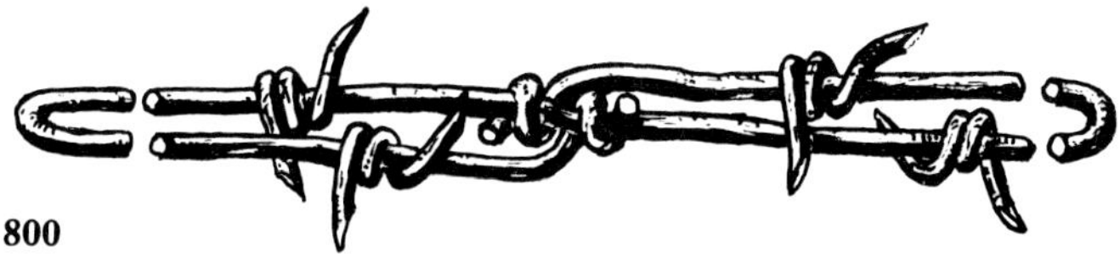

800

GLIDDEN LINK WIRE
PAT J.F. GLIDDEN NOV. 24, 1874

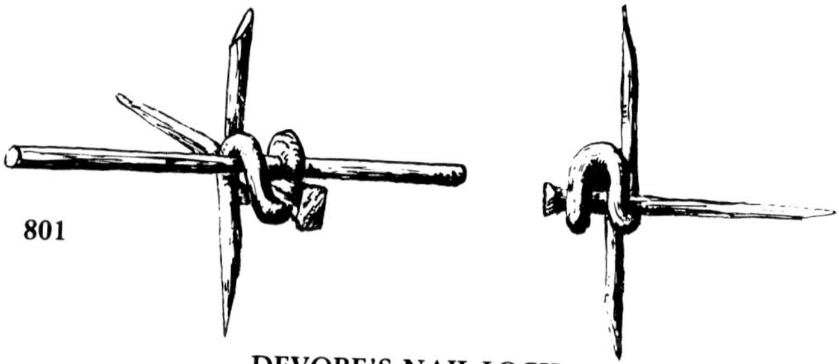

801

DEVORE'S NAIL LOCK
PAT. L.M. DEVORE OCT. 19, 1875

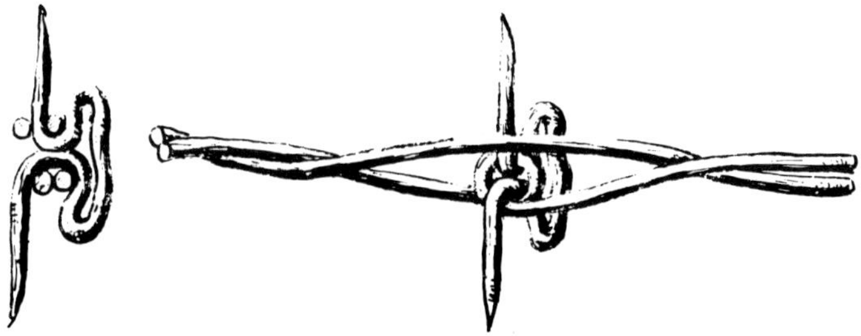

802

PITNEYS LOOP LINE
PAT. A.L. PITNEY OCT. 1, 1878

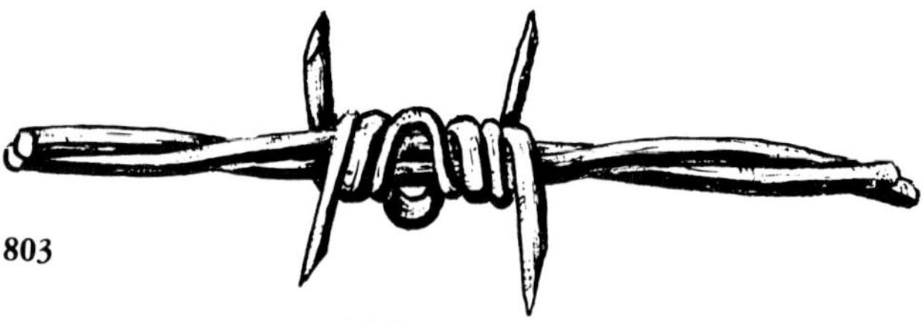

803

MERRILLS HALFROUND
HALF OF BARB IS HALF ROUND
PAT. J.C. MERRILL SEPT. 29, 1874

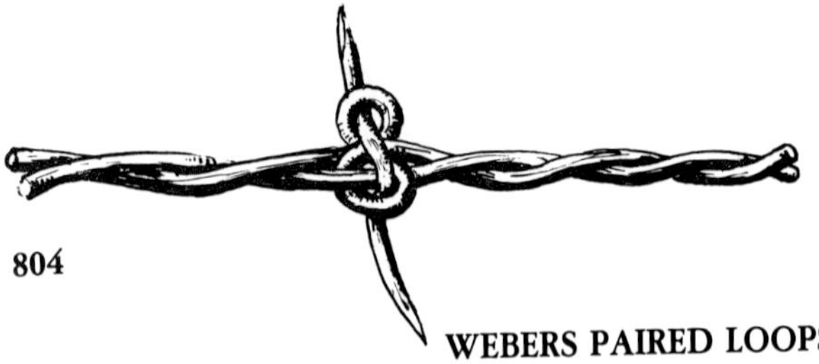

804

WEBERS PAIRED LOOPS
PAT. T.A. WEBER OCT. 9, 1883

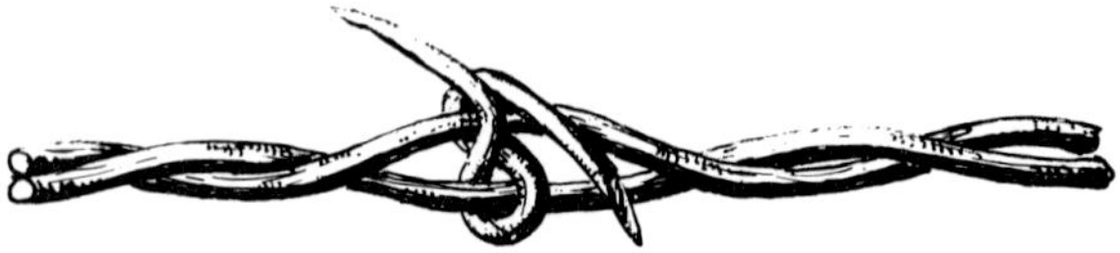

805

NADELHOFFER'S WING
PAT. JOHN NADELHOFFER NOV. 4, 1884

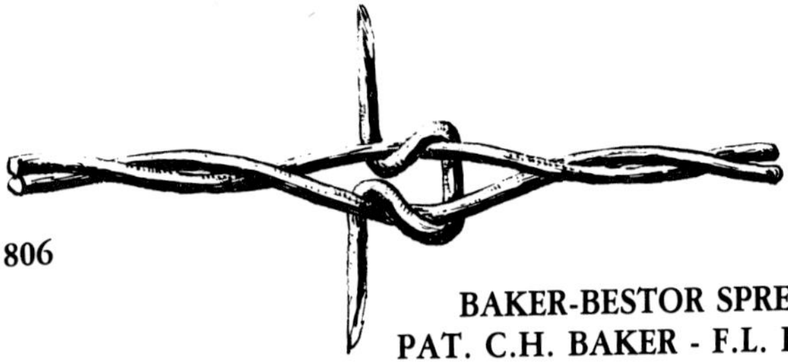

806

BAKER-BESTOR SPREAD
PAT. C.H. BAKER - F.L. BESTOR
SEPT. 17, 1878

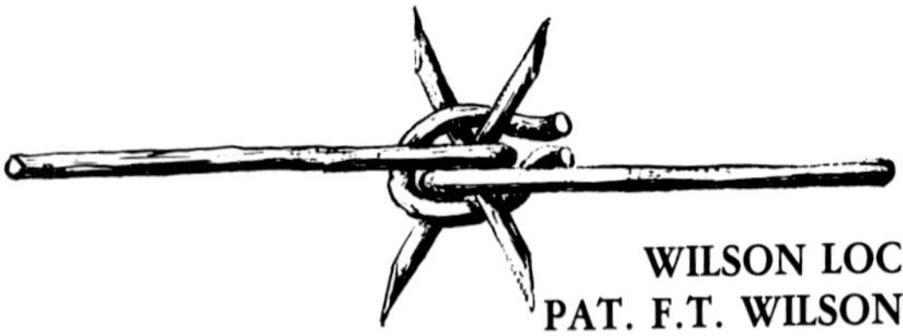

807

WILSON LOCK BARB
PAT. F.T. WILSON JAN. 5, 1875

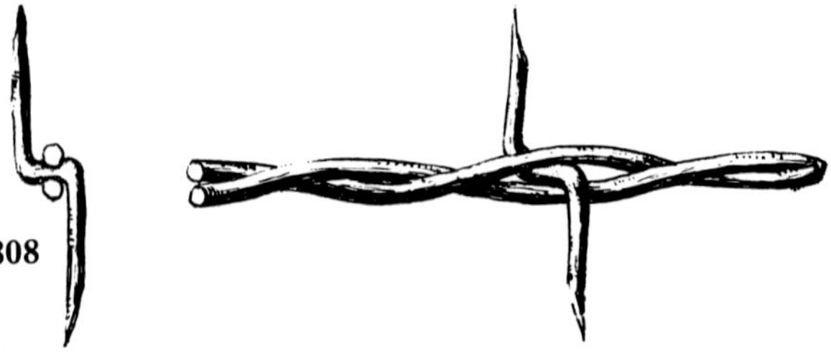

808

BEERBOWERS FLAT CENTER
PAT. G.M. BEERBOWER OCT. 6, 1885

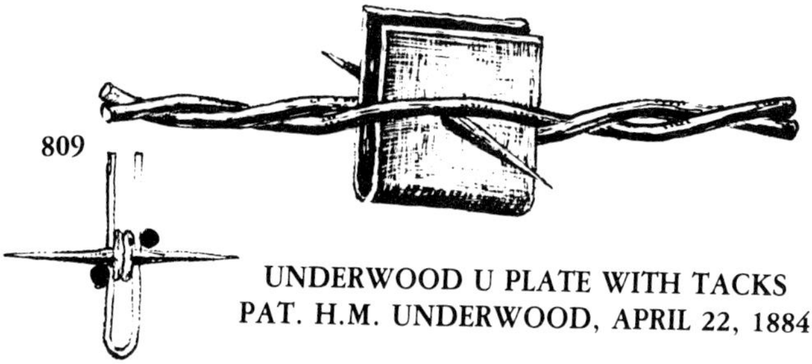

809

UNDERWOOD U PLATE WITH TACKS
PAT. H.M. UNDERWOOD, APRIL 22, 1884

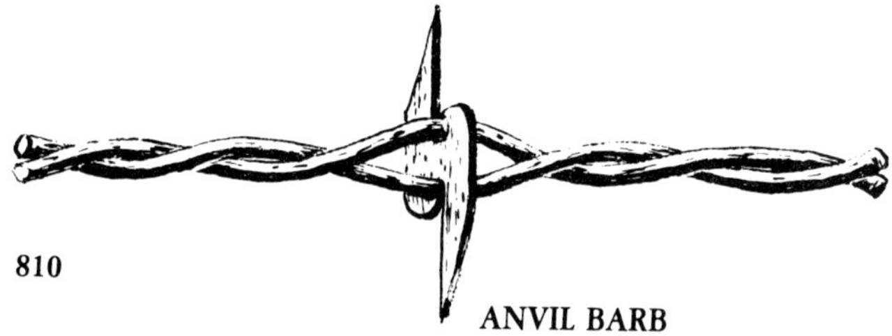

810

ANVIL BARB
PAT. JACOB HAISH DEC. 15, 1885

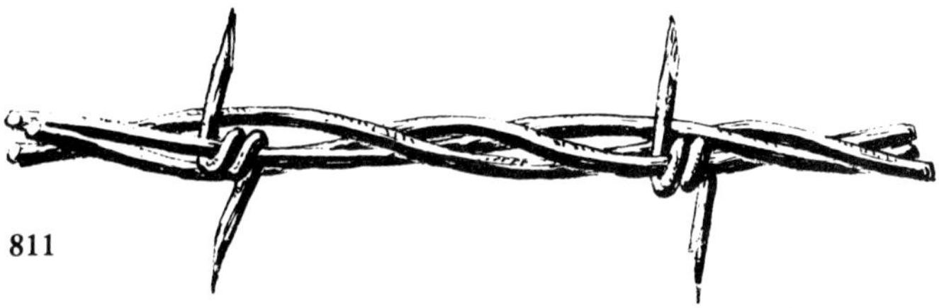

811

GLIDDEN 3 STRAND, BARBS ON
2 LINES PAT. J.F. GLIDDEN NOV. 24, 1874

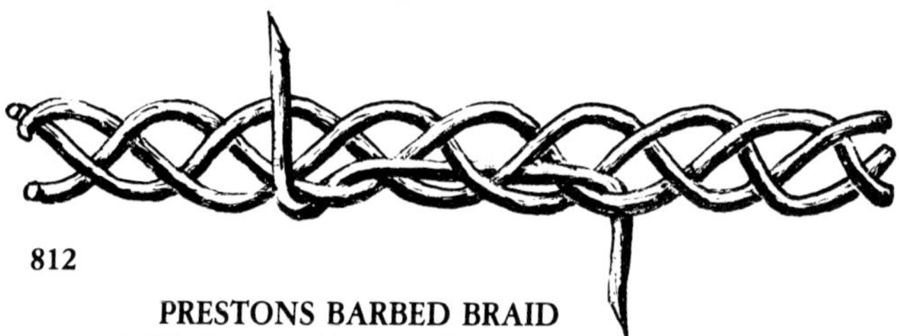

812

PRESTONS BARBED BRAID
PAT. O. PRESTON OCT. 18, 1881

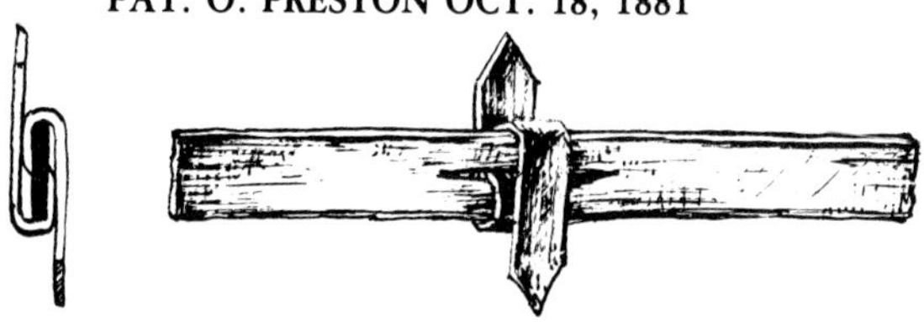

813

SLIT RIBBON WITH BARB
PAT. A.J. UPHAM SEPT. 16, 1884

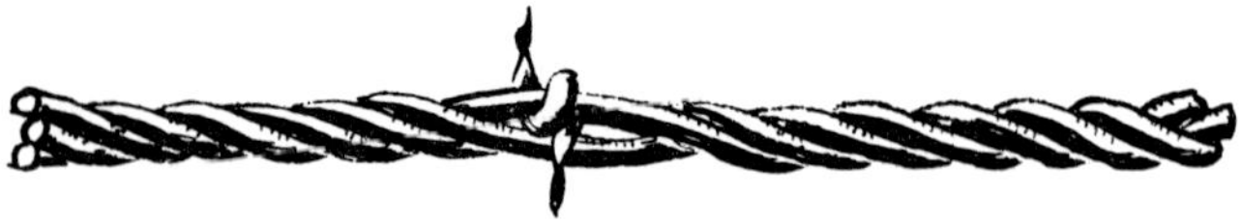

814 CURTIS FAST TWIST THREE LINE
PAT. J.D. CURTIS MARCH 15, 1892

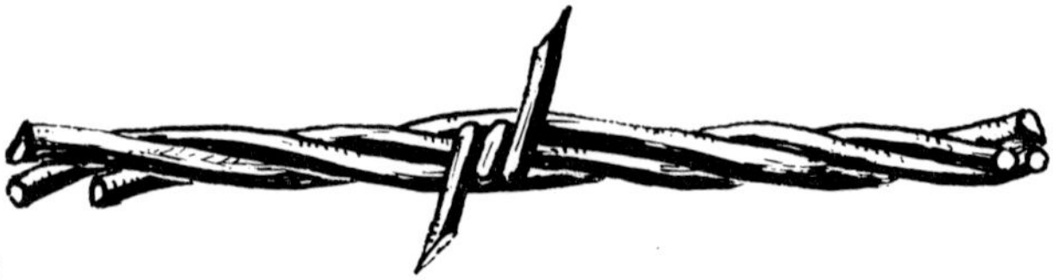

815

GLIDDEN TEAR DROP 3 LINE
PAT. J.F. GLIDDEN NOV. 24, 1874

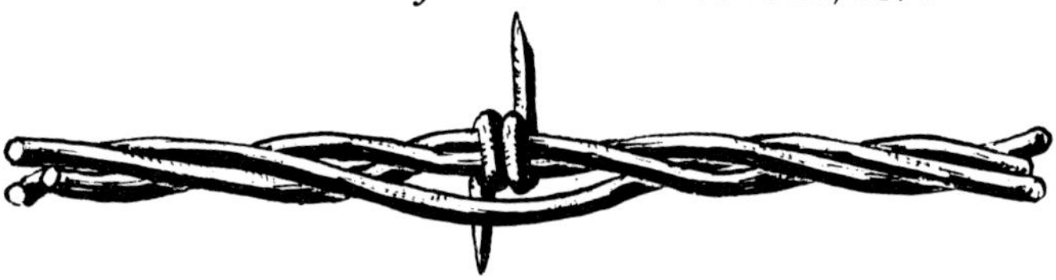

816 GLIDDEN 3 LINE BARBAROUND 2
PAT. J.F. GLIDDEN NOV. 24, 1874

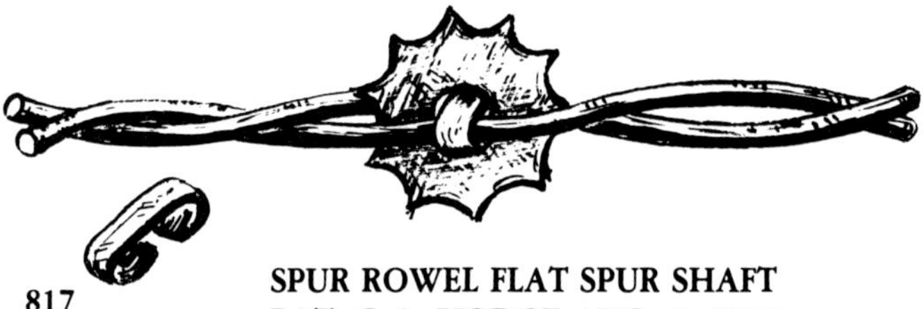

817

SPUR ROWEL FLAT SPUR SHAFT
PAT. C.A. HODGE AUG. 2, 1887

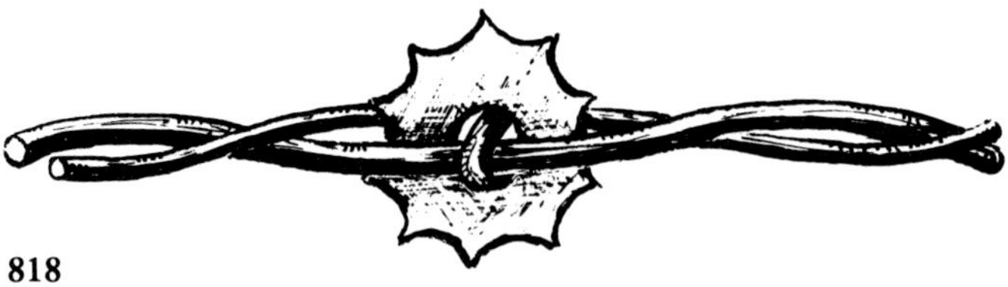

818

SPUR ROWEL LARGE & SMALL LINES
PAT. C.A. HODGE AUG. 2, 1887

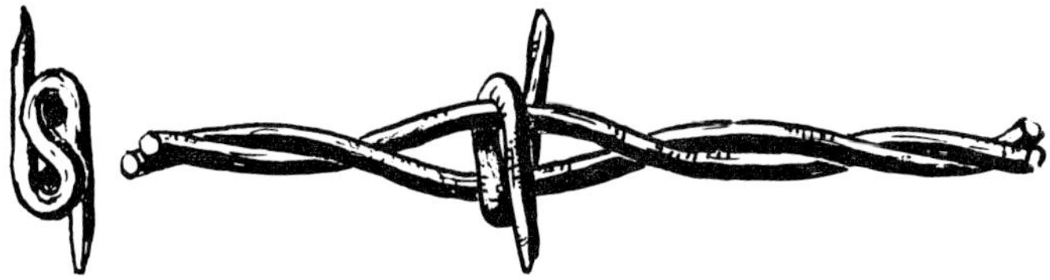

819 VARIATION HAISH'S ORIG. "S"
PAT. JACOB HAISH AUG. 31, 1875

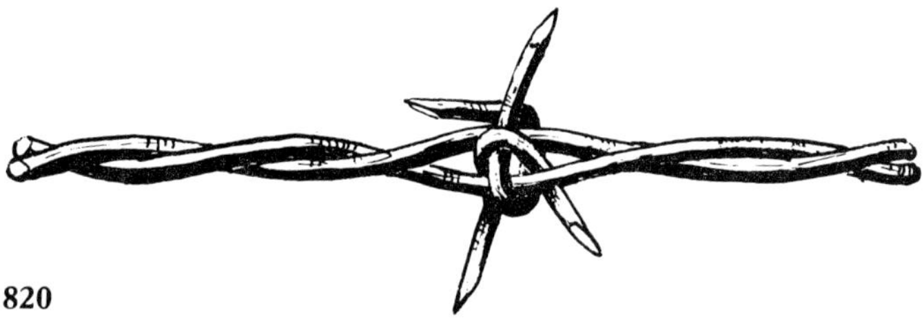

820 AUSTRALIAN BENT STAPLE
BARB SM. GA.

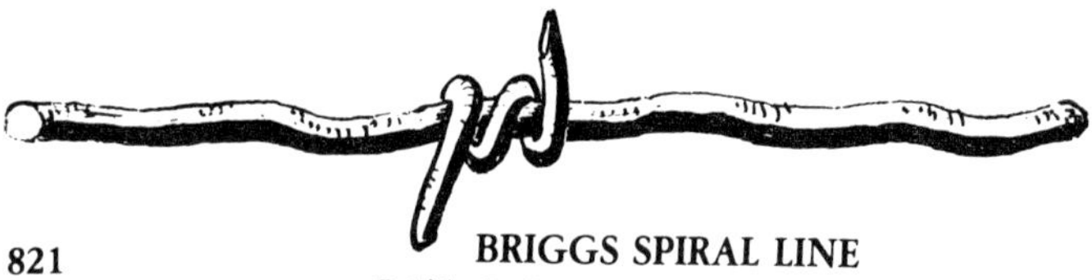

821 BRIGGS SPIRAL LINE
PAT. O.P. BRIGGS JULY 1, 1884

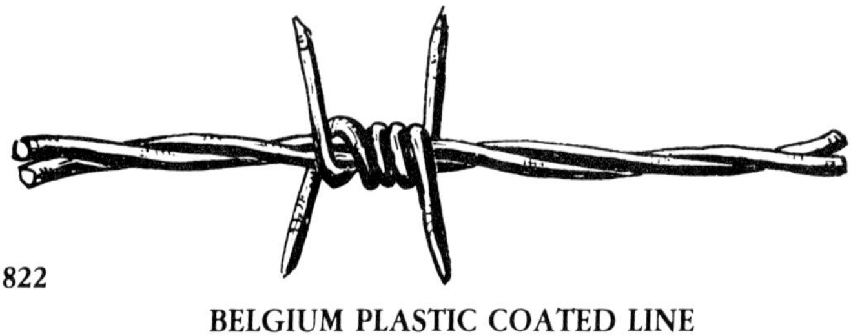

822 BELGIUM PLASTIC COATED LINE
WIRE ALUMINUM BARBS

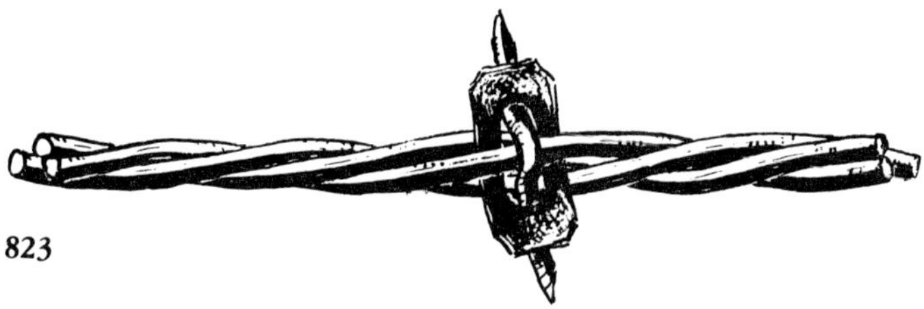

823 THREE LINE VOSHBURGH
PAT. C.A. VOSBURGH OCT. 3, 1876

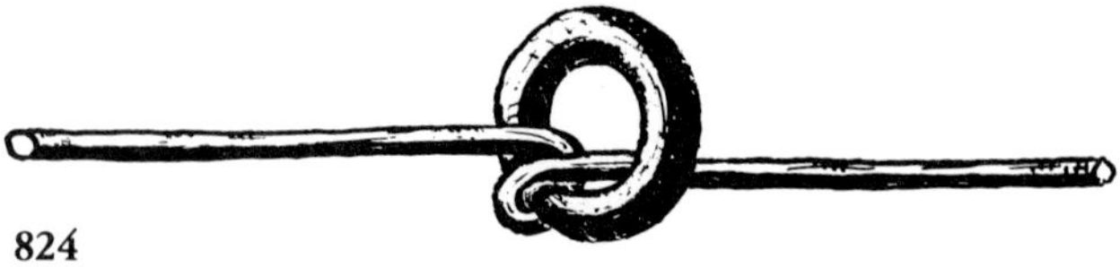

824

PLANTER LOOP
PAT. CAVINESS & McCORMICK JAN. 30, 1883

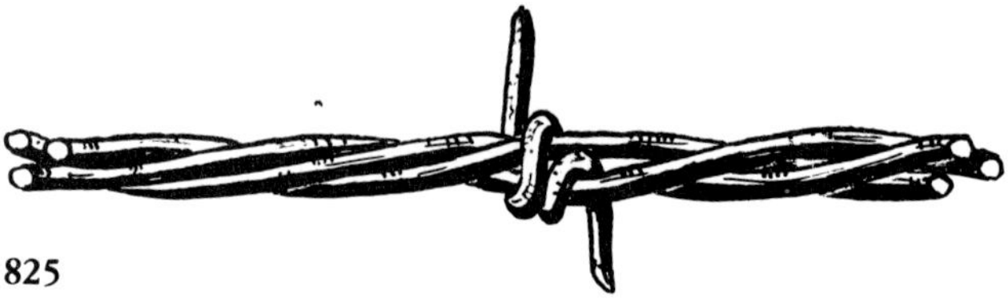

825

BROTHERTON BARB AROUND 3 LINES
PAT. J. BROTHERTON SEPT. 3, 1878

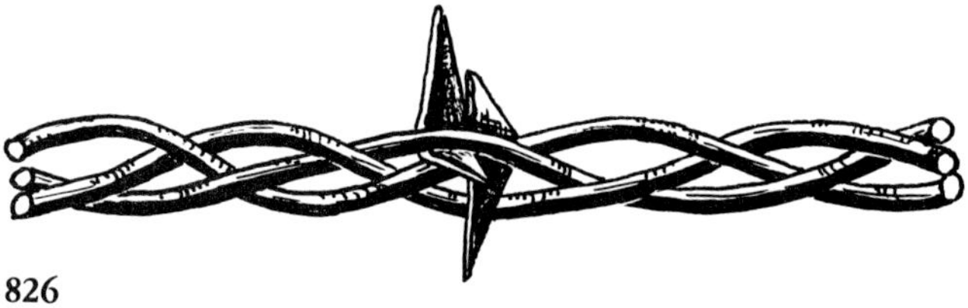

826

PRESTON BRAID WITH ARROWPLATE
O. PRESTON OCT. 18, 1881

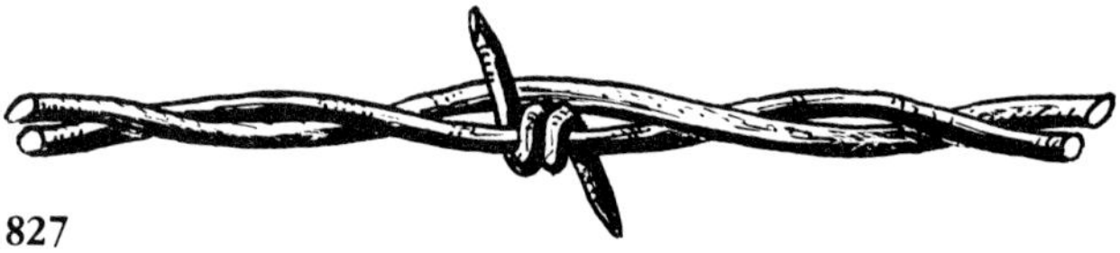

827

GLIDDEN SMALL LINE ONE OVAL
PAT. J.F. GLIDDEN NOV. 24, 1874

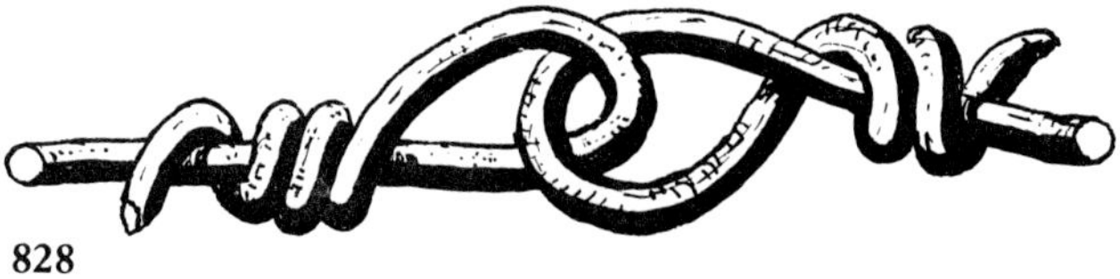

828

CRANDALS GATE LINK
PAT. E.M. CRANDAL MAR. 14, 1875

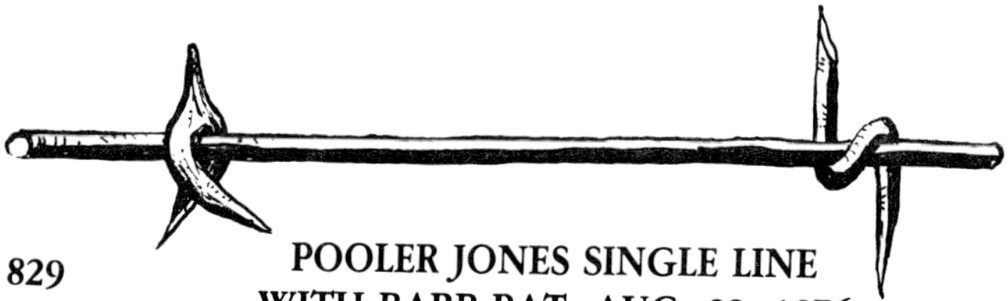

829 POOLER JONES SINGLE LINE
WITH BARB PAT. AUG. 29, 1876

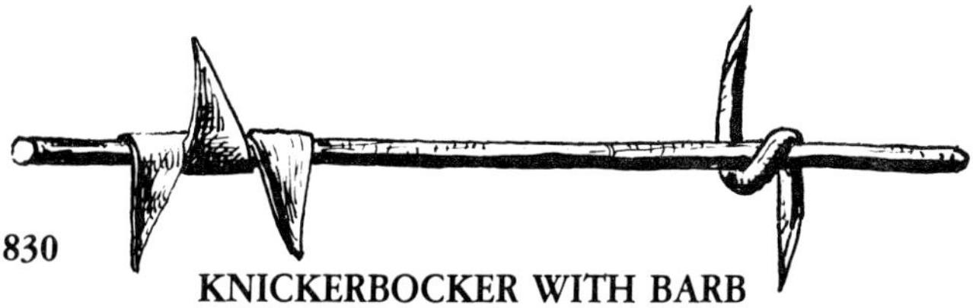

830 KNICKERBOCKER WITH BARB
PAT. DEC. 12, 1876

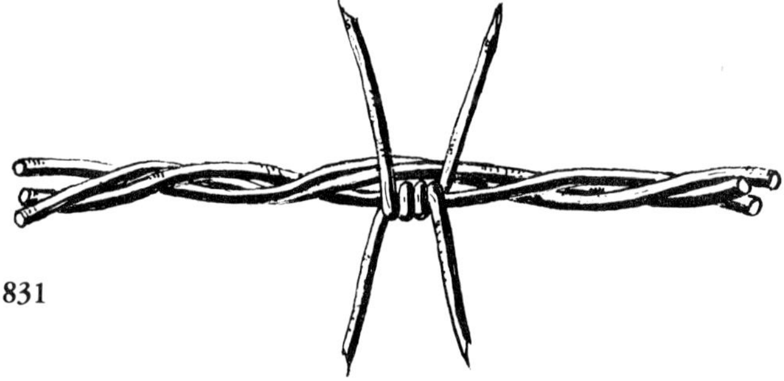

831 THREE STRAND ITALIAN
MILITARY WIRE W.W. II

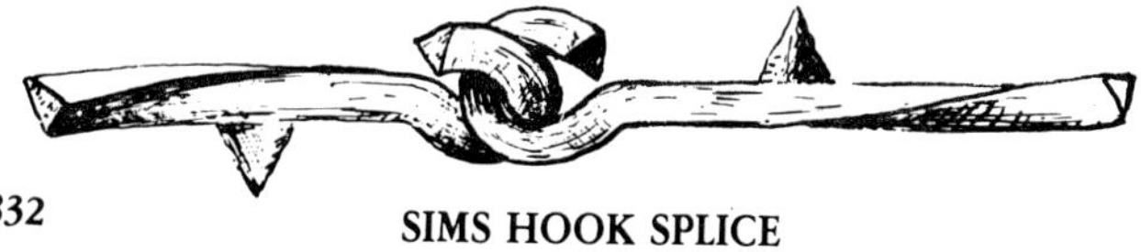

832 SIMS HOOK SPLICE
PAT. E. SIMS MAY 30, 1876

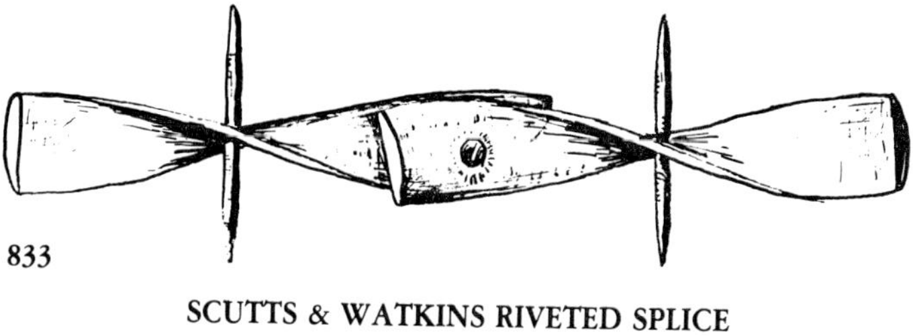

833 SCUTTS & WATKINS RIVETED SPLICE
PAT. JUNE 1, 1875

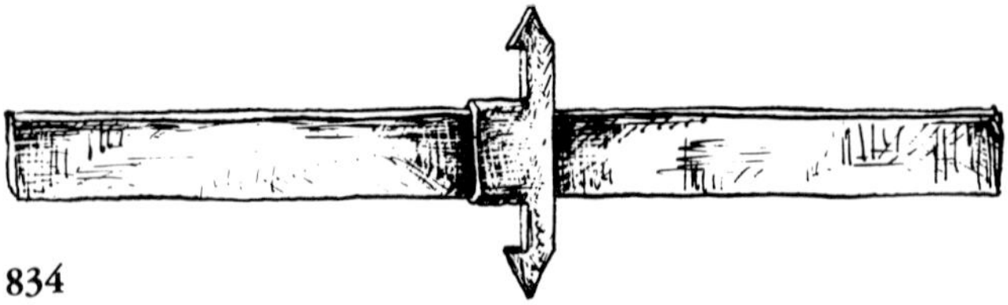

834

BRINKERHOFFS DOUBLE HOOK
PAT. J. BRINKERHOFF APR. 8, 1879

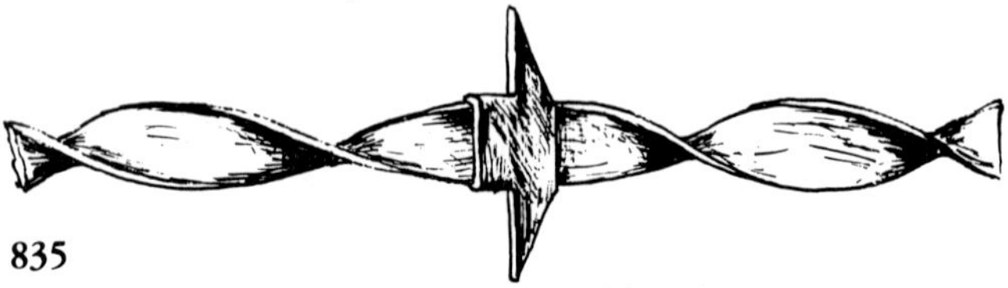

835

BRINKERHOFFS TWISTED LINE
PAT. J. BRINKERHOFF APR. 8, 1879

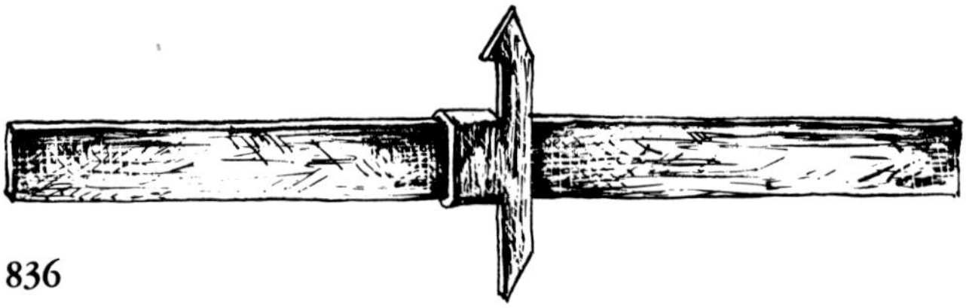

836

BRINKERHOFFS SINGLE HOOK
PAT. J. BRINKERHOFF APR. 8, 1879

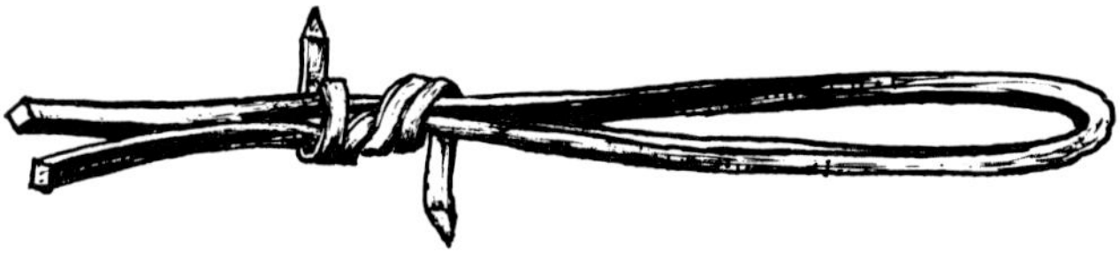

837 CRANDALS SQUARE LINE LINK
PAT. MAY 24, 1881

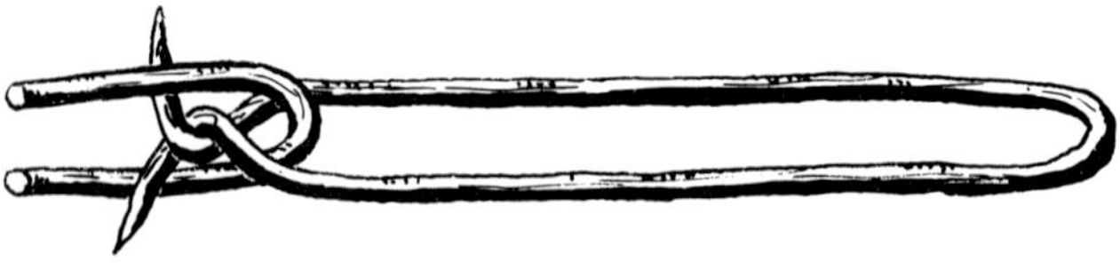

838 CRANDALS "T" LINK
PAT. E.M. CRANDAL NOV. 28, 1876

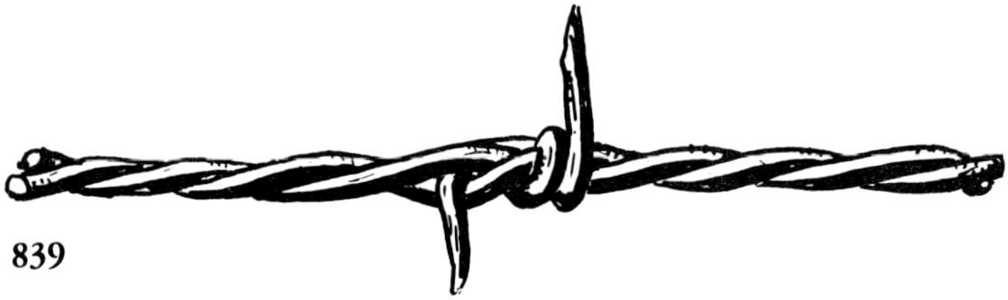

839

BROTHERTON 2 PT. ARCHED BARB
PAT. J. BROTHERTON SEPT. 3, 1878

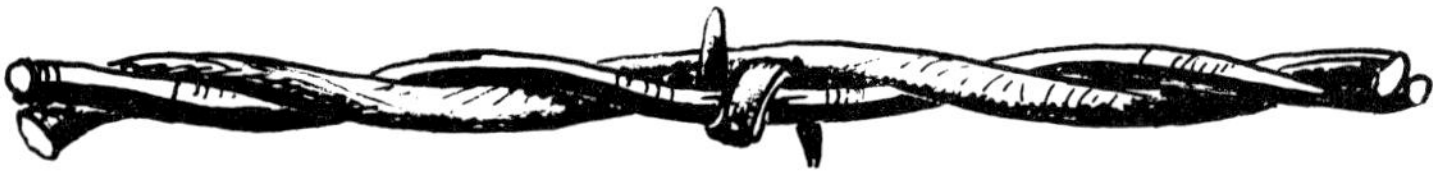

840

BAKER DRAWN EXTRA LARGE OVAL LINE
PAT. FEB. 27, 1883

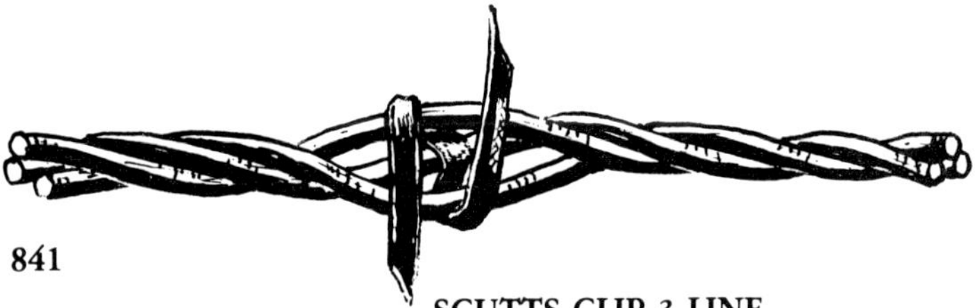

841

SCUTTS CLIP 3 LINE
PAT. H.B. SCUTT DEC. 22, 1885

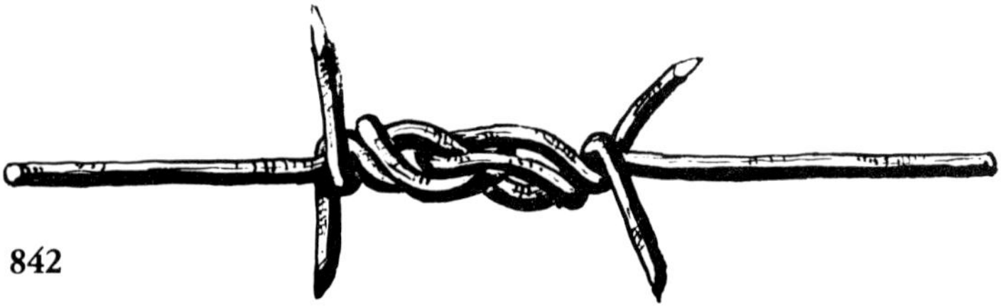

842

ABBOTTS ALL ALUMINUM
PAT. JAN. 19, 1943

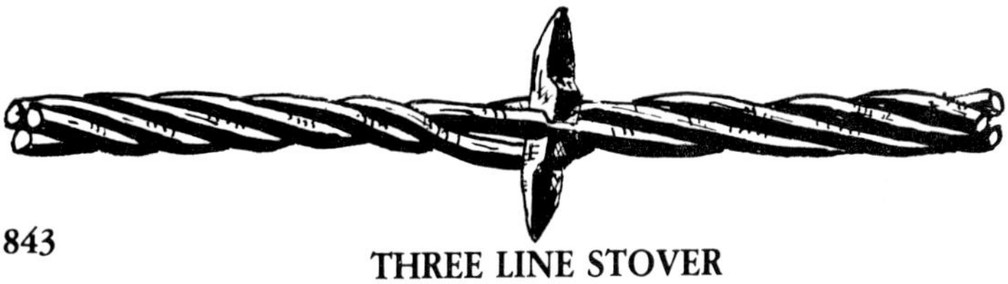

843

THREE LINE STOVER
PAT. D.C. STOVER JUNE 29, 1875

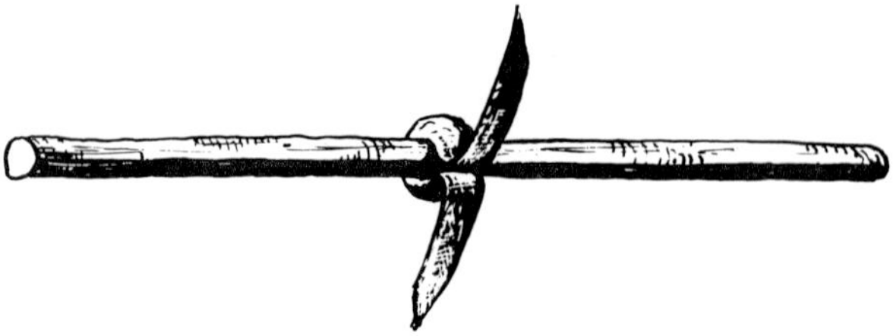

844

HAISH STICK TITE HALF ROUND BARB
AUSTRALIAN

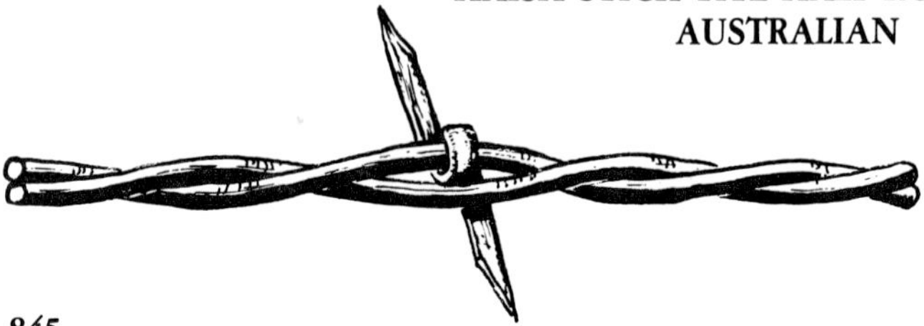

845

GERMAN MADE BAKER LARGE BARB

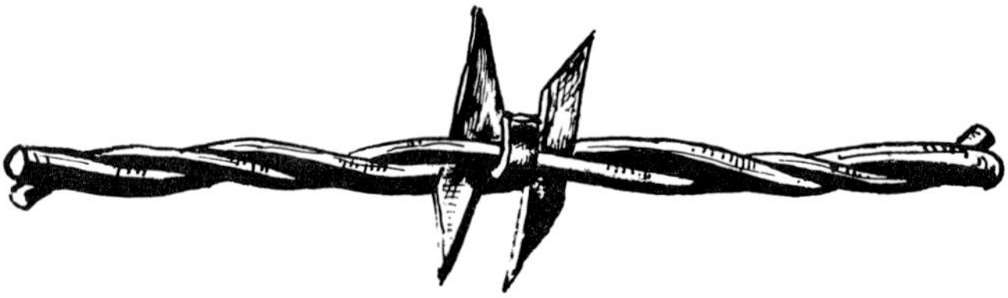

846

SCUTTS EARLY CLIP TWO SIZE
LINE WIRES PAT. JUNE 18, 1878

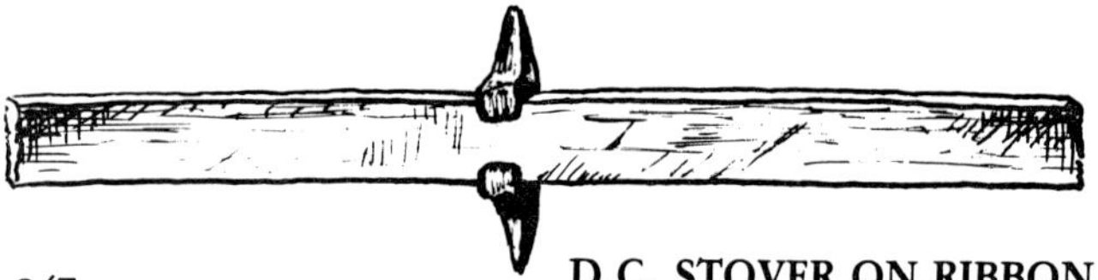

847

D.C. STOVER ON RIBBON LINE
PAT. JUNE 29, 1875

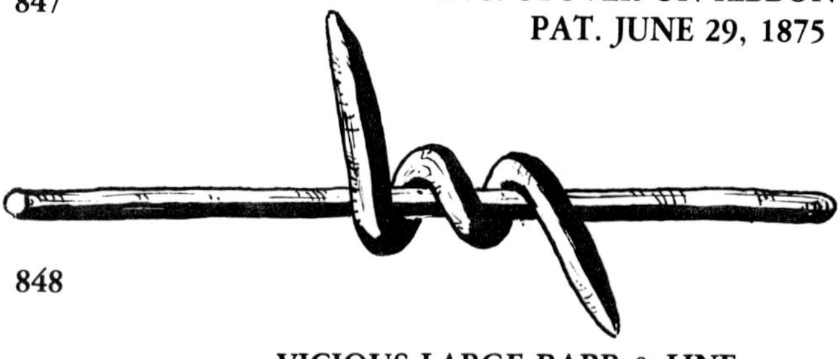

848

VICIOUS LARGE BARB & LINE

849

WOODEN PICKET WIRE COMES
IN FOUR GA. LINE WIRES

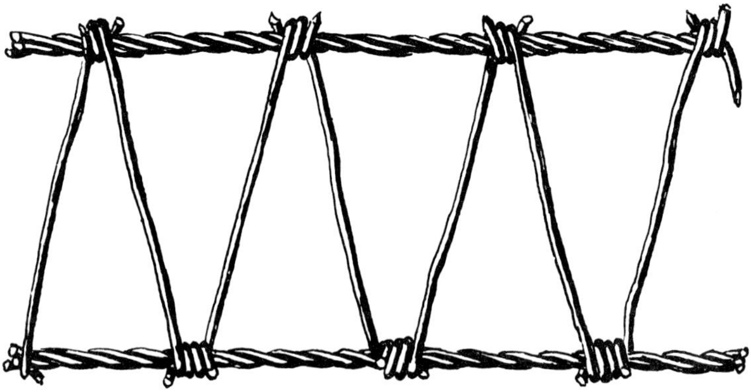

850

TOP RUNNER MESH

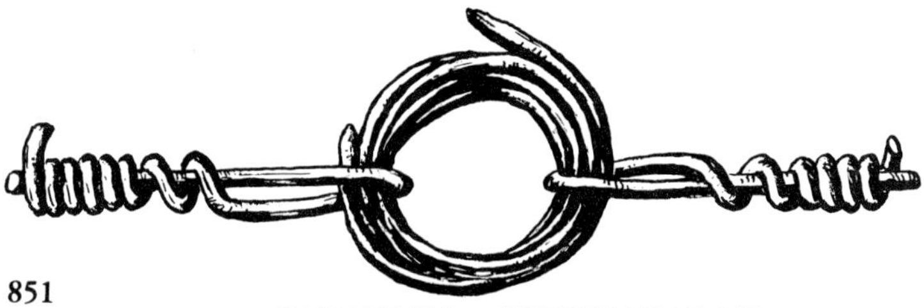

851

CANADIAN & AMERICAN HIGH
TENSILE SPLICE

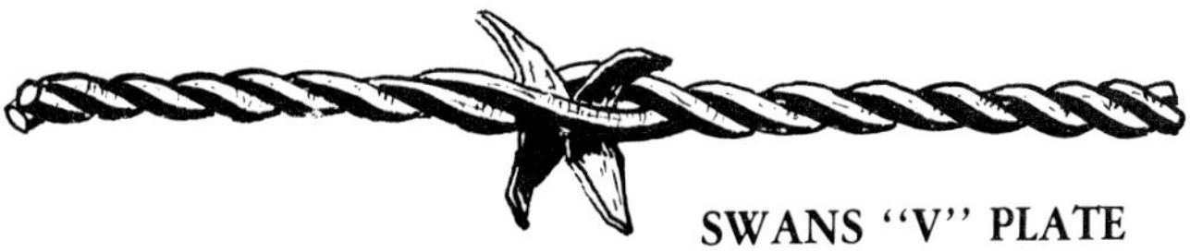

852

SWANS "V" PLATE
PAT. JUNE 10, 1879

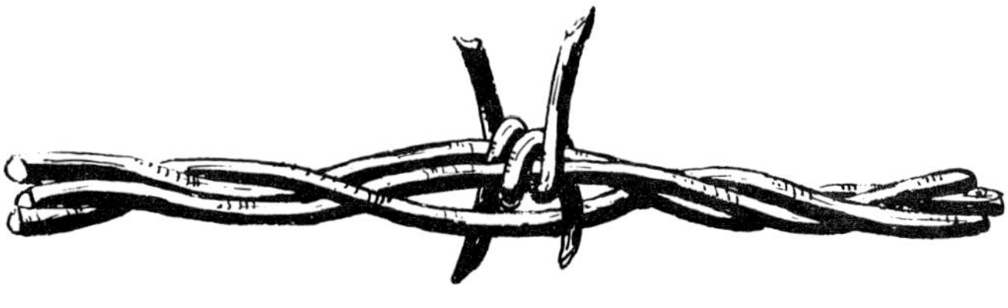

853

N.B. ROSS THREE LINE
PAT. JUNE 10, 1879

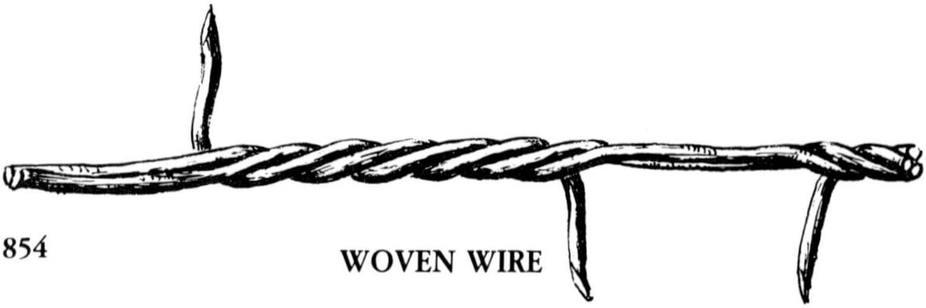

854 WOVEN WIRE

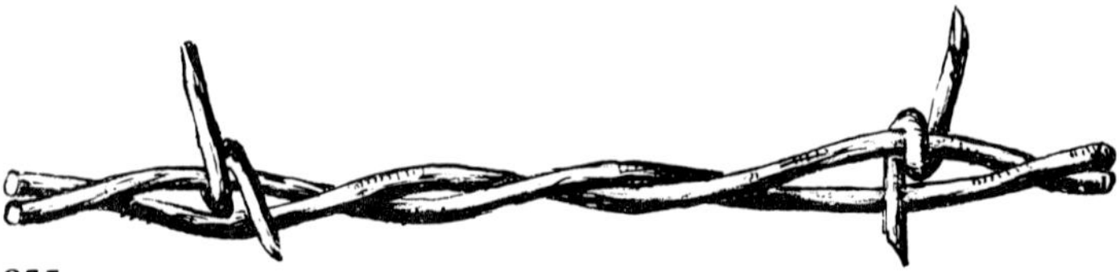

855 PURCHES MILES REVERSE BARB
PAT. OCT. 8, 1878

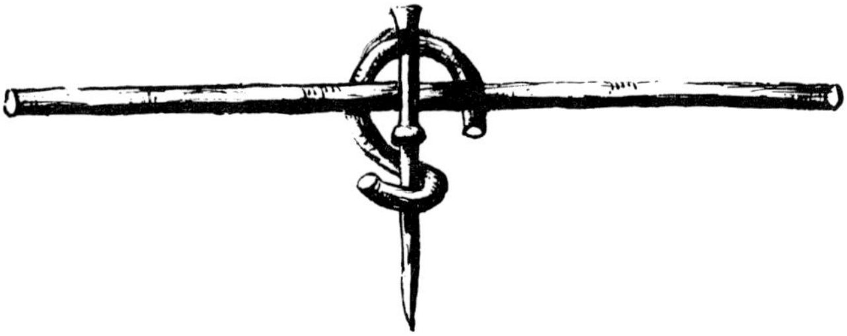

856 CLAMP & NAIL STAPLE

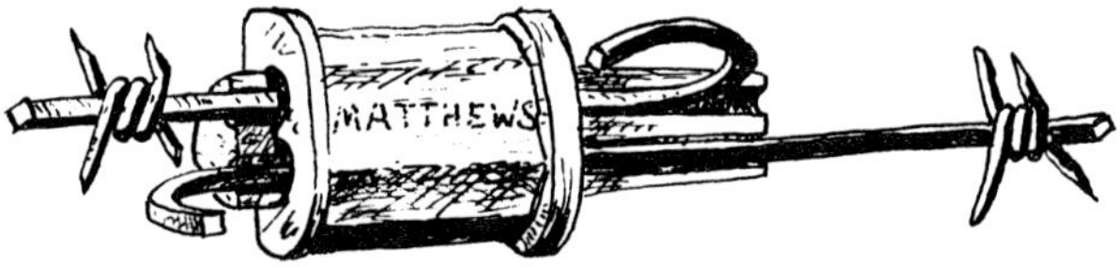

857 CAST SPLICER

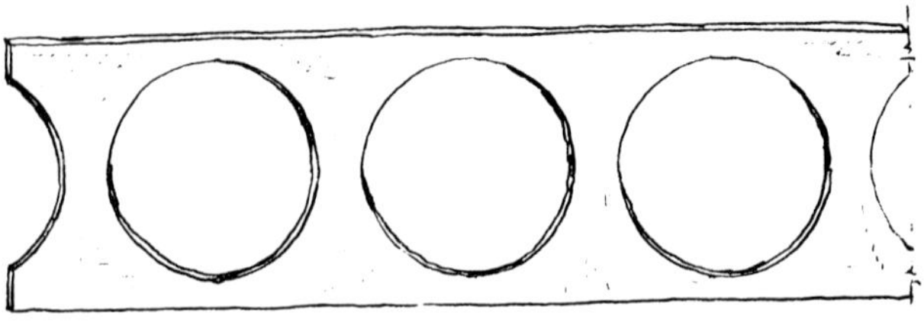

858 BRASS BAWDY HOUSE TOKEN
SCRAP FENCING

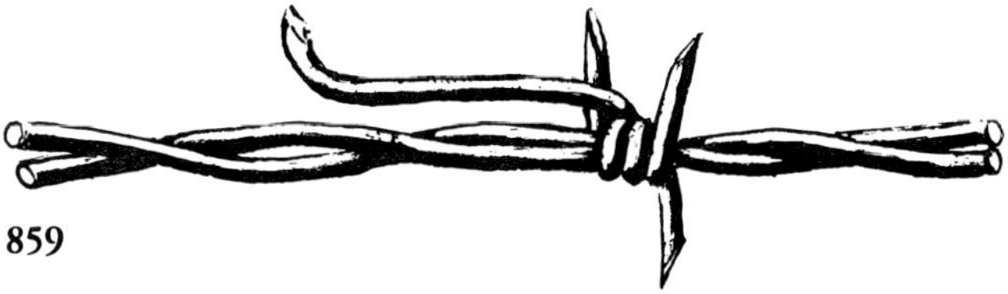

859

DODGE & WASHBURN MACHINE ERROR
PAT. JAN. 4, 1872

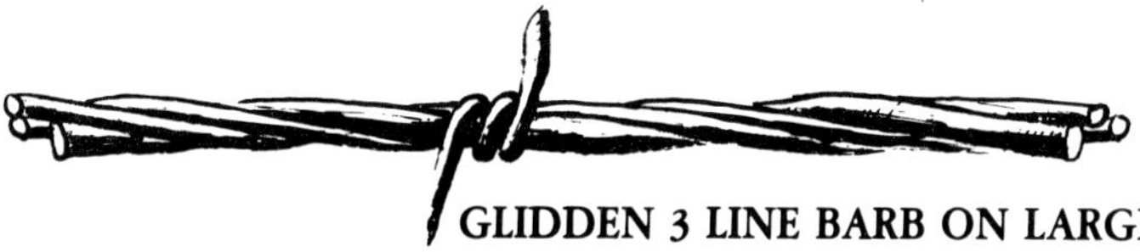

860

GLIDDEN 3 LINE BARB ON LARGE OVAL LINE
PAT. NOV. 4, 1874

861

GLIDDEN R.R. WIRE WITH EXTRA LARGE
SQ. LINE. PAT. NOV. 24, 1874

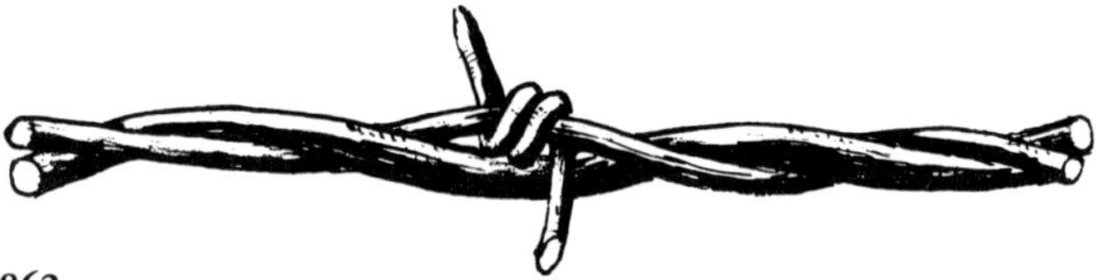

862

GLIDDEN LARGE & SMALL LINE, LARGE LINE
GALV. PAT. NOV. 24, 1874

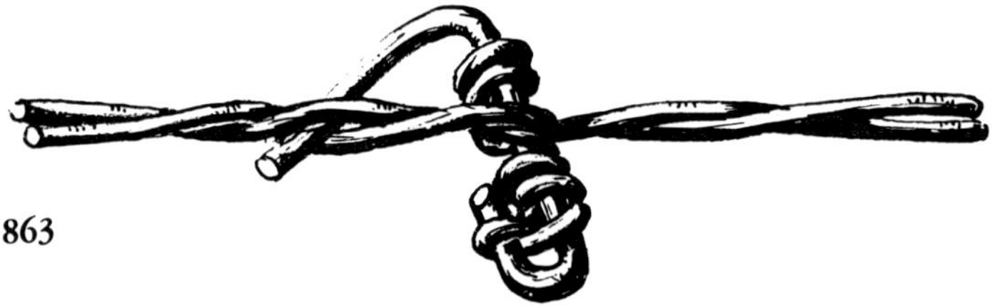

863

IN LINE TIGHTENER, TOOL APPLIED

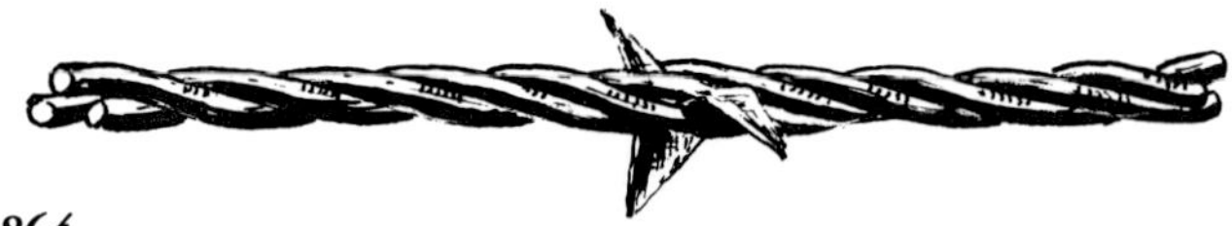

864

SMALL WORMLEY THREE STRAND
Y BARB PAT. NOV. 2, 1875

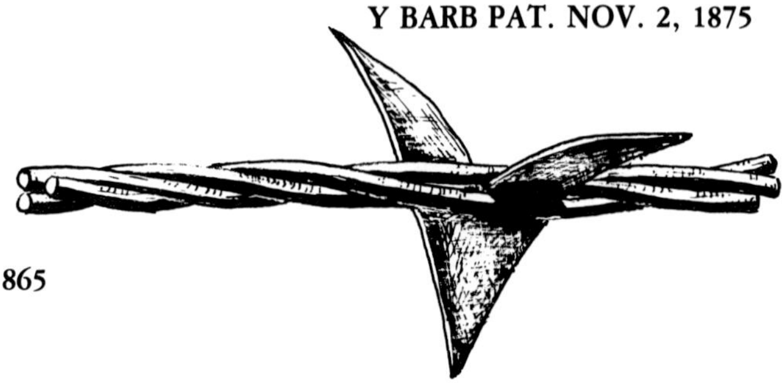

865

GIANT WORMLEY THREE STRAND
Y BARB PAT. NOV. 2, 1875

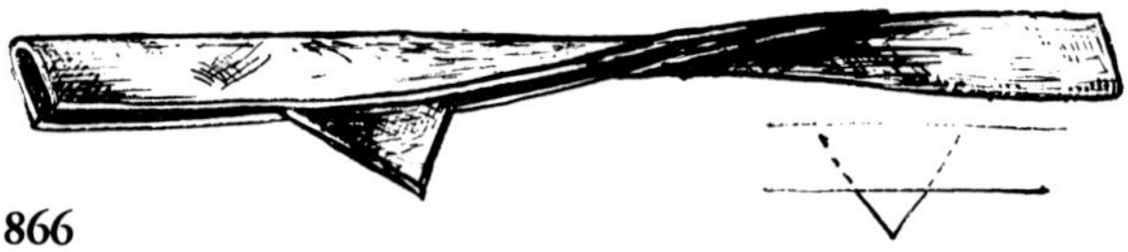

866

W.E. BROCK FOLDED STRIP
PAT. SEPT. 16, 1884

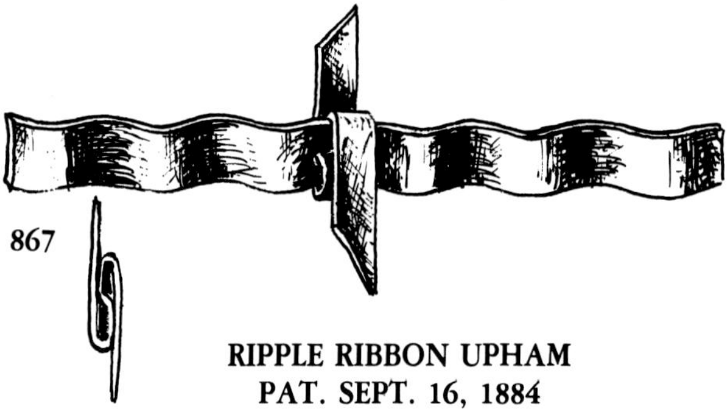

867

RIPPLE RIBBON UPHAM
PAT. SEPT. 16, 1884

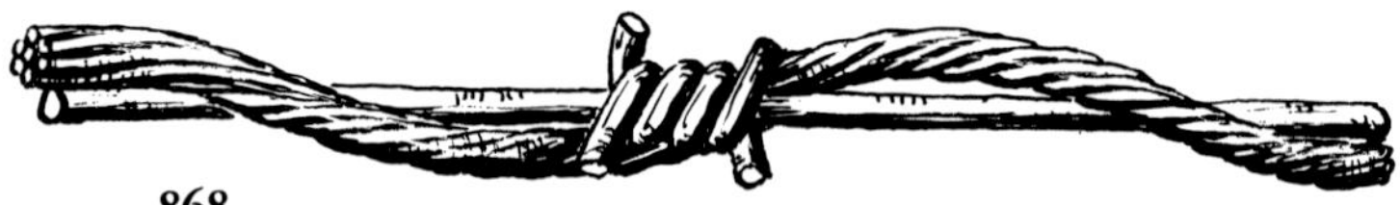

868

7 STRAND CABLE ON FOUR POINT
BLUNT CUT BARB ON STEEL LINE

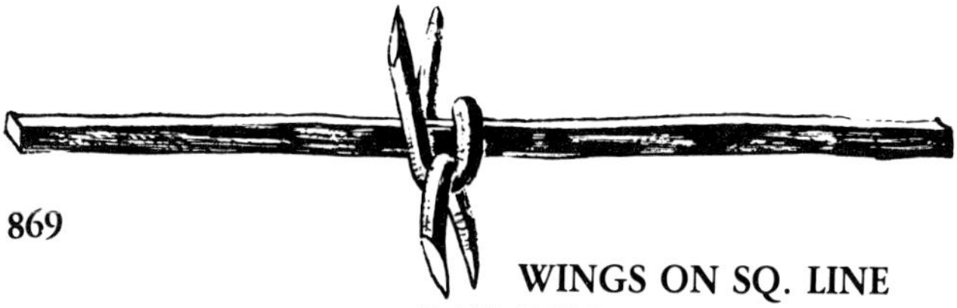

869

WINGS ON SQ. LINE
PAT. L.T.L. WING FEB. 26, 1878

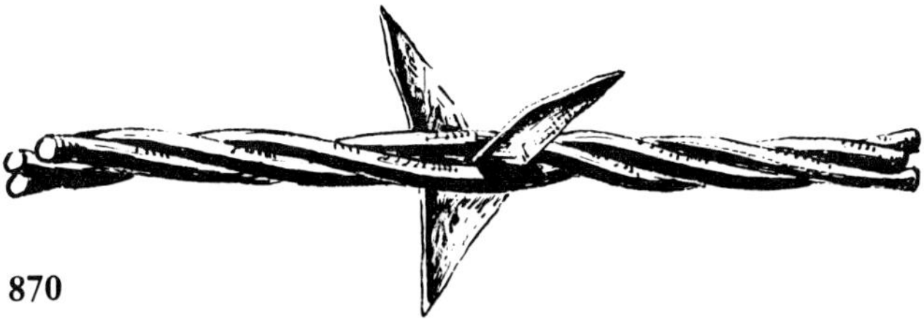

870

MEDIUM WORMLEY 3 STRAND Y BARB
PAT. NOV. 2, 1875

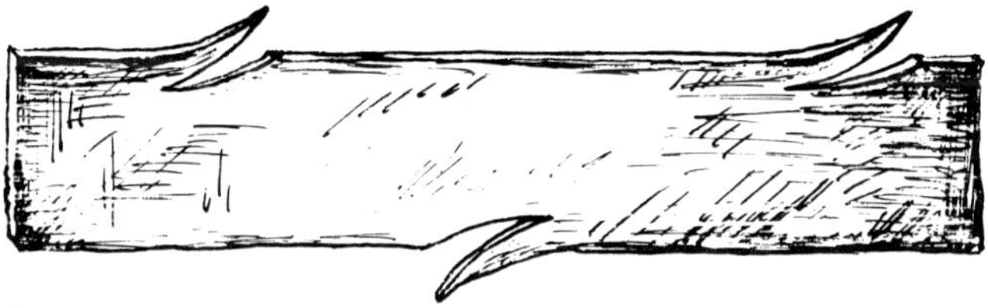

871

JUDSON MEDIUM WIDTH RIBBON
PAT. AUG. 15, 1871

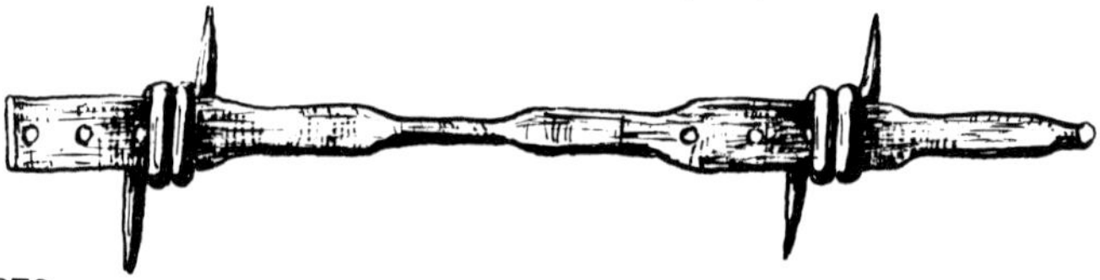

872

LENOX VARIABLE LINE ROUND,
OVAL & FLAT WITH DOTS PAT. JUNE 24, 1884

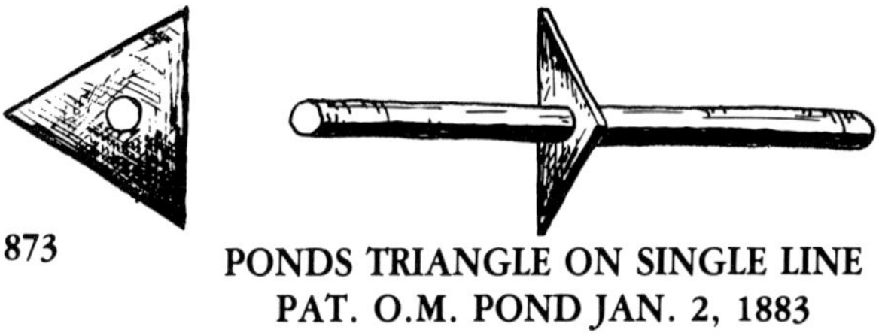

873

PONDS TRIANGLE ON SINGLE LINE
PAT. O.M. POND JAN. 2, 1883

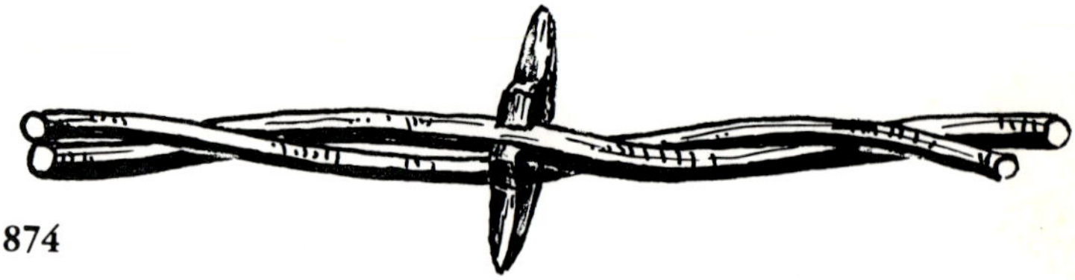

874

D.C. STOVER BARB ON LARGE LINE
PAT. JUNE 29, 1875

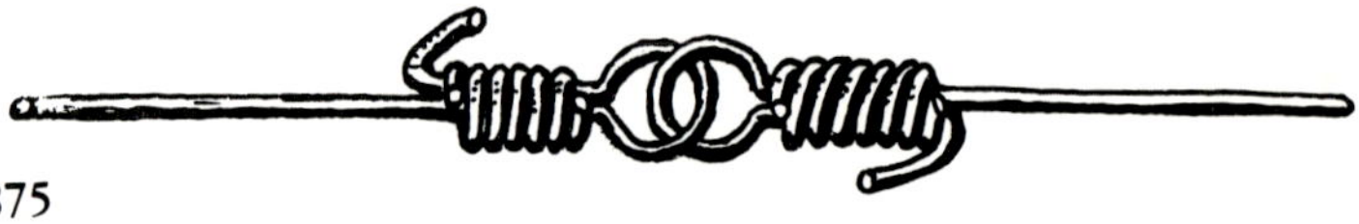

875

T.C. LORD PLANTER WIRE PAT. SEPT. 14, 1880

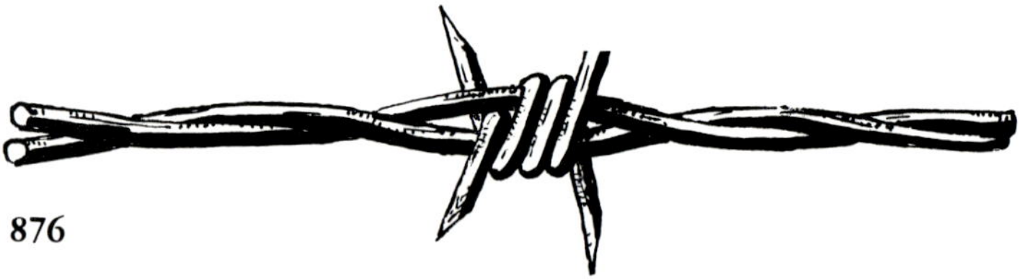

876

ROSS FOUR POINT STEEL LINES ALUMINUM
BARBS PAT. JUNE 10, 1879

877

SCUTTS "V" NOTCH BARB HAS THICK
CENTER PAT. AUG. 1, 1876

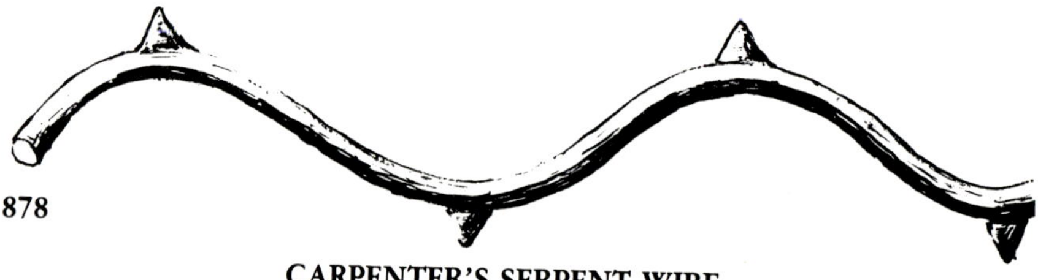

878

CARPENTER'S SERPENT WIRE
PAT. JUNE 6, 1882

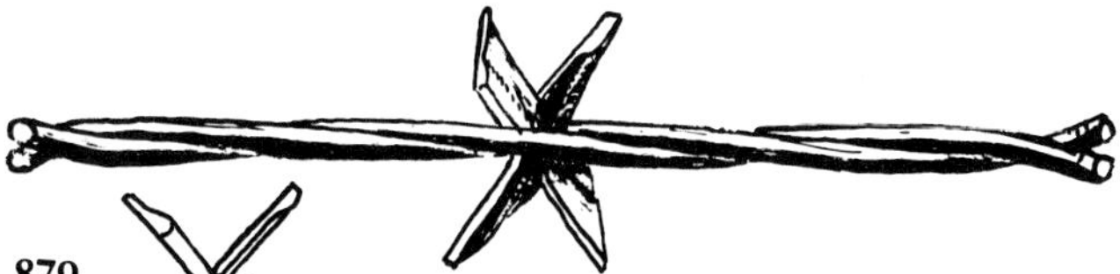

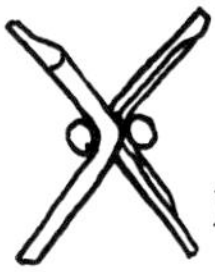

879

M & W WITH RIBS ON BARB
PAT. JUNE 18, 1878

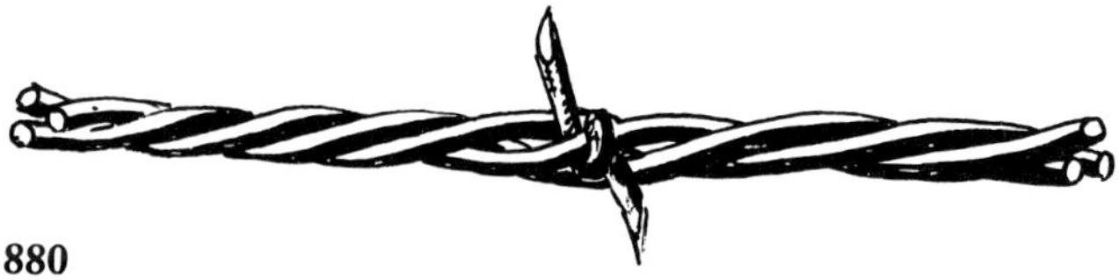

880

MILES 3 LINE PAT. OCT. 8, 1878

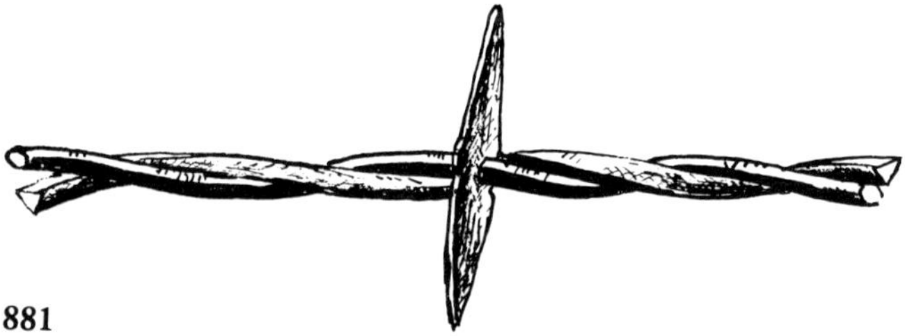

881

KELLY TWO LINE. ONE TRIANGLE
PAT. FEB. 11, 1868

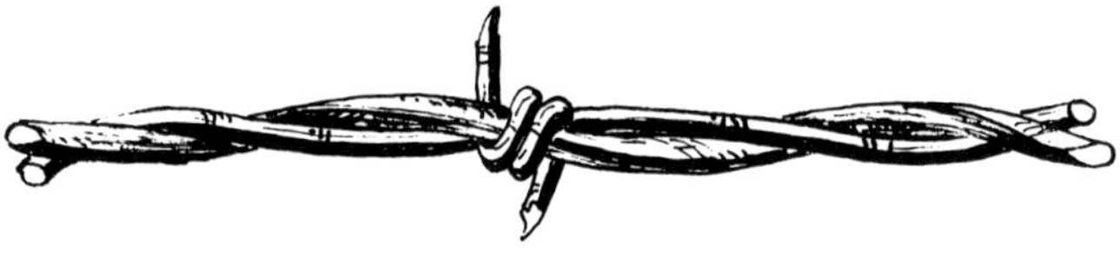

882

GLIDDEN BARB ON RIBBED LINE
PAT. NOV. 24, 1874

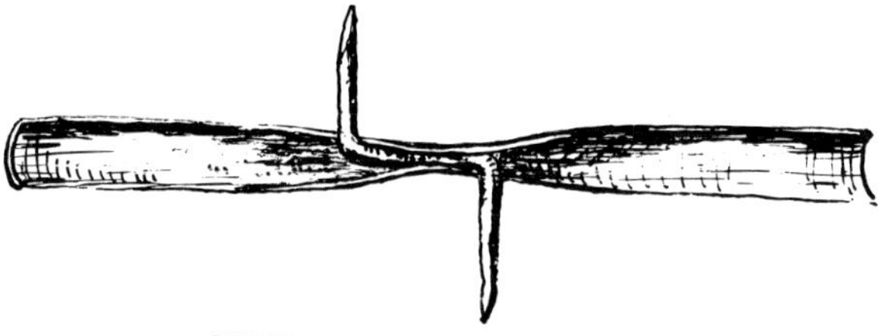

883

HAISH'S CONCAVE RAIL BARB
CRIMPED IN. PAT. DEC. 15, 1885

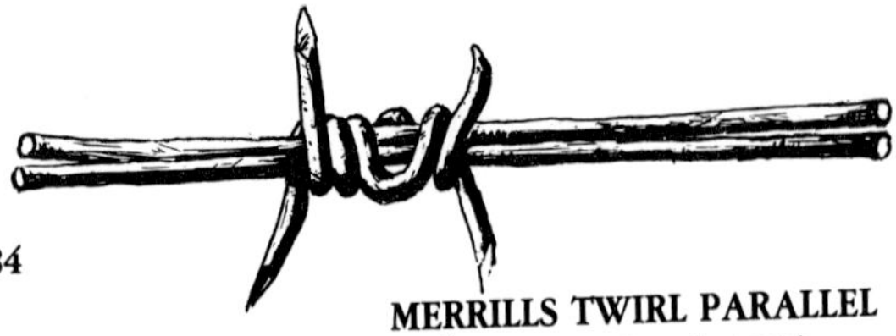

884

MERRILLS TWIRL PARALLEL
PAT. DEC. 26, 1876

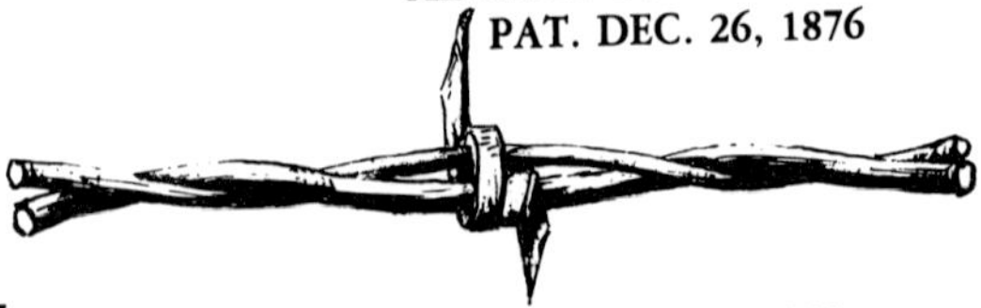

885

BROTHERTON FLAT BARB ON
DIFF. GA. LINE WIRES PAT. SEPT. 13, 1887

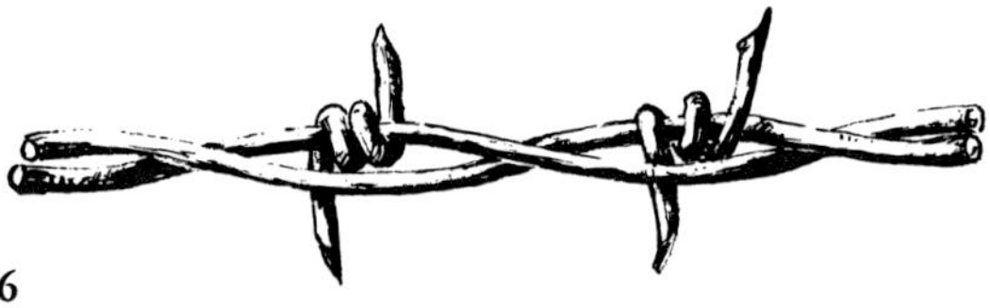

886

GLIDDEN & ROSE ON TWO LINES
PAT. NOV. 24, 1874

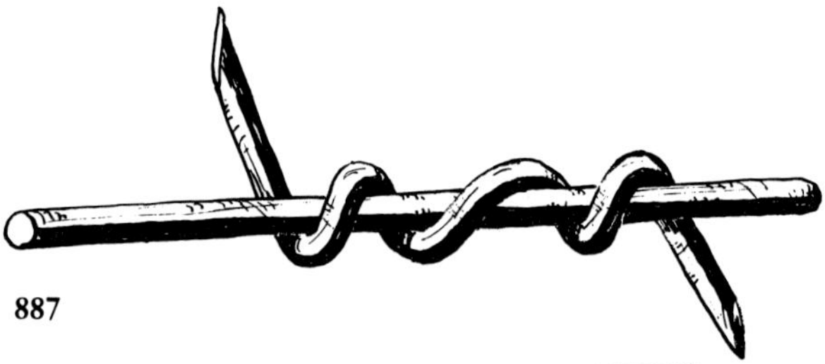

887

AUSTRALIAN VICIOUS WAMEGO
PAT. AUSTRALIA

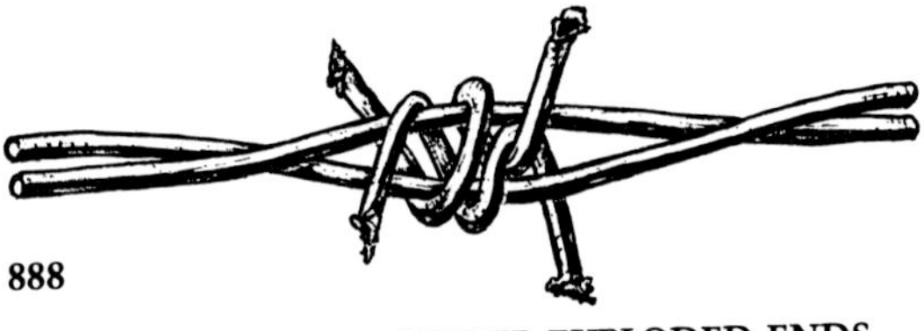

888

BURNELL COPPER EXPLODED ENDS
PAT. JUNE 19, 1877

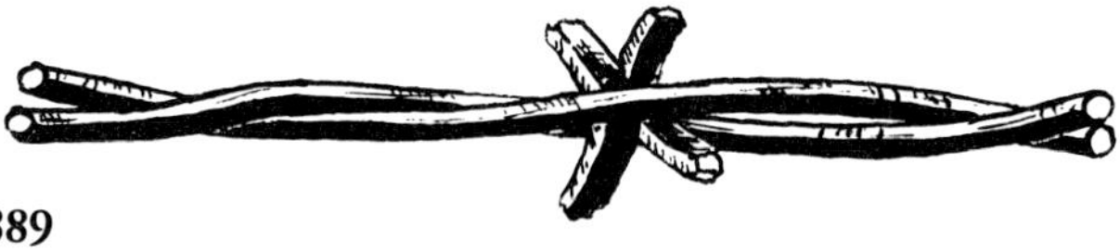

889

T.C. LORD SMALL BARB
PAT. FEB. 12, 1884

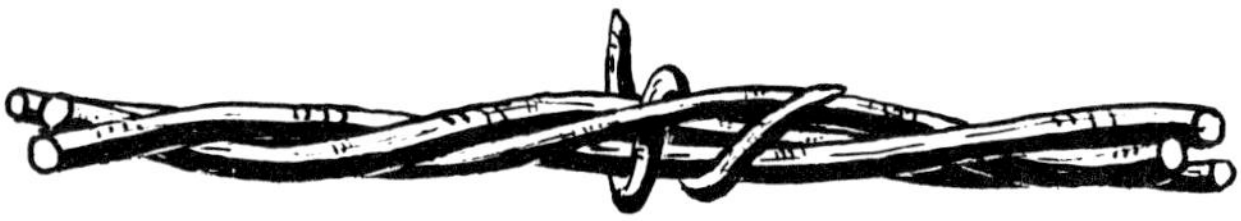

890

EDEMBARN 3 LINE 1 LINE SMALL
PAT. MAR. 17, 1885

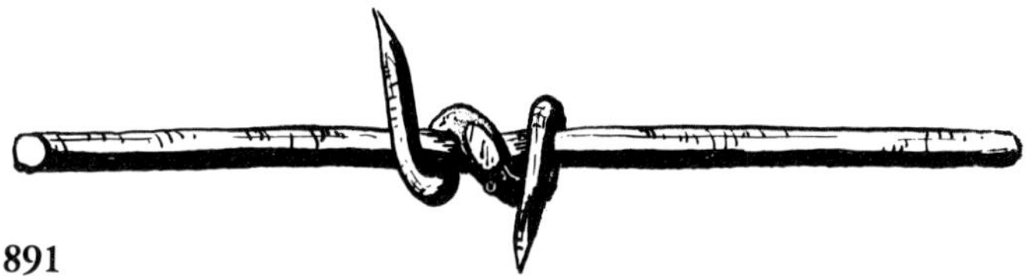

891

SUNDERLAND HAMMERED BARB
PAT. AUG. 12, 1884

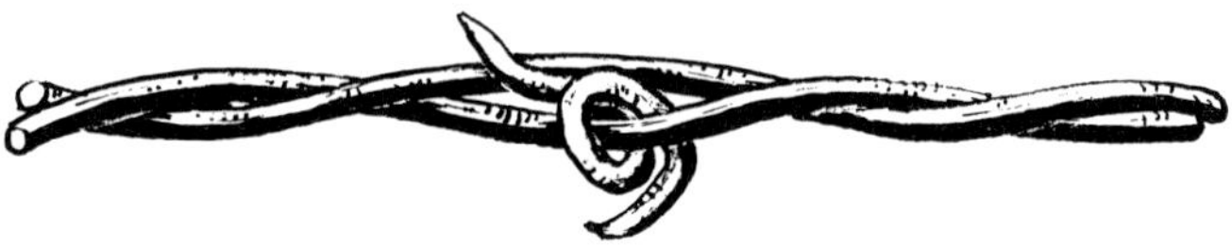

892

BRAINMARDS "DROP LOOP"
LOOSE BARB PAT. DEC. 5, 1882

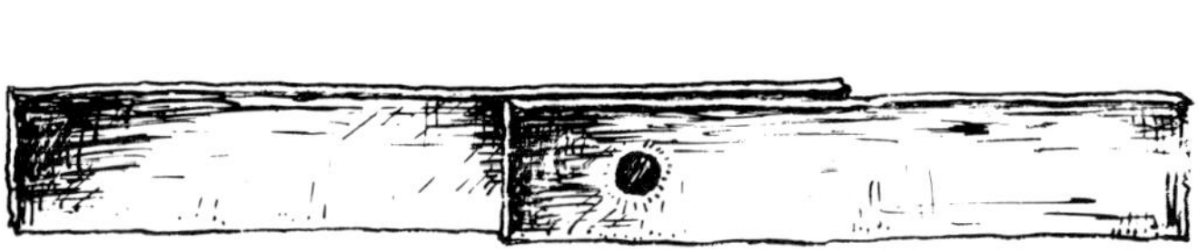

893

SPANISH LAND GRANT RIBBON RIVET
SPLICE 1880 PERIOD

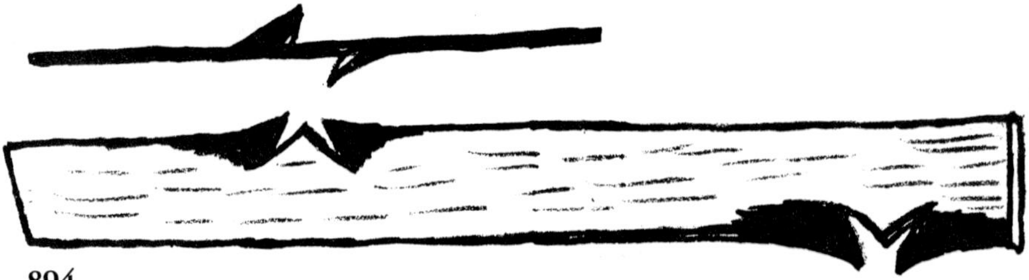

894

BOW CUT RIBBON PAT. F.D. FORD
JAN. 13, 1885

895

(ENLARGED CUT SKETCH)

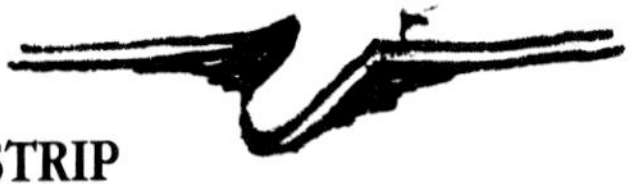

STAMPED RIBBON STRIP
PAT. E. JORDAN JULY 22, 1884

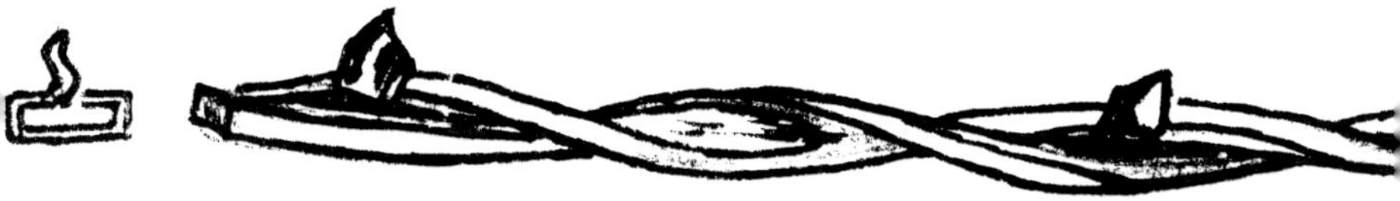

896

HIGH TINSEL ALUMINUM BOX LINE

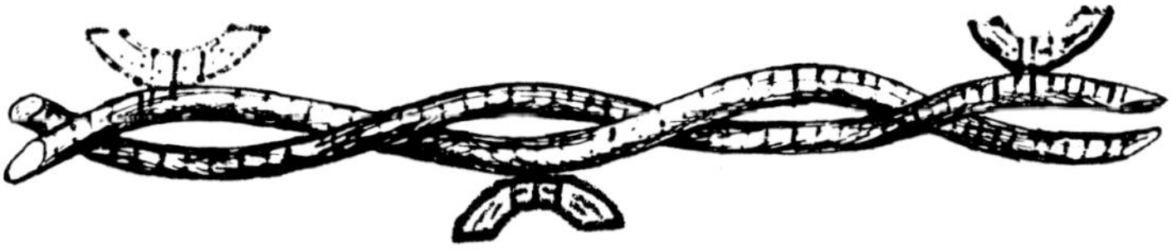

897

"ROCKING BARB" PAT. PERRY & GINN
AUG. 24, 1897

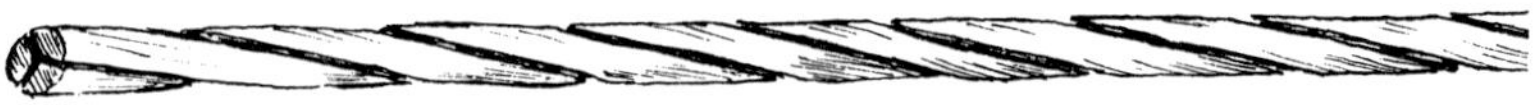

898

THREE LINE ROCK SAW WIRE

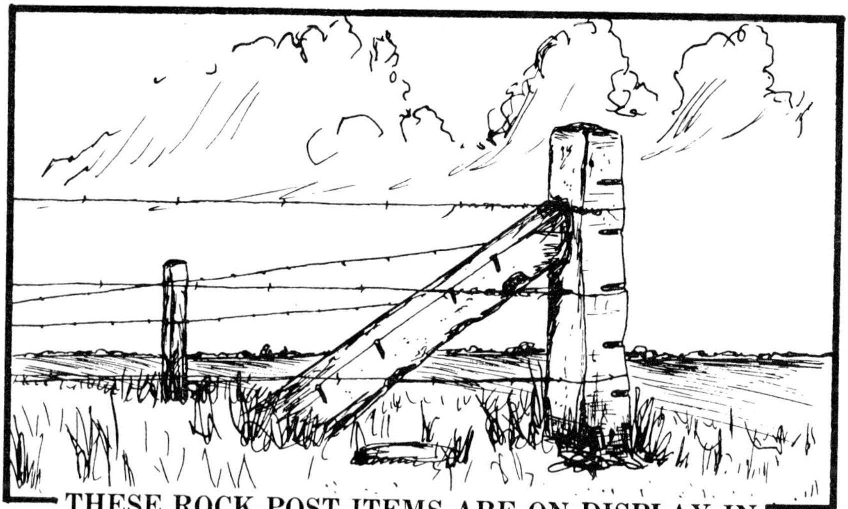

THESE ROCK POST ITEMS ARE ON DISPLAY IN THE ROCK POST MUSEUM AT LA CROSSE, KAN.

T1 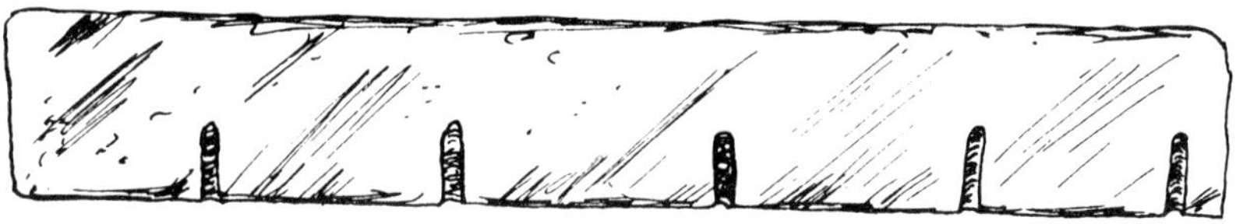

LIMESTONE POST

T2 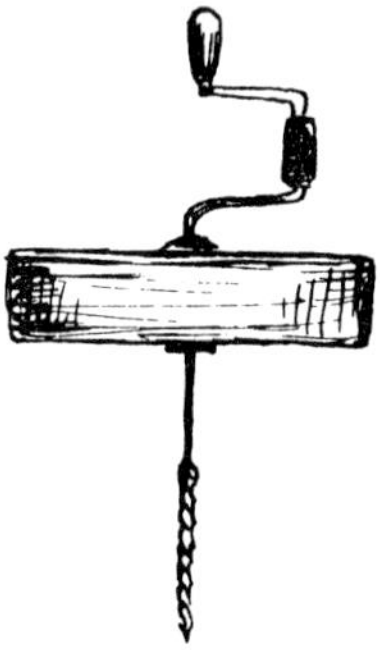

DRILL

T3 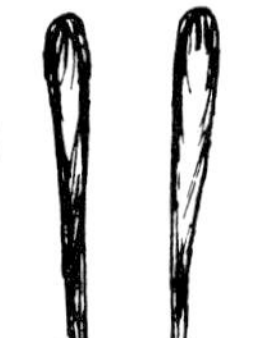

FEATHERS

T4 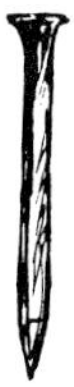

WEDGE

T5 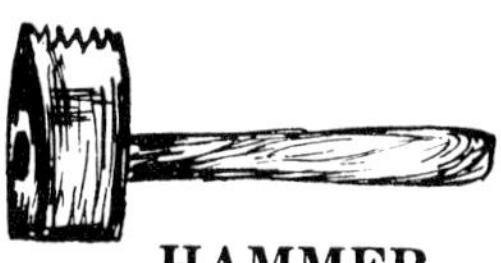

HAMMER

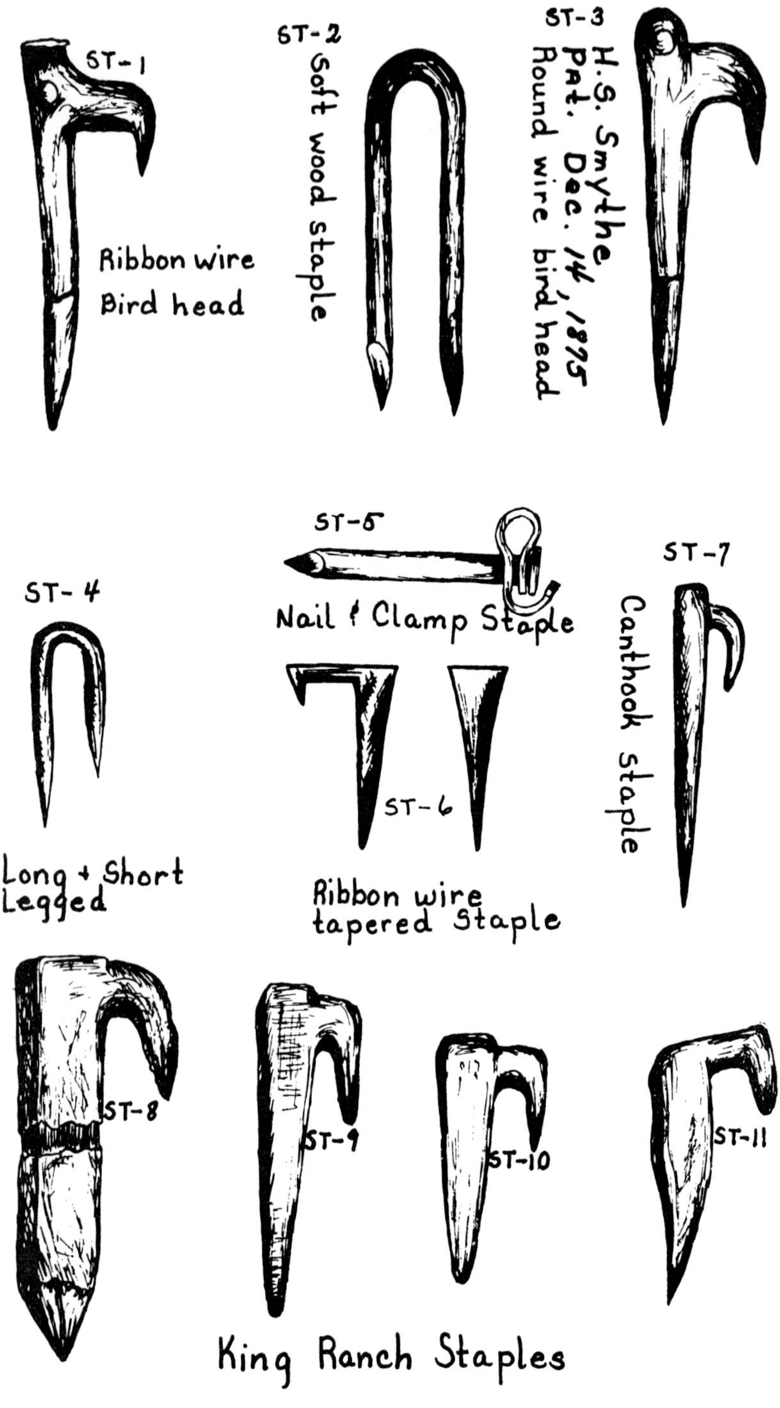
ST-1
Ribbon wire
Bird head
ST-2
Soft wood staple
ST-3
H.S. Smythe
Pat. Dec. 14, 1875
Round wire bird head
ST-5
Nail & Clamp Staple
ST-4
Long + Short
Legged
ST-6
Ribbon wire
tapered Staple
ST-7
Canthook staple
ST-8
ST-9
ST-10
ST-11
King Ranch Staples

ST-12

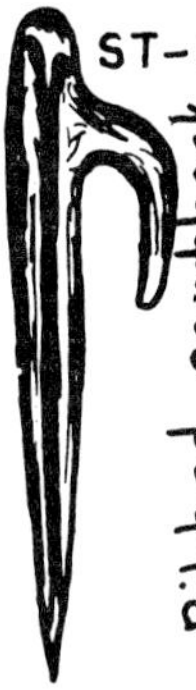

Ribbed canthook

ST-13

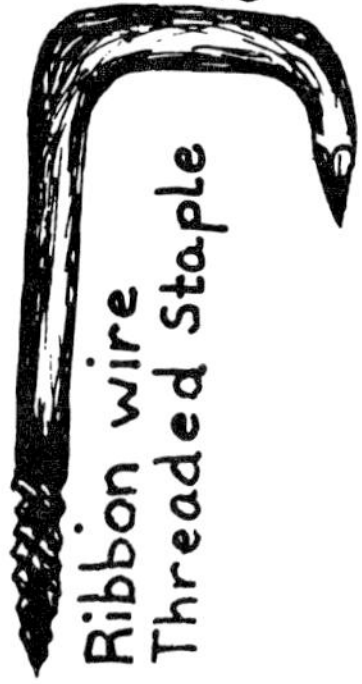

Ribbon wire
Threaded Staple

ST-14

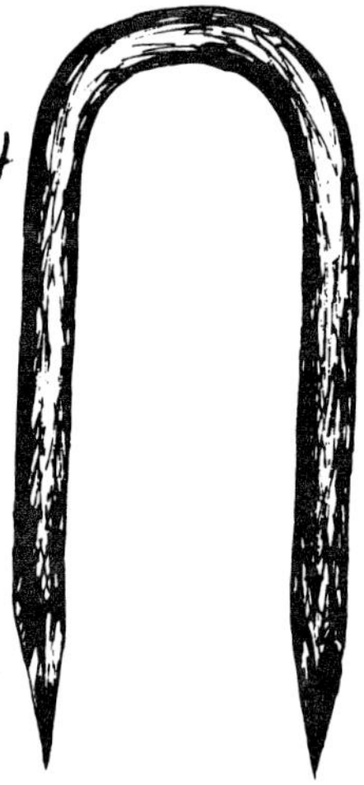

Gate and barn door staple

ST-15

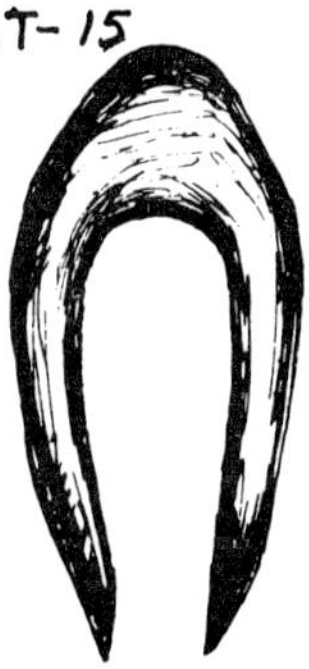

Punched Sheet metal

ST-16

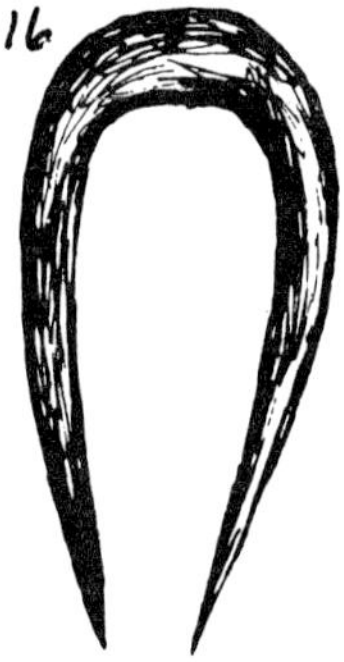

Handmade

ST-17

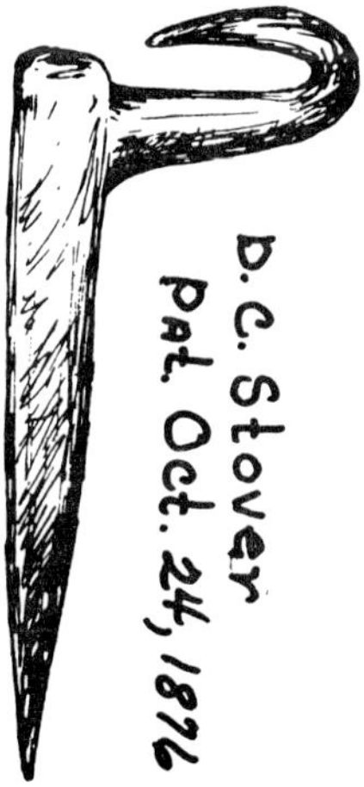

D.C. Stover
Pat. Oct. 24, 1876

ST-18

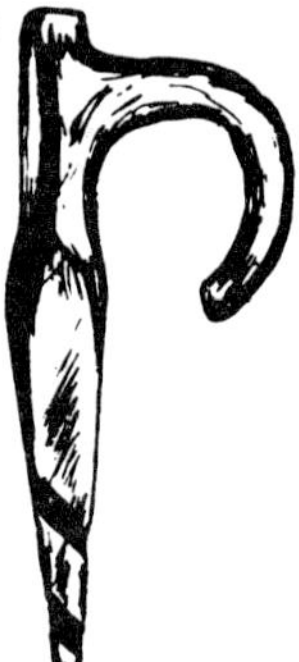

Had big ring so
wire could slide

ST-19

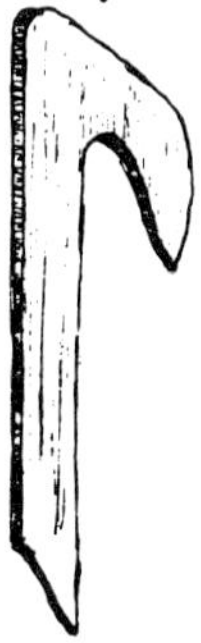

Z.I. Pratt
Pat. June 2, 1874

ST-20

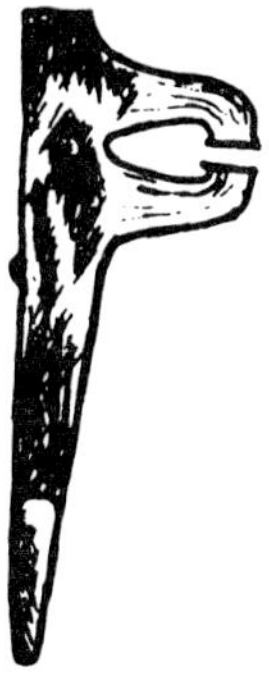

J. & W. Brinkerhoff
Pat. June 24, 1879

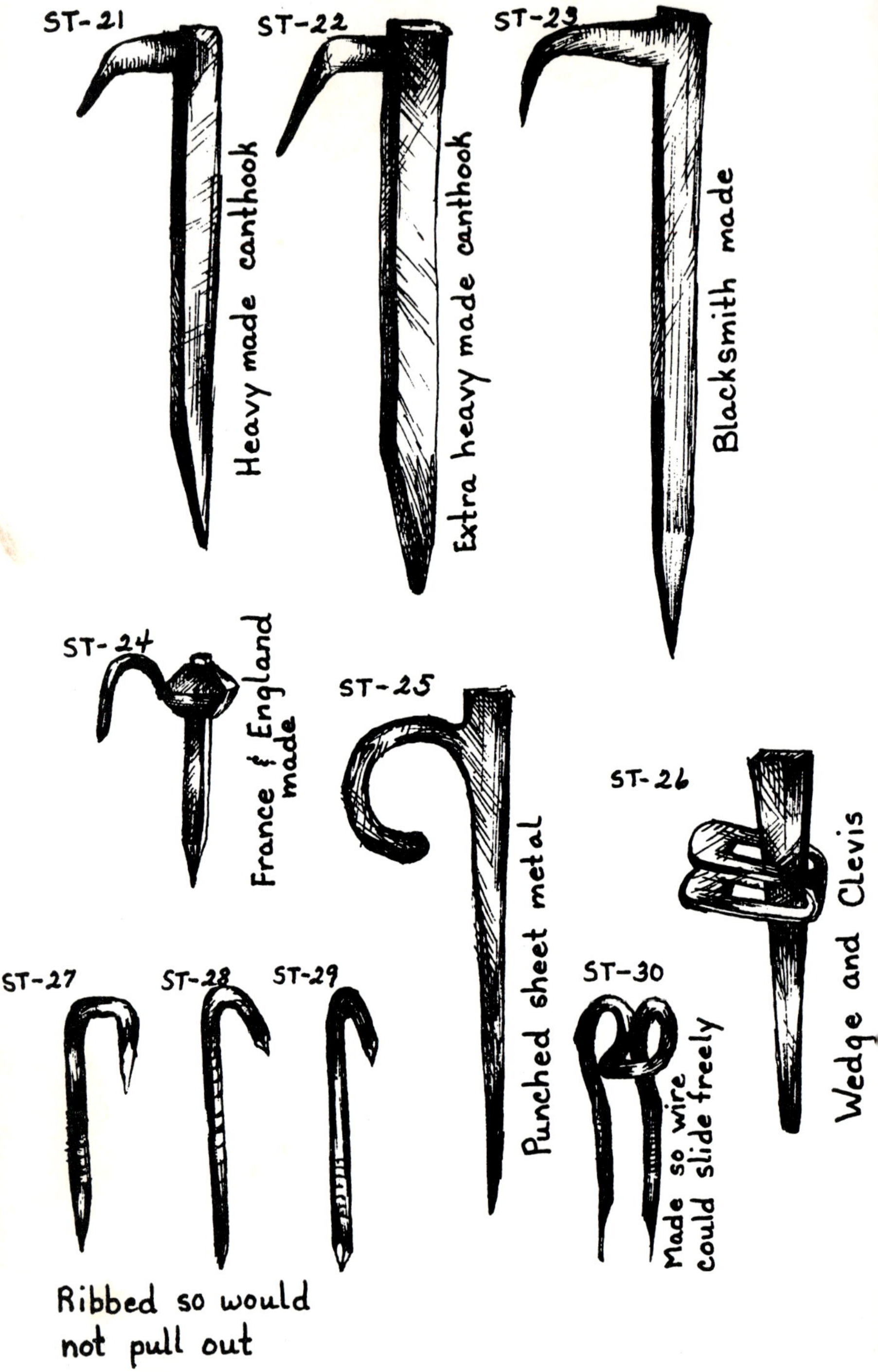
ST-21
Heavy made canthook
ST-22
Extra heavy made canthook
ST-23
Blacksmith made
ST-24
France & England made
ST-25
Punched sheet metal
ST-26
Wedge and Clevis
ST-27
ST-28
ST-29
Ribbed so would not pull out
ST-30
Made so wire could slide freely

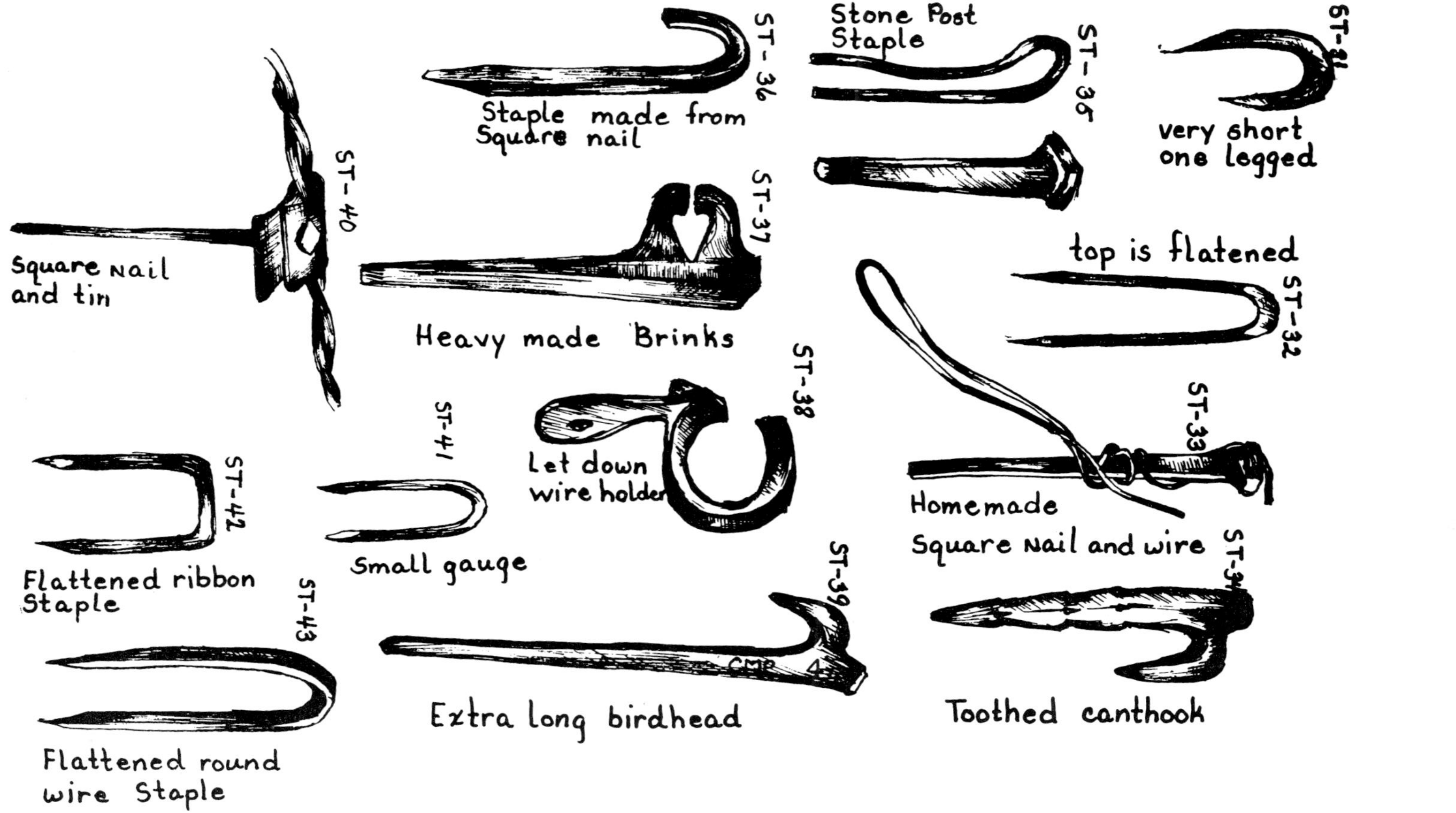
ST-31
very short
one legged
ST-35
Stone Post
Staple
ST-32
top is flatened
ST-33
Homemade
Square Nail and wire
ST-34
Toothed canthook
ST-36
Staple made from
Square nail
ST-37
Heavy made Brinks
ST-38
Let down
wire holder
ST-39
CMP
4
Extra long birdhead
ST-40
Square Nail
and tin
ST-41
Small gauge
ST-42
Flattened ribbon
Staple
ST-43
Flattened round
wire Staple

INDEX

NOTE: A Section Consisting Of Fencing Tools And Related Items May Be Found At The End Of The Wire Section

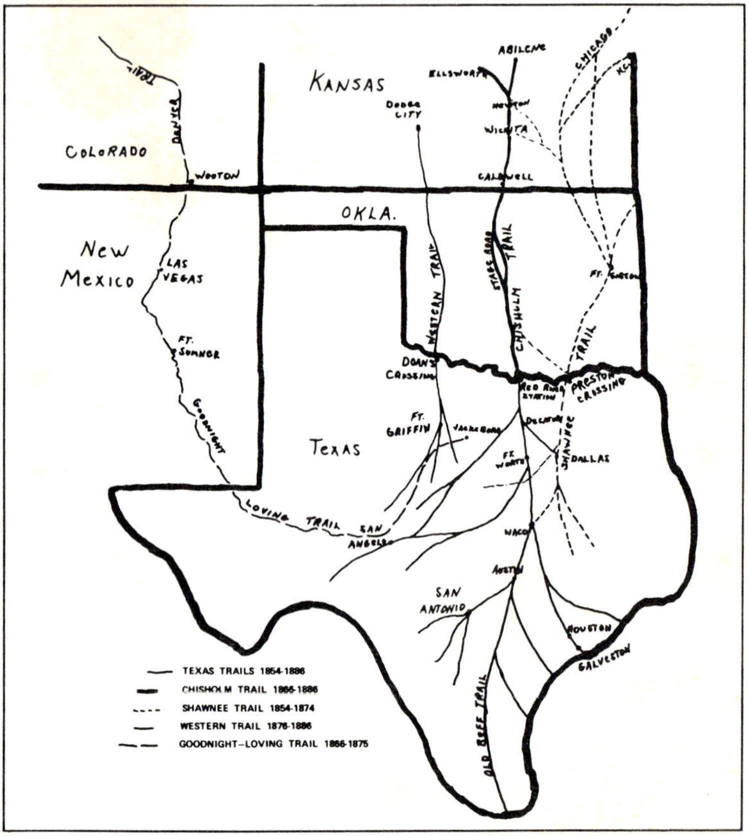

CATTLE TRAILS